S0-BCT-342

EARLY COMMENTS ON *CLIMATE PERIL*

"A brilliant book, and one that might just change the world. By far the best overview of climate science and its implications for our planet that I've ever read."

> TIM FLANNERY
> Chief Councilor, Australian Climate Council
> Author of *The Weathermakers*

"John Berger has produced a critically needed and eloquent compendium of not only the crisis of global warming, but of the many interdependent shortages and crises—of financial risks, and of land, water, cultural preservation, and of economic opportunity for the poor—that all exist today and that conspire to challenge us to be better stewards of the planet."

> DANIEL M. KAMMEN
> Distinguished Professor of Energy
> University of California, Berkeley

"I applaud *Climate Peril* for showing so clearly that climatic disruption is the consuming issue of our time and our response in the next few years will determine the fate of this civilization. At some point not far away, the corporate and political venality *Climate Peril* characterizes so clearly becomes criminal neglect of duty as costs of the disruption soar and effects become irreversible."

> GEORGE M. WOODWELL
> Founder, Director Emeritus, and Senior Scientist
> Woods Hole Research Center

"*Climate Peril* should be required reading for the remaining climate skeptics—if, indeed, there remain any whose views are not shaped purely by the sources of their paychecks. In the second volume of his trilogy, John Berger continues to probe the climate issue with unusual insight and clarity."

> DENIS HAYES
> President, The Bullitt Foundation
> Director, Solar Energy Research Institute (1979–1981)
> Coordinator, Earth Day 1970

"*Climate Peril* provides a comprehensive and accurate, yet very readable account of the climate change threat. If you are aware of the problem, but not yet convinced that it presents a near and present danger, read this call to arms. You'll understand why it is that our ongoing burning of fossil fuels, left unchecked, imperils our health, access to food and fresh water, national security, and the stability of our environment. And you'll see why there is still time to act to avert a global catastrophe."

MICHAEL MANN
Distinguished Professor of Meteorology, Penn State University
Director of the Penn State Earth System Science Center
Author of *The Hockey Stick and the Climate Wars*

"*Climate Peril* is precisely what is needed: a clear-eyed, accessible account of climate change and its implications for the future of life on Earth and for that matter, of civilization itself. This should be required reading for all who care about future generations."

THOMAS E. LOVEJOY
University Professor of Environmental Science and Policy
George Mason University
Senior Fellow, The United Nations Foundation

"*Climate Peril* is a lucid primer on what may be the greatest challenge human beings have ever faced. If we make no further moves against climate change than the timid steps taken so far, the world is due for something potentially as catastrophic as a slow-motion nuclear war. Even if we do do everything we possibly can, we're still in for big trouble. John J. Berger has done a great service by describing what confronts us in a thorough, careful way that a lay person can understand."

ADAM HOCHSCHILD
Author and journalist
Cofounder, *Mother Jones*
Lecturer, Graduate School of Journalism
University of California, Berkeley

"*Climate Peril* is a compelling, accessible, and accurate summary of the climate crisis. Highly recommended!"

JONATHAN G. KOOMEY
Research Fellow, Steyer-Taylor Center for Energy Policy
and Finance, Stanford University
Author of *Cold Cash, Cool Climate: Science-Based Advice
for Ecological Entrepreneurs*

"Never has a book been so vitally important. Politicians are looking the other way while global warming is staring them in the face. We the people urgently need to become fully informed, rise up as one to challenge, educate, and pressure our elected representatives so they will exert ultimate responsibility for the future of our children, grandchildren, and 30 million species that cohabit the planet. Read *Climate Peril*, then act."

DR. HELEN CALDICOTT, MD
President, The Helen Caldicott Foundation
Physicians for Social Responsibility (former president)
Author of *Loving this Planet* and other books

"Seven billion of us—growing by 200,000 net per day, and increasing their demands for consumption at an even faster rate—are driving the climate upward rapidly and unpredictably. This book, and the work of which it is a part, clearly illuminate the dilemma that we face together, its severe effects on biodiversity and the functioning of our planet. *Climate Peril* thus lays the foundation for understanding the actions we must take to begin building a sustainable world for the future. Highly recommended!"

PETER H. RAVEN
President Emeritus, Missouri Botanical Garden

"*Climate Peril* is an excellent primer on the causes and effects of climate change, which, as John Berger notes, imperils our very existence and that of all natural systems on which we depend."

LESTER R. BROWN
President, Earth Policy Institute
Author of *Full Planet, Empty Plates*

CLIMATE PERIL

ALSO BY JOHN J. BERGER

CLIMATE PERIL

The Intelligent Reader's Guide to Understanding the Climate Crisis

John J. Berger

Northbrae
Books

Climate Peril: The Intelligent Reader's Guide to Understanding the Climate Crisis
Copyright © John J Berger, 2014
All rights reserved.

First published in 2014 in the United States by Northbrae Books.

Northbrae Books, Publishers
941 The Alameda
Berkeley, California 94707
Northbraebooks@gmail.com

Library of Congress Preassigned Control Number 2014904435

Publisher's Cataloging-In-Publication Data

Berger, John J., 1945-
   Climate peril : the intelligent reader's guide to understanding the climate crisis / John J. Berger ; introduction by Paul R. and Anne H. Ehrlich. -- First edition.

   pages : illustrations ; cm

   Issued also as an ebook.
   Includes bibliographical references and index.
   ISBN: 978-0-9859092-3-9 (paperback)
   ISBN: 978-0-9859092-4-6 (hardcover)

   1. Climatic changes. 2. Global warming. 3. Greenhouse gases--Environmental aspects.
I. Ehrlich, Paul R. II. Ehrlich, Anne H. III. Title.

QC903 .B47 2014
363.738/74   2014904435

ISBN: 978-0-9859092-2-2 (electronic book)

Cover design by Shannon Bodie, Lightbourne, Inc.
Interior design by Nancy Austin
Author photo by Phil Saltonstall

Copies produced by Thomson-Shore of Dexter, MI, are printed on 100 percent recycled
Rolland Enviro Natural paper. Paper consumption by other printers has been offset
through donations to support tree planting by Sempervirens Fund of Los Altos, California,
an organization dediated to restoration and protection of forest and parkland.

Manufactured in the United States of America
First edition June 2014
10 9 8 7 6 5 4 3 2 1

Dedicated to everyone raising public awareness of climate change and acting to protect the climate.

The climate crisis is the mirror in which we see reflected the combined ecological impact of our industrialised age.

HIS HIGHNESS CHARLES, THE PRINCE OF WALES
ADDRESS TO THE 2009 COPENHAGEN CLIMATE CHANGE CONFERENCE

CONTENTS

FOREWORD

BEN SANTER

IN THE MID-1990S, I was the convening lead author of the eighth chapter of the *Second Assessment Report* of the Intergovernmental Panel on Climate Change (IPCC), ["Detection of Climate Change and Attribution of Causes"]. After years of careful evaluation of the available scientific evidence, my scientific coauthors and I concluded in 1995 that: "the balance of evidence suggests a discernible human influence on global climate."

Subsequent IPCC Scientific Assessment Reports in 2001, 2007, and 2013 confirmed our 1995 finding. The most recent *2013 Assessment Report* concluded that: "It is *extremely likely* that human influence has been the dominant cause of the observed warming since the mid-20th century." The words "extremely likely" had a specific probabilistic meaning: greater than 95 percent probability of occurrence.

This dramatic evolution in our scientific understanding—from the cautious "balance of evidence" statement to the more definitive finding that humans have been "the dominant cause" of the observed warming since 1950—has occurred in less than two decades. In parallel with this evolution in understanding, scientists have observed global-scale increases in the temperatures of the land and ocean surfaces, in the temperature of the lowest layer of the atmosphere, and in the heat content of the ocean. Global-mean sea level has risen. Arctic sea ice extent and thickness has decreased, along with Northern Hemisphere snow coverage. Large-scale climate change is also evident in the water cycle—in zonal-mean rainfall patterns, surface humidity, the amount of water vapor in the atmosphere, and runoff from major river basins. These extraordinary transformations are occurring over the span of one human lifetime.

From the careful detective work performed by many hundreds of climate scientists, it is clear that natural factors alone cannot explain these distinctive

changes in the atmosphere and oceans, in the snow and ice, and in the distributions and abundances of many plant and animal species around the world.
While the climate can and does vary naturally in response to changes in
the sun's energy output, large volcanic eruptions, or internal oscillations in
the climate system (phenomena such as El Niños and La Niñas), a "natural
causation" diagnosis does not fit the incredibly rich array of observational evidence. In order to best explain the unusual observed behavior, a strong human
influence is required. It is immutable fact that this strong human influence is
primarily due to fossil fuel burning, and the resulting changes in atmospheric
levels of heat-trapping greenhouse gases.

As I write these words in March 2014, scientists have just discovered the
gravity wave "fingerprints" from the Big Bang. This profoundly important
discovery—a marvelous feat of technology and scientific imagination—may
help us to understand the very beginnings of space and time. And yet, even as
we are deciphering the origins of the universe, we are fundamentally and dramatically changing our own planet. Humanity's "fingerprints" are now iden-
tifiable in many different aspects of the climate system, and human-caused
climate change appears destined to impact life on Earth for millennia to come.

But even as we look outward to understand the universe, many people ·
still refuse to look at our own home planet, and are unable to recognize
how we are changing our fragile atmosphere. Dr. John Berger's *Climate Peril*
performs an extraordinary public service—it helps all of us clearly see that
we are no longer just innocent bystanders in the climate system. Dr. Berger
lucidly describes our likely climatic future, past and present-day climate, and
the human and natural drivers of climate change. He then shows some of the
projected impacts of human-caused climate change, such as changes in the
properties of extreme events, progressive acidification of the world's oceans,
and species extinctions. These perils are not future hypothetical events. We are
experiencing them now, in our lifetimes.

To make informed decisions on how to respond to human-caused climate
change, we need an informed, scientifically savvy global citizenry. Climate science needs to be accessible to the many, not just to a select few. *Climate Peril*
fulfills this goal of making the science accessible. The book is a comprehensive,
plain-English introduction to complex scientific issues. It should be read by
anyone who has children or grandchildren and cares about the kind of world
with which we leave them.

While a sobering book, *Climate Peril* also offers a hopeful perspective on
humanity's future, and on our ability to provide cheap, low-carbon energy. A

species that is clever enough to peer more than 13 billion years into the past, and identify the gravity wave signature of the Big Bang, can surely figure out ways of providing cheap, low-carbon energy for the billions of citizens of planet Earth. But as *Climate Peril* so starkly reveals, that is a global enterprise of greatest urgency—a looming test of whether our species can graduate from troubled adolescence to maturity. Passing this test will require the development and deployment of advanced energy technology, engaging the best scientific minds, and mustering the political will to solve a problem that affects every one of Earth's inhabitants.

Ben Santer, PhD, is a climate research scientist at the Program for Climate Model Diagnosis and Intercomparison, Lawrence Livermore National Laboratory, Livermore, CA 94550. For more information, see page 317.

PREFACE

During some 40 years of professional work I have been deeply concerned about the effects of human activities on the environment. My focus was originally on the environmental impacts of energy technologies. After studying the energy options globally available, I was convinced by the mid-1970s that the most economically and environmentally sound way forward required the vigorous adoption of renewable energy and efficiency technologies. In the decades and books that followed, I strongly advocated for those technologies and for more intensive use of environmental restoration technologies to repair damaged natural resource systems.

Through all these concerns, I came to the study of climate change in the 1980s and now view it as the paramount issue of our time. It affects all humanity as well as the entire Earth, together with all its ecosystems.

Climate Peril: The Intelligent Reader's Guide to Understanding the Climate Crisis was written not only to explain the causes of climate change and the natural science needed to understand their effects, but most importantly, to convey the catastrophic impacts on plants, animals, human health, and economic welfare to which a continuation of rapid climate change is currently leading.

To bring this enormous story to life, *Climate Peril* ranges from the Himalayan glaciers to the ocean depths; from the melting Greenland Ice Cap and thawing permafrost to the tropical rainforests of the Amazon; from the African Sahara to the melting methane ice crystals beneath the polar seas; and from America's mountain snowpacks to its seashores and coastal cities. I hope the book conveys a clear understanding of how and why climate change occurs; its human origins; its extraordinary speed; and most of all, that catastrophic events have already begun.

To set the stage for this epic story, *Climate Peril* explains key scientific ideas and concepts, such as the role that carbon dioxide and other long-lived, heat-trapping gases play in the climate system, as well as positive feedback processes and tipping point risks. But even if you are a newcomer to climate change issues, you will not need any prior knowledge to read the book. You can do so quickly for general understanding or more intensively for in-depth knowledge.

My goal in *Climate Peril* is to show why climate change is the most critical threat to the planet today, and why we need to muster an emergency response. *Climate Peril* is the second in a series of three books. The first was *Climate Myths: The Campaign Against Climate Science* (Northbrae Books, 2013). The discussion in *Climate Peril* of the processes that cause climate change and its physical, biological, and economic impacts is, I believe, vital for understanding the need for the solutions proposed in the third book of this series, *Climate Solutions: Turning Climate Crisis Into Jobs, Prosperity, and a Sustainable Future* (forthcoming). That final volume provides an overview of the technologies available for mitigating climate change and the strategies plus tactics required to ensure that essential climate policies are implemented, despite even the most formidable obstacles.[a] Before proceeding further, a note about temperature: All temperatures in the US edition of *Climate Peril* are in Fahrenheit; information for converting temperature to Celsius is provided on page xx.

The Perils We Face

Tragically, it is now apparent that human activity has already triggered massive and rapid global climate change. The reason is clear: we have raised the concentration of heat-trapping gases in the atmosphere to levels not seen on the planet for millions of years. Chapter 3 explains how these gases serve as a planetary thermostat that influences its temperature and how long they will persist in the atmosphere.

If we continue business as usual and go on increasing those emissions, the Earth's average temperature is likely to rise by 7 to 11°F or more by 2100, with far greater heating on land than at sea, in the continental continental interiors, and in the Arctic.[b] Moreover, this transitory 2100 temperature response will ultimately be greatly exceeded over carbon dioxide's millennial

---

[a] The forerunner of the current three-book series is an earlier climate book of mine, *Beating the Heat: Why and How We Must Combat Global Warming* (Berkeley Hills Books, 2000). Where relevant, portions of *Beating the Heat* have been integrated into the current volumes, which greatly expand, extend, and update the earlier material.

[b] Some sophisticated climate models project that 4°C could be reached as early as 2060 if emissions continue their rapid increase.

atmospheric lifetime as Earth's temperature rises to a final equilibrium dictated by the enduring atmospheric presence of this, and other, long-lived greenhouse gases.

World-renowned climate scientist James Hansen, lately of NASA's Goddard Institute for Space Studies, warns that, "even 2°C [3.6°F] warming is a disaster scenario, as warm as the Pliocene a few million years ago, when sea level was 50 feet higher. It would lead to a different planet, not immediately, but with ongoing change out of humanity's control."[1]

An 11°F rise would be *three times hotter* than the 3.6°F of additional heating that Dr. Hansen warns would already be a "disaster scenario." Some experts, however, have projected that if we continue with our current carbon emissions policies, we are on track to reach 11°F not by 2100 but far sooner.

While we don't know exactly how much more global heating gases the atmosphere can hold before the world's average temperature rises beyond 3.6°F, we are apparently very close, if not already past, the point at which it might already be too late to avoid a 3.6°F increase[2] by 2100 or earlier.

The effects of overheating the planet would not be limited to extreme weather, droughts and widespread water shortages, along with inevitable famine. As explained in chapter 11, it would also upset the chemistry of the oceans with disastrous consequences for marine life and all who depend on it for food and livelihood. Meanwhile, people living by the ocean would face rapidly rising sea levels and flooding, leading to the abandonment of much of what we today consider our coastline.

Professor Corinne Le Quéré, lead scientist of the British Antarctic Survey, headed a recent study published in *Nature Geoscience* that chronicled the world's rapidly rising emissions. She warned that the billions of tons of carbon dioxide we are now discharging to the air each year may be gradually reducing the ocean's critically important ability to absorb excess carbon dioxide.[3] The ocean currently takes up about half of the carbon dioxide that human activity has discharged to the atmosphere. Therefore, if the ocean's ability to moderate the effects of additional carbon dioxide emissions is impaired, the climate will heat up even more quickly than projected.

With global temperature rising steeply, mass extinctions of both land and water species are inevitable. In fact, extinction rates are already at exceptionally high levels. As described in chapter 10, wonders of nature—some as yet unknown, many still little-studied and poorly understood—are vanishing from the planet forever. Entire regional ecosystems will be destroyed, along with the life-support functions they provide us.

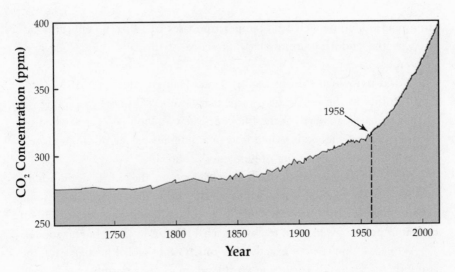

FIGURE P-1. The concentration of carbon dioxide (CO_2) in the atmosphere has been rising since the start of the Industrial Revolution. The increase began almost imperceptibly in the late eighteenth century but accelerated sharply in the twentieth century to reach its current peak near 400 parts per million (ppm). The collection of data from 1958 to the present was begun by climate researcher Charles David Keeling and is known as the Keeling Curve. By sampling the atmosphere at Hawaii's Mauna Loa Observatory on the flanks of the Mauna Loa volcano high above sea level, scientists are able to get accurate CO_2 levels in the east-west trade winds that blow across Hawaii, relatively unaffected by local "hot spots" of ground-level CO_2 contamination. The data prior to 1958 is derived from ice core analysis. Courtesy of Scripps Institute of Oceanography and The Keeling Curve website at the University of California, San Diego, http://keelingcurve.ucsd.edu.

As chapter 8 relates, in the hotter world that is developing, tropical diseases will spread northward from the tropics. Weeds and other invasive pest species will become more prevalent. Insufferable heat waves that prove fatal to the young, the old, the weak, and the infirm will become far more common.

These effects, compounded together, will eventually but inexorably translate into thirst, hunger, starvation, disease, displacement, despair, and death for hundreds of millions of people.[5] An intensification of global resource conflicts and heightened risk of major wars is also expected as swelling populations with increased resource demands encounter a shrinking, climate-devastated resource base. This has already been observed in places as diverse as Nigeria and Syria.[6,7] Survivors will lament that timely measures were not taken to head off the consequences of climate change that for decades loomed large but virtually neglected on our horizon.

It still *may* be possible to avoid the worst impacts of climate change by rapidly scaling back carbon dioxide and other emissions. We need to reach for

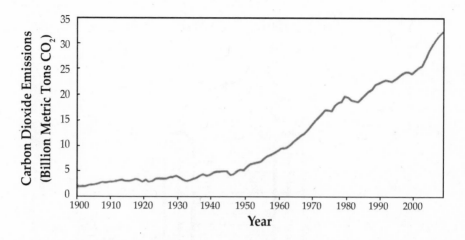

FIGURE P-2. The sixteen-fold growth in global carbon dioxide emissions shown above is largely responsible for the increased concentrations of carbon dioxide observed in the atmosphere from 1900 to 2000. (See preceding figure.) Courtesy of US EPA. Source: T.A. Boden et al., 2010.[4]

that possibility with all our might and deliberate speed by reducing carbon dioxide and other heat-trapping gas emissions as quickly as possible.

Fortunately, the peoples of the Earth do collectively have the knowledge, skill, wealth, and natural resources needed to shift the planet from carbon-based fuels to carbon-free (and carbon-neutral) renewable energy. Moreover, the climate-safe technologies the world needs are genuinely available, and the clean, safe energy flows they utilize—from the sun, wind, water, and Earth—are timeless and abundant.

Thus, from a technological standpoint, all the ingredients needed to protect the climate are at hand. But the educational, political, and economic obstacles are daunting and unprecedented. Underscoring that, the United States and other nations have so far lacked the will and political leadership to really launch this process.

Let's get started now to deepen our understanding of climate change's causes and effects. Armed with this knowledge, those who choose to do so will be in a better position to work for climate protection. As noted earlier, detailed recommendations and analysis about what should be done about climate change can be found in the final volume of this series, *Climate Solutions: Turning Global Crisis Into Jobs, Prosperity, and a Sustainable Future.*

Temperature Conversion Tool

$$°C = \frac{5}{9}(°F - 32)$$

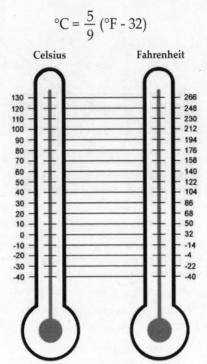

Equivalent Temperature Changes

° Celsius	° Fahrenheit
1	1.8
2	3.6
3	5.4
4	7.2
5	9
6	10.8
7	12.6
8	14.4
9	16.2
10	18

INTRODUCTION

PAUL R. EHRLICH AND ANNE H. EHRLICH

C LIMATE DISRUPTION IS the most widely discussed element of the "perfect storm" of environmental problems now confronting humanity.[a] And
well it might be. While other problems are ominous—such as the increasing
chance of vast epidemics, the global spread of toxic chemicals, or the accelerating extermination of the plants, animals, and microorganisms that support
us—climate disruption amplifies all these major environmental problems. For
example, it exposes people to more diseases by increasing the geographic
range of nasty tropical pathogens; it changes the distribution (and perhaps
increases the toxicity) of persistent organic pollutants, and it accelerates the
global rate of extinctions. That leads to the breakdown of natural ecosystems
and disruption of human food production and water supply systems.

Because climate disruption is caused by human-generated processes and
land use changes that civilization heavily depends upon, and because it has
such far-reaching effects and connections, climate change is almost certainly
the most complicated issue ever to confront the modern world. *Climate Peril*
is an excellent introduction to the complexities of both the causes and the
consequences. Dr. Berger explains the connections between industrialization,
habitat destruction, population growth, personal consumption, and climate
change. The book covers the wide range of climate change issues that we have
been concerned about for many years, especially the often overlooked interactions between human population growth and the climate system.

Consider that each person added to the planet generally will require
more resources, food, and energy and will emit more greenhouse gases
(GHGs) into the atmosphere than the last person. That's because people pick
the low-hanging fruit first, so newcomers on average must be fed from more
marginal land, use water transported further or purified more, and get metals

---

[a] http://www.theguardian.com/environment/2012/feb/20/climate-change-overconsumption

from poorer ores. The acquisition and manufacture of virtually every material object that person uses, every trip in an automobile or airplane, and every meal that he or she eats will lead to the release of additional carbon dioxide, worsening climate change. Our food system itself is responsible for roughly a quarter of all greenhouse gas emissions. Increasing food production will cause more greenhouse gases to be emitted, but they, in turn, will likely have a malign impact on human beings by reducing harvests. Atmospheric warming and increased adverse weather are already reducing crop yields in many parts of the world, and higher temperatures and CO_2 acidifying the oceans is endangering what's left of their bounty.

Typhoon Haiyan in 2013 was likely so huge and powerful because the roughly 1°C global temperature rise since industrialization made the western Pacific Ocean warmer, and warmer oceans make storms more intense. High population density and poverty increase people's vulnerability to extreme weather events generated by climate change. The high population density of the Philippines (321 people per km^2; nearly equal to Japan's) and its poverty (a ninth of Japan's per capita income) made the impact of Haiyan disproportionately worse than it might have been in another society. Large numbers of people were living in exposed areas, often in fragile wooden shacks that were blown away by high winds or washed away by storm surges. The result was thousands of deaths and hundreds of thousands left homeless.

Similar effects were seen when Hurricane Mitch struck Honduras and Nicaragua in 1998; more than 20,000 people died or were missing. Many if not most of them in those less developed nations were poor people living in exposed and hazardous situations, such as river valleys below steep overgrazed hillsides. Of course, as economic, technological, and social forces, along with demographic pressures, have caused more and more people to move from rural to urban situations, almost half of the human population lives in coastal areas.[b] And as John Berger explains, global warming will lead to sea-level rises that will force large numbers of coastal dwellers to become climate refugees in the future.

These issues do not impinge only on the poor citizens of developing nations. Climate change is currently affecting all of us in terms of the harsher weather and changing climates we are already experiencing. This is reflected in the higher costs we are paying, or will soon be paying, for food and insurance, and the taxes we will have to pay to cover the costs of damage

---

[b] http://coastalchallenges.com/2010/01/31/un-atlas-60-of-us-live-in-the-coastal-areas

relief and rebuilding in the wake of severe storms and floods, among other climate-related disasters. Citizens of developed nations are not immune to such calamities. As *Climate Peril* clearly demonstrates, taxpayers will have to pay for huge investments in infrastructure repair and new adaptive development to reduce vulnerability to future disasters. We will also have to live with the knowledge that we will not be leaving this planet in good condition for our sons and daughters, grandchildren and great-grandchildren, and that that will adversely affect the quality of their lives.

So read *Climate Peril* to become well informed about what probably is the greatest threat ever faced by civilization. By the time you are done, perhaps you will agree with us that the only long-term solution is not only to curb greenhouse emissions and find alternative ways to provide energy and produce food and other goods, but also to deal with the rapidly rising global population and the desperate poverty that are making both climate disruption and resource destruction ever worse. In addition to making family planning and health services readily available to all throughout the world and improving girls' access to education in developing nations, the solution to the human dilemma requires that equity issues also be addressed. We must find ways to reduce wasteful consumption by the rich while increasing needed consumption by the poor. Both people and the environment must be treated justly and ethically if we are to create the conditions for a safe climate and a sustainable civilization. People must recognize that the human enterprise must ultimately be scaled to fit the resources of the planet rather than being allowed to destroy both the planet's climate and its resource base.

Paul R. Ehrlich is the Bing Professor of Population Studies and president of the Center for Conservation Biology at Stanford University. Anne E. Ehrlich is senior research scientist and associate director for policy of the Center for Conservation Biology at Stanford University and coauthor of several books on population and environmental issues. For more information, see pages 318–319.

CHAPTER I

Global Climate, 2100 AD

*First, I worry about climate change. It's the only thing
that I believe has the power to fundamentally end the march
of civilization as we know it, and make a lot of the other
efforts that we're making irrelevant and impossible.*

—PRESIDENT BILL CLINTON, WORLD ECONOMIC FORUM AT
DAVOS, SWITZERLAND (JANUARY 31, 2006)

This chapter depicts the dangerous and often terrifying ways in which
a new and hotter climate will manifest itself, from North America to
Africa, Asia, and the islands of Oceania.

Where We're Heading, Once and for All

PREPARE YOURSELF FOR SOME BAD NEWS. I promise I'll try to make it
brief. I'm going to take you on a whirlwind tour to see what our world
will be like in less than 90 years, if present climate abuse continues and greenhouse gas emissions thus continue their current growth. This scenario not
only could really happen; in some respects it has already begun, and we are
now in its early stages.

Some may view this forecast as alarmist, because the temperature increases
that underlie it are toward the high end of the range of those expected, but as
you will learn later in the book (chapters 4 and 6), some temperature projections are even higher, and the climate impacts may well be compounded by
rapid global population growth as well as by the human health effects, wars,
and other social conflicts discussed in chapter 8.

So your life may still seem normal now, or not, but rapid climate change is already altering the world in ways that are truly alarming. Long before 2100 AD, it will profoundly affect our health, our homes, our businesses, and our farms, as well as our water, power, and transportation systems. Calling attention to the real dangers now is both a moral obligation and an essential part of the struggle to avert these consequences (not to be confused with alarmism).

The impacts of climate channge are not limited to easily recognized extreme weather events, such as floods, hurricanes, tornadoes, or droughts—deadly and expensive as those events are. The impacts also include other dire consequences:

- Heat waves
- Dying forests
- Abnormally large wildfires
- Habitat destruction
- Accelerating rates of extinction
- Altered seasons and disruption of normal seasonal ecological relationships
- Invasive species encroaching deeper into once-intact ecosystems
- Lethal diseases fanning out from the tropics
- Island nations about to be obliterated
- Disappearing sea ice and glaciers
- Rising seas
- Acidifying oceans
- Declining ocean plankton
- Melting permafrost and Arctic wetlands

These phenomena are undeniable, although their causes are still disputed by climate science deniers. During our imaginary "visit to the future," we will glimpse how these dangerous and seemingly isolated climate changes collectively threaten our future welfare and survival. By the time you complete this book, I hope you will understand why and how all these effects are indeed a result of "manmade" climate change.

Current climate change trends reveal that the world is almost certainly going to surpass an average warming of 3.6°F (2°C), probably in about 40 years or so, on its way to *much* higher temperatures. While at first 3.6°F may not sound like much, it is nonetheless about two and a half times the warming that the Earth has already experienced since preindustrial times.[a] For the past two decades, this 3.6°F has been the amount of warming that most scientists and policymakers had regarded as the upper safe boundary between acceptable and dangerous climate change.

In a colossal global policy failure, however, the leading nations of the world over those same 20 years have failed in many rounds of international negotiations to stabilize or bring down global carbon dioxide emissions. Now, however, over the past decade, astonishing and alarming global climate changes have already begun in response to only 1.4°F of average global surface temperature warming. Thus, 3.6°F of warming—rather than a safety threshold—is a nebulous transition zone between highly dangerous and extremely dangerous climate change.[1]

Carbon Dioxide, a Tenacious Gas

To those who are new to climate studies, it may come as a shock that the major climate change that has now begun is largely irreversible with current technology. Contrary to popular belief, we cannot just "overshoot" a safe temperature and then, as if with a magic global thermostat, somehow turn the heat down to normal again. Once we "overshoot," we are in effect stranded at the new equilibrium temperature. The climate would remain overheated because of the long-lived heat-trapping gases with which we have overloaded the atmosphere.

Contrary to common popular belief, carbon dioxide, the principal long-lived heat-trapping gas, does not disappear within hundreds of years, which would be bad enough. In fact, a substantial portion of the carbon dioxide remains in the atmosphere for thousands of years.[2] Through our emissions from industrial pollution, transportation, agriculture, and deforestation, we have in effect created a new world atmosphere. It will take millennia for Earth's natural carbon-removal processes to fully reabsorb those gases from the atmosphere.

Only if we could extract the extra heat-trapping gas from the air could we restore a semblance of our previous climate. However, we would have to extract a great deal (hundreds of billions of tons), and the climate would

---

[a] The onset of the modern industrial era is generally viewed as coinciding with the start of the Industrial Revolution in England at about 1760. (It was marked by a shift from the use of wood to greater reliance on coal and by the use of machinery in manufacturing, as well as by improvements in the efficiency of steam engines, and by technological advances in important industries.)

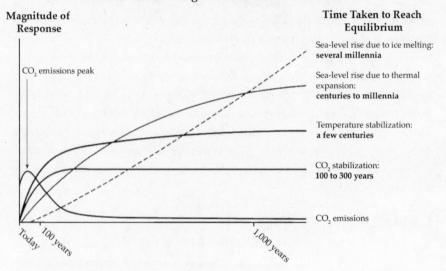

CO₂ Concentration, Temperature, and Sea Level
Continue to Rise Long After Emissions Are Reduced

FIGURE 1-1. This diagram illustrates that even after carbon dioxide emissions peak and begin declining, carbon dioxide concentrations, global temperatures, and sea level all continue rising toward new elevated equilibrium conditions over periods ranging from centuries to millennia. Note that although carbon dioxide emissions fall quickly and steeply on the diagram, the excess carbon dioxide humanity added to the atmosphere remains there and thus temperature remains elevated instead of returning to normal preindustrial levels. Source: Taroh Matsunos et al., 2012. © 2012 The Japan Academy.[3]

respond very slowly.[b] But unfortunately, we currently have no quick, affordable way to scrub vast amounts of carbon out of the atmosphere. So the really bad news is that, because of the properties of long-lived global heating gases, whatever peak temperature we reach as a result of having added them to the atmosphere will be Earth's new temperature for a long, long time. That makes it imperative for us to reduce our emissions as much as possible and as quickly as possible. The longer we delay this process, the larger the emission reductions we'll have to make in the future to achieve the same result.

Now you understand the severity and finality of climate change. It has no quick fix. A realistic understanding of the problem is necessary, however, if we are to limit climate damage. Realism is preferable to illusions and false optimism.

The good news is that knowledge about the risks created by our present climate policies, or more properly, by their inadequacy, is spreading rapidly

---

[b] See www.trillionthton.org for current cumulative human-induced releases of carbon dioxide.

throughout the world. If enough people understand them, there is a chance that we may yet adopt climate policies that bring prompt, deep reductions in emissions.

The technology to constrain emissions is definitely here, and the world definitely has the resources to tackle the challenge. Many cities and states, for example, have already set ambitious carbon dioxide reduction targets, and some regions have set up multistate and international carbon exchanges. The Federal government is tightening energy efficiency standards and fuel efficiency standards for vehicles while moving to regulate carbon pollution from power plants for the first time.[c] To see why we must do far more, however, join me now for an imaginary journey into the future to see the havoc that will be produced by the current laissez-faire global climate policy. In the absence of an enforceable international climate treaty with a commonly agreed-upon global emissions reduction target, each nation will decide for itself how much carbon pollution it wants to release.

2100 AD

Suppose today is August 1, 2100. The world's average temperature has risen more than 10°F.[4] That's an awful lot. Some climate models project that, with business as usual (that is, with no meaningful international effort to curb the rise in global emissions), temperature might be up only 7°F by 2100. But that would be extremely damaging, too.

Because most of the planet is covered by oceans, which on the average are cooler than the land, when the world's average temperature goes up 1°F, the average temperature over land will rise about 1.5°F in the early stages of warming. Toward the interior of continents, it will go up about 2°F.

The tremendous 10°F average planetary temperature increase, or the very large temperature increase of 7°F, would also be magnified two or three times in the Arctic. Temperatures in the high Arctic, including parts of Alaska, Canada, and Greenland, might thus have risen as much as 14 to 20°F. Moreover, these temperatures would not be final but transitional (as will be further explained in chapter 3) on their way to even higher temperatures. Again, that's because of the long-lived atmospheric residence time of carbon dioxide and some other greenhouse gases.

---

[c] See President Obama's climate change speech at Georgetown University on June 26, 2013, and the Administration's June 2013 Climate Action Plan.

The Great Melting

Let's now consider climate changes that we could see by 2100 if the warming continues to accelerate as forecast in response to continued increases in carbon pollution. The changes may initially seem remote to your daily life if you live in a city in the temperate zone, but as we proceed from considering the remote polar regions of the Earth to the middle latitudes, you will see how these changes will adversely impact us all.

By 2100, the Arctic Ocean is virtually ice-free. That amplifies global warming, because the reflective ice is replaced by darker water, which absorbs more heat. Without the ice, walruses are virtually gone. The huge shoals of shrimp-like krill that swarmed and bred beneath the margins of the ice shelves are gone. The whales that strained tons of krill have starved. So have the krill-eating fish and seals that had eaten the fish—and that needed ice during breeding and pupping season. Polar bears that depended on the seals and that bore their young on ice have become very scarce. A small population remains on land where they interbred with grizzlies.

FIGURE 1-2. Polar bears on a melting ice floe in Canada's Beaufort Sea in August 2004. Credit: Dan Crosbie, Environment Canada.

The Marine Food Web

Floating summer sea ice is coated underneath by a biologically important layer of marine algae. Tiny floating aquatic organisms called zooplankton feed on the algae. But with the ice gone, they have now lost an important food supply. The summer sea ice also shielded the water from the mixing effects of wind. It thus had protected a layer of warmer surface water that had nurtured microscopic floating plants known as phytoplankton.

These unpresumptuous little organisms didn't get much attention back in 2013. But as they formed the base of the ocean's food web, they were essential to the whole marine food chain. When the ice and its coating of marine algae disappeared, the spring phytoplankton bloom was delayed. That led to a scarcity of zooplankton. In turn, that led to a shortage or disappearance of the young krill and other plankton feeders, such as juvenile marine organisms, including baby fish.

Thus, with the collapse of the Arctic sea ice, the Arctic ecosystem went into a tailspin. The seemingly boundless fisheries became a distant memory. Arctic commercial fishing ended. The Inuit and other native subsistence hunters and fishermen had to give up their traditional lifestyles.

On the other end of the world, the planet's warming affected Antarctica. It had not only melted the floating Arctic sea ice, but as sea level rose, it had elevated the ice shelves attached to the shore, while warmer ocean currents melted away at their base. The hinge-like connections between the ice shelves and the shore weakened and then ... but I'm getting ahead of the story.

Back in 2010, the Antarctic Peninsula had already warmed 4.5°F in only 60 years—more than three times the Earth's average warming.[5] Thus it was far from surprising when, with continued warming of the air and undersea currents, much of the great Ross Ice Shelf—an area the size of France—collapsed in the late twenty-first century.

Although loss of the Ross Ice Shelf didn't raise sea level (its weight was already supported by the sea, which displaced an equivalent mass of water), the Ross Ice Shelf nonetheless had served as an ice dam. The dam had helped to hold the Antarctic Ice Cap and its glaciers firmly in place on land. Once the ice shelf collapsed, the flow of ice from Antarctica accelerated. This icy new infusion did raise sea level.

The Responsive Ocean

The trend was ominous because Antarctica contains about nine times as much ice as Greenland. The last time that the Earth had warmed by over

3.6°F—during the Eemian interglacial period, which ended 114,000 years ago—enough ice had melted to lift the sea 13 to 20 feet (4 to 6 meters) higher than at present.[6]

The Greenland Ice Sheet, meanwhile, had not been stable between 2000 and 2100.[d] In 2000 AD, Greenland was losing about 180 billion tons of ice a year.[7] The melting was adding nearly three hundredths of an inch a year to sea level by 2013. Over the next 87 years, however, the loss accelerated greatly. Thousands of billions of tons of Greenland ice melted and cascaded into the ocean. Along with contributions from the West Antarctic Peninsula, further buoyed by the expansion of warmed seawater, average sea level had thus risen four feet by 2100.[8] The elevated oceans were now also rising very quickly—almost half a foot every ten years.[9,10]

Sea-level rise is not uniform everywhere. By 2100, some coastal land had subsided, due to the effects of the continuous extraction of groundwater for irrigation.[11] The seas along these coasts were thus even higher, relative to the land, than the global average. The overall resulting net average rise in sea level by 2100 had a catastrophic impact on coastal and near-coastal areas around the world. Simultaneously, the infusion of freshwater from melted ice reduced the ocean's salt content and seawater density.

These changes in seawater chemistry were compounded by a major increase in the concentration of dissolved carbon dioxide in the ocean, increasing an acidification of seawater that had begun in the twentieth century. Whereas the population of the ocean's pervasive floating algae (phytoplankton) had fallen by 40 percent in the late twentieth century, it was now experiencing an even more serious crash. (See chapter 11.) No one really knew at what point their population would stabilize, or how their species composition would change across the ocean in response to the new conditions.

Since, as noted earlier, plankton populations form the base of the ocean's food web, their decline and population disturbance directly jeopardized the ocean's productivity. All plankton feeders were thus adversely affected. Simultaneously, the water's increased acidity interfered with the ability of many plankton feeders, such as oysters, clams, mussels, and crustaceans (like lobsters and shrimp), to accumulate enough calcium to form their shells or, in the case of coral, their skeletons. Thus, the Dungeness crab—once hauled by the millions from the Pacific Ocean of North America—the blue crabs of the Chesapeake

---

[d] For a current report on the extent of the melting of the Greenland Ice Sheet, see http://nsidc.org/greenland-today.

Bay, and the lobsters of Maine all had now vanished or were rare. Restaurants and seafood consumers noticed. About a billion people worldwide depend directly or indirectly on seafood as an important source of protein, and many millions more depend on seafood or products for their livelihoods.

The same salinity and density changes that devastated plankton and the ocean's food chain also slowed the vast system of currents across the oceans known as the oceanic conveyor belt circulation. That slowdown then reduced the northward transport of heat from the equator by the Gulf Stream. So paradoxically, as the world warmed, Western Europe's climate began to more closely resemble Iceland's. Its agriculture suffered. People noticed that, too, but it was too late.

The Great Parching

In the vast permafrost region of Canada in 2100, scrawny trees tilted at crazy angles in the soft earth. The permafrost that used to underlie their relatively shallow root systems had melted. Some of the northern forests were dry. Some had burned many years ago during exceptionally hot wildfires and were now dead. Others looked tired and sickly. In the highest latitudes, however, more rain and snow had fallen as temperatures rose.

At lower latitudes, most of North and South America, including the entire Amazon Basin and Brazil, however, had become much, much drier. Throughout the world, dry areas in general had become drier. In the United States, droughts had become more frequent in the Midwest, Southwest, and heavily populated parts of the East.[12] Heat waves had become more frequent, prolonged, and ferociously hot. Some areas where a heat wave previously meant a temperature of 100 to 102°F now had heat waves in which the mercury hit 114°F or above.

In the southern and blistering southwestern United States, important rivers had simply vanished in the intense summer heat, unable to keep pace with greater evaporation and ever-increasing demands from farms, suburbs, and cities. Backup water supplies over-pumped from the ground had by now failed. Long before the wells went dry, however, groundwater tables had dropped far beneath the root zones for many plants, creating patches of desert within arid landscapes or expanding existing deserts.

Drought, Food, and Hunger

Desperate municipalities, meanwhile, had applied relentless pressure on agricultural water users to give up their scant irrigation water supplies.

Agriculture in early twenty-first-century America had typically consumed about 90 percent of all water used. Agricultural counties in 2100 regretted not having set up groundwater management districts in the early twenty-first century to apportion their scarce water resources in a more rational manner. Instead, the famous "Tragedy of the Commons," described by twentieth-century ecologist Garrett Hardin, had come to pass.[13] Each groundwater user had basically grabbed for as much water access as he could. That seemed to maximize each person's share, short-term, but hastened the demise of the unmanaged common groundwater basins. So everyone lost out.

With water so scarce and costly, many farms over the past decades had first fallowed their fields and, when the rains failed, had finally gone out of business. Then food prices had shot up. Farm workers had been thrown out of work. Manufacturers and farm equipment dealers had had to cut back. At the same time, businesses built around commerce in feed, seed, fertilizer, agricultural chemicals, and crops had all withered. Real estate prices had slumped. Many people simply left the region. Farm economies unraveled.

As portions of the South had become too hot or dry for corn and soybeans, the remaining farmers had shifted to crops that could still survive. But pastures and rangeland dried out. A lot of livestock died. The United States was now forced to bid for corn and soybeans on world markets to keep food affordable. But importing food had become difficult. Few nations had crop surpluses. Climate change had made the global climate more variable and thus harder for farmers to anticipate. Moreover, when enough rain fell to produce crops, the heat reduced yields.

China, India, and other ever-hungrier and more desperate nations bid heavily against the United States for food. World grain prices skyrocketed. Malnutrition increased in hard-hit, drought-stricken areas, including southern and central Africa, Southeast Asia, Australia, and southern Europe. Due to the high food prices, water shortages, and widespread poverty, food and water riots were common. Governments now rose or fell in tandem with their ability to keep people fed and quench their thirst. Roughly 30 percent of the world's land area was afflicted with some degree of drought at any given time in 2100. Livestock and chickens that once gorged on cheap corn, sorghum, and soy had become luxuries. The price of milk and baked goods, too, was far beyond the means of many low-income people.

In the United States, drier summers and scarce runoff in the Great Lakes Basin had brought water levels down. Sunken deeply, the Great Lakes were discolored by greenish-yellow algal blooms. Inflow to the St. Lawrence

Seaway had dwindled, making navigation difficult. The twentieth century's ambitious and expensive efforts to restore the Great Lakes had been abandoned years earlier because of droughts and because the nation was coping with even more serious problems.

Urban Defense

Tremendous changes had occurred along the coastlines of Connecticut, Delaware, the District of Columbia, Maryland, New York, New Jersey, Virginia, and Rhode Island. By 2100, massive new seawalls and breakwaters had sprouted around many urban centers in that northeastern corridor. Long convoys of large trucks were still supplying these huge construction sites on a daily basis with many tons of rock and cement. Many less important populated areas had simply been abandoned to the ocean and were now under water.

Superstorm Sandy in 2012 proved to be a foretaste of the large hurricanes that came with increasing frequency in the decades that followed.[e] With average sea level elevated by more than three feet,[f] 3 percent of Boston was below sea level along with 7 percent of both New York City and Jacksonville, Florida.[14] Nine percent of Norfolk, Virginia—subsiding anyway[g]—was now under water.[15] The situation was much worse in Florida where 15 percent of Tampa and 18 percent of Miami were submerged.[16]

Miami's plight was complicated by the fact that the city reposes on porous carbonate rock. As the Atlantic Ocean rose ever-higher, the carbonate filled with seawater below ground like a sponge. The ocean then bubbled up through basements, streets, and parks.[17] It was, of course, impossible and prohibitively expensive to entirely protect all these areas from the ocean.[h] The tip of Florida, for example, was entirely submerged. More than nine-tenths of New Orleans was below sea level.

---

[e] Hurricane Sandy, in the fall of 2012, took more than a hundred American lives, destroyed homes and businesses, crippled transportation systems, caused tens of billions of dollars in damage, and brought power outages and fuel shortages to millions. The storm was remarkable for its size, extending from the Carolinas to Maine and from New York to Michigan.

[f] Some experts project five feet or more (see reference 12).

[g] It sits on the impact crater that formed Chesapeake Bay 35 million years ago.

[h] To eliminate storm surges, the wall would need to be close to 20 feet high, and it would cut people off from sight of and access to the ocean, something few residents would countenance.

Blemishes on the Big Apple

In 2012, New York was one of the world's wealthiest cities, a showcase of urban luxury and sophistication, an entertainment and media powerhouse, a center of global finance and trade. Thus, the city had a lot to lose from sea-level rise, including two trillion dollars in coastal property.[18] It tried its best to fend off the sea.

For decades, despite considerable seawall construction, parts of lower Manhattan and the financial district were repeatedly subjected to major flooding by a series of twnety-first-century hurricanes and floods. Every two years or so it seemed the city would experience severe flooding from what used to be considered "storms of the century."[19]

It wasn't the storms themselves that caused the most havoc, but the towering storm surges that often accompanied them. During Superstorm Sandy in 2012, a storm surge of 14 feet had hit Staten Island. Roughly a century later, the storms were even more powerful and frequent and were rising above greatly elevated seas.

Tunnels for subways and roads connecting Manhattan with Queens, Brooklyn, and New Jersey had proven very vulnerable to flooding. The subways had suffered more than $5 billion in damages from the millions of gallons of corrosive seawater that flooded the system during Superstorm Sandy. Billions more were required in an effort to keep pace with continuing damage from subsequent storms and to stormproof the system. Additional billions were needed to protect low-lying John F. Kennedy International and LaGuardia airports, parts of which had to be abandoned due to flooding.

When mass transit periodically had had to shut down in New York in the mid-twenty-first century, millions were unable to get to work or find fresh food. Many were not able to reach hospitals. Power shortages had complicated the situation. Certain low-income neighborhoods got more than their share of storm surges and transit cutoffs. They deteriorated.

It was gradual at first. Anyone able to leave did so. Finally, with businesses and residents moving out, the remnant neighborhoods degenerated into squalid, impoverished no-man's lands, plagued with crime, decay, and unemployment, sweltering in humid summer temperatures that rarely dipped beneath 100°F for weeks on end. Heavily armed National Guard units were called out frequently, not just to provide disaster relief but to patrol the streets of blighted neighborhoods.

Unfortunately, not all the East Coast's vulnerable fuel depots, chemical plants, nuclear power plants, toxic waste dumps, and landfills had been successfully protected or moved from the coast. So it was in the New York–New Jersey area. With the US economy struggling, industrial facilities in New Jersey and elsewhere were unable to cope with the multibillion dollar costs of decommissioning and relocating or protecting these facilities. Leaks of toxic chemicals at some of these sites became a chronic problem as the seas rose. Caustic and carcinogenic fumes wafted on the wind, and liquids leaked into floodwaters and seeped into the soil.

Finally, in 2070, New York completed its flood-protection system. Years in planning and vastly over budget, the five-mile-long massive outer harbor barrier had huge movable gates that could permit navigation and tidal action when opened but could be shut against onrushing storm surges. The massive barriers did a lot to moderate large storm surges, but some areas still flooded and ghost factories still dotted the shoreline.

Similar scenarios played out in other coastal cities throughout the nation, exhausting their disaster relief funds and emergency response capabilities. Few urban areas, however, could equal the financial resources that New York had mustered to cope with the ocean. Many low-lying urban regions of the United States in 2100 therefore resembled "Third World" countries. The inhospitable new climate was thus not just a stress on natural systems, like forests and wetlands, but also an enemy of social justice, public order, and human progress. Climate change became the consummate destroyer of human hope for a better future.

Of course, the degradation of New York and other great American cities had heavily burdened the US economy. The nation was beset by climate-related woes and was no longer on the path to greater prosperity. Instead, it had grown preoccupied with warding off a climate-related economic contraction.

The national debt had been contracting in the 2050s but, by the later part of the century, it had swollen to unprecedented levels as the nation reeled from climate-related disasters and the ensuing fiscal strain. Simultaneously, the United States was at last belatedly spending heavily on new energy and transportation infrastructure while also trying to extend economic aid to climate-battered developing nations. These "aid" payments, of course, were not exactly altruistic. They resembled climate ransoms paid to recipient nations in exchange for their cooperation in reducing their carbon pollution of the global atmosphere.

Thus, instead of economic optimism and global prosperity, the economic pundits of 2100 talked endlessly and apprehensively about the world's economic future. Their concerns were understandable. Every day the news brought stories of increasing global strife, hunger, instability, and worsening ecological collapse on an overpopulated planet where billions depended precariously at best on a shrinking natural resource base.

Rural Shores

On the outskirts of a typical small town on Nantucket Sound—which had not been protected from the ocean due to the expense—an old road led straight into the water in 2100. Waves covered what used to be the beach, nearby fields, and small near-shore freshwater ponds.

Skeletal remains of shorefront buildings and walls protruded from the surf. A congealed mass of plastic flotsam and jetsam identified the high-water mark. (With a global population of more than 13 billion still using disposable plastic containers and packaging, ocean pollution had by this time reached crisis proportions. But the world now had more pressing concerns than enforcing recycling regulations.)

Over long stretches of the rural New England and Middle Atlantic coast, new high-water marks attested to savage storm surges that had pounded inland across roads and breakwaters far from what was once the coast. Many coastal roads were now elevated onto a system of dikes to withstand further flooding. Utility and rail lines had also been reconstructed farther inland away from treacherous seas. Power poles and pylons leaned crookedly, the ground beneath them waterlogged by intruding seawater.

Wetlands, Bays, and Barrier Isles

At least half of the nation's coastal wetlands had also been flooded by the rising seas. Normally, when seas rise slowly over long time periods, wetland plants in a natural intertidal habitat simply spread inland onto higher ground as their original habitat is drowned by the sea. But in urban areas, roads, levees, and landfills fence wetlands in and trap them in place. The wetlands therefore had shrunk or disappeared beneath the waves.

The wetland loss in turn had helped devastate the fishing industry. More than half of the commercial fish species use wetlands during their life cycle. Fish populations had plummeted by 2100 because of the wetland loss compounded by overfishing and the climate-driven changes in the ocean's salinity, acidity, temperature, and currents.

In Chesapeake Bay, one of America's most important estuaries, the water in 2100 was warm and murky with sediment and algae. Chesapeake Bay oysters, famous for centuries, had dwindled in number due to the dirty, acidic water, and oyster farms had closed. In towns and farms along the bay shores, freshwater wells had been ruined by rising seas. Urban water intake pipes had been moved farther up the Potomac River. The bay's smallest islands were totally gone; larger ones had shrunk.

Farther south, waves lapped where islands once buffered Virginia from heavy seas. The Outer Banks of North Carolina, including the Cape Hatteras barrier, were virtually gone. Albemarle-Pamlico Sound was exposed to the full force of Atlantic storms that once would have spent their fury on the lost barrier beaches. North Carolina by 2100 had lost more than 1,000 square miles to the sea. Whereas wars historically have broken out when one nation has taken even a few miles of land from another, the United States by the twenty-first century had given up thousands of square miles to the sea without putting up a fight—until it was all too late.

The map of the southeastern United States had been redrawn. Thousands of square miles of once-dry coastal land are under saltwater here by 2100, including the famous multibillion dollar Miami shoreline, the Florida Keys with its crystalline coves, and much of Everglades National Park.

Malaria, yellow fever, and bonebreak (dengue) fever are now prevalent in Florida as pesticide-resistant, disease-carrying mosquito populations thrive in the warmer weather. Mosquitoes that transmit malaria survive only where winter temperatures exceed 61°F. Chillier temperatures had kept them in check during the twentieth century. During the twenty-first century, however, the area of potential infection had expanded with rising temperatures, spreading the disease north and to higher elevations. Tourists and retirees who could afford to leave shunned the infested areas of Florida.

The Delta and Gulf

By 2100, much of the low-lying Mississippi Delta area and parts of Louisiana were also flooded. Many coastal pipelines and much other important infrastructure were underwater. The commercially important spotted sea trout, oyster larvae, and flounder had lost most of their habitat. The commercial fish and shellfish industries not damaged by Gulf oil spills were gone. Few edible fish were left to catch, and many of those contained dangerous toxic chemicals.

Flooding had become a huge problem everywhere in bayou country by 2100. Following an incomplete recovery after Hurricane Katrina in 2005, New Orleans, which was then six and a half feet below sea level, sank another several feet due to subsidence and sea-level rise. With so many localities all begging for help at once, federal flood disaster relief funds were soon over-committed. Federally subsidized flood insurance was slashed. A strange pervasive numbness—a fatalistic kind of national "compassion fatigue"—seemed to set in on those not directly affected by each new disaster.

On the Gulf Coast of Texas, Galveston spent billions reinforcing its seawall. Sunken pipelines and abandoned oil pumping stations nonetheless remained visible from the air. Here and there, greenish coppery slicks coated the coastal water above leaking submerged landfills. As the seas had risen along the Texas Gulf Coast, much of the wetlands and wetland-dependent birds, fish, and shellfish had disappeared here, too. The state's once economically important brown shrimp catch was devastated. Just as on the Louisiana Coast, most fishing people here lost their livelihoods.

Scorched Earth and Sweltering Valleys

It was sad to contemplate how much the country had changed west of the Mississippi by 2100. Ever since 2030, vast wildfires on drought-stricken land in the Southwest had blackened large swathes of Arizona, New Mexico, and Colorado, destroying thousands of homes and killing unlucky residents and firefighters. And instead of a summer fire season, many parts of California now had a year-round fire season.

On the day that our time traveler entered California in 2100, a couple of large wildfires were still actively burning out of control in brushy hill areas east of the populated southern coast. Thick ash-laden grey smoke from these fires had hung for days in low valleys as the blazes consumed vegetation and homes built along bone-dry canyons. People with asthma or chronic obstructive pulmonary diseases, such as bronchitis and emphysema, were seeking help with their breathing at overcrowded local emergency rooms.

The San Francisco area—not known for high temperatures—on the average had had only 12 days of extreme heat in the early twenty-first century. Most residents had not even owned air conditioners. However, with the 10°F average global increase in temperature that had occurred by 2100, even once-temperate San Francisco was having 70 to 94 days of extreme heat per year.

In hotter areas of the state, temperatures approached 120°F in bad heat waves. Surges in heat-related illnesses and deaths would ensue, particularly among the very young, the elderly, and those suffering from heart disease or other chronic ailments.

Whereas the state had added tens of millions of people during the twenty-first century, it was nonetheless having to make do with a lot less water. Miles of once-productive farmland in California's Imperial and Coachella Valleys lay hot, dry, and fallow. Parched by decades of drought, the Colorado River basin was no longer able to provide the one million acre-feet of water it used to send to California's Metropolitan Water District. Stringent water rationing was now common during the state's very dry summer and fall.

While California's winter runoff had increased by 2100—charged by heavy winter rains—the winter snowpack had greatly diminished and melted earlier each year. Spring and summer runoff therefore had fallen to critically low levels. Although reservoirs in the state had been enlarged in multibillion dollar retrofits, they still weren't able to hold enough of the larger winter flow to make up for the summer shortages. Water was thus scarcest in California when farms needed it most for irrigation. Wild California salmon populations and other cold-water species that needed high spring flows had long since been decimated by the chronic low flows and the warmer temperatures.

Along the Pacific Coast, the medium-density residential developments built over coastal sand dunes right up to the beaches had been destroyed by the rising seas and storm surges. Expensive property near San Francisco Bay was also flooded by rising seas and intense winter storms.

Snowless Mountains and Failing Dikes

The Sacramento-San Joaquin River Delta—the largest freshwater estuary on the West Coast in 2013—had provided drinking water to two-thirds of California.[20] It was thus crucial to the state's water supply system. However, parts of the Delta that were already some 25 feet below sea level behind clay dikes did not fare well with the passage of the twenty-first century.[21]

The Delta's aging, unstable system of fragile dikes, weakened by storm surges and saltwater intrusion, failed after a major earthquake on the San Andreas Fault in 2085. Some of the delta's freshwater marshes turned salty as seawater pressed inexorably inland. Submerged roads and bridges throughout the Delta were still visible in 2100. Much of the freshwater supplies once stored by the Delta were gone.[22] Some of the area's water conveyance structures had also failed in the quake.

In the Pacific Northwest, because the diminished mountain snowpack had melted earlier each year, little water was available in the summer and fall of 2100 to produce hydropower, on which much of the region's economy depended. Stream flows in many waterways also often were too low for fish; as in California, the region's magnificent wild salmon stocks were nearly extinct. On Mount Rainier, where the penstemon, primrose, and white heather used to grow, trees had replaced the splendid alpine wild-flower meadows.

Asia

Across the Pacific, the rapid climate change of the past century had spread disease in many areas. Malaria now threatened 60 percent of the world's residents, many of whom lived in Asia. Hundreds of millions of people were infected each year, and millions were dying. More Asian people were also falling victim to bonebreak fever, river blindness, encephalitis, cholera, yellow fever, and waterborne intestinal illnesses. Rising temperatures, which speed up spoilage, had also caused increases in food poisoning from salmonella in contaminated food.

In the Pacific Ocean, the nations of Kiribati, Tuvalu, and the Marshall Islands had disappeared into the sea. The same thing had happened to other "stepping stones" across the Pacific. Some of the larger Solomon Islands had survived, but were now smaller and partially evacuated.

Most of the world's tropical coral had died or was dying by 2100, due to a combination of higher ocean temperatures, severe storms, freshwater and sediment from heavy downpours, and contamination. The once-spectacular coral reef between the Kyushu and the Ryukyu Islands, with its brilliantly colored fishes and anemone, was now desolate, bleached an unnatural white.

The Yangtze Flood

A colossal flood on the Yangtze River had left 250 million Chinese homeless in 2091. Massive refugee camps stretched as far as the eye could see along the edges of the enormous flood zone. Although some 220 million people returned to the floodplain after the waters receded, 30 million had nothing to go back to. It was as if a mass of people equal to 70 percent of the US population in 2010 had been driven from their homes and then a population larger than that of Texas in 2010 had been left destitute indefinitely. Ironically, despite the flooding, the north of China was at the same time parched by extraordinarily dry weather.

Between 2070 and 2095, several million people were also displaced by sea-level rise and flooding in coastal areas of Myanmar, Thailand, and Vietnam, as well as Indonesia, the Philippines, and Malaysia. During the same interval, two-thirds of Bangladesh went underwater from an extraordinarily severe monsoon that halted its already impoverished economy.

Remembering Africa

In the twentieth century, Africa was not exactly a model of fine resource stewardship. The continent's tropical forests were being rapidly destroyed. Wildlife was being brutally exterminated. Species were going extinct. Water resources were being depleted. Soils were eroding, and grasslands were turning into deserts. The carnage had continued and in some places had intensified during the twenty-first century.

Africa's population in the twenty-first century had continued expanding faster than its economic growth, while its natural resources deteriorated. Even in 2012, tens of millions of Africans had been living in shacks and sheds with few comforts. By 2100, Africans had even fewer natural resources left per capita to support themselves. Wars and lawlessness over large areas had created millions of miserable refugees. Meanwhile, the continent had remained largely dependent on rain-fed agriculture to raise food for domestic consumption and export, while drought had stalked through the countryside. Needless to say, this was not a great business model. Millions of starving farmers were forced to leave their dying herds and drying fields in search of food and water in the cities.

Political instability still plagued Egypt. The situation had worsened when a large part of the Nile Delta was lost to rising seas in the twenty-first century through the combined effects of flooding and erosion. Rising sea levels actually had delivered Egypt a two-fisted blow, reducing economic production while creating a crush of environmental refugees. Meanwhile, food prices had risen sharply.

Large numbers of impoverished Egyptians—no one even knew exactly how many—were suffering from hunger, malnutrition, preventable disease, and, yes, starvation in the twenty-first century. Revenue from tourism, an important part of the twentieth century Egyptian economy, had virtually disappeared in response to the widespread misery and menacing instability.

Along West Africa's coast, few nations had the money to defend themselves against the sea, though many countries had large, low-lying coastal cities. Severe flooding and heavy economic and human losses were the norm in these cities by 2100. And with less rainfall, crop failures and famine were even more frequent than in the twentieth century.

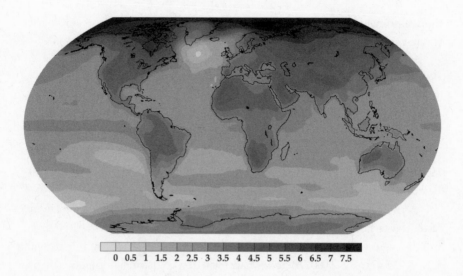

0 0.5 1 1.5 2 2.5 3 3.5 4 4.5 5 5.5 6 6.5 7 7.5

FIGURE 1-3. Projected surface temperature changes in the late twenty-first century (2090–2099) relative to the average surface temperature from 1980–1999, in degrees Celsius (1°C = 1.8°F) over different regions of the globe. Source: IPCC, Summary for Policymakers, *Fifth Assessment Report* of the Intergovernmental Panel on Climate Change, 2013.[23]

The national parks and wildlife refuges on which West Africa's tourism depended had been gravely damaged by poaching, abuse, neglect, and droughts. Tourists no longer arrived at coastal resorts, where the beaches had disappeared and flood dangers had risen. Much of West Africa's economy was thus crippled, along with the Sahelian lands to the south of the Sahara.

Dog Days in Europe

After having endured a severe economic crisis in the early part of the twenty-first century, Spain, Italy, Greece, and parts of Turkey were especially barren and dry in 2100. Chronic water shortages, especially in the more southerly areas, were aggravated by higher temperatures, heat waves, and drought. Many reservoirs and wells dried up.

In Italy's low-lying Po River plain, construction crews were once again trying to elevate already towering flood barriers. In Venice, which was built on a lagoon, whole neighborhoods and world-famous tourist attractions had been abandoned due to the repeated flooding.

In the Alps, the imposing glaciers were gone by 2100, save for a few small pockets of ice in cool, sheltered areas. They seemed almost a mockery of the majestic mountain glaciers that once drew millions of tourists to Austria, Italy,

France, and Switzerland. Mountain slopes had now become unstable due to the melting of the glaciers and high-elevation permafrost. Landslides had become common after heavy rains. Some downslope villages had to be abandoned. Average alpine temperatures had shot up as the century had progressed and, as a result, mountain snowpacks had declined across Europe. Water levels in the Rhine had dropped significantly, and runoff declined sharply in places like Hungary.

The extensive interior wetlands of the Netherlands were quite dry in 2100, due to higher temperatures and increased evaporation. All along their shorelines, coastal marshes had shrunk. Long accustomed to battling the sea, the Dutch had had to erect ever more massive flood barriers. Upgraded shoreline defenses had also been built around Hamburg, London, and other wealthy Western cities.

The Uncertain Future

The scenarios in this imaginary round-the-world journey are not absolutely certain. They are merely reasonable projections, rooted in ever-improving climate science and daily evidence of accelerating climate change. Nonetheless, it is always possible to make more optimistic assumptions, as we will see in the next chapter.

CHAPTER 2

Current Climate Impacts

Civilizations collapse when their environments are ruined.
The obliteration of Nature is a dangerous strategy.

—E. O. Wilson

By altering temperature, rainfall, wind, and weather patterns, sea levels, ocean chemistry, and ocean currents, a rapidly changing climate affects all natural resources, all species, all people, everywhere. This chapter provides far-ranging, irrefutable evidence of several major climate changes that are already profoundly affecting the Earth, its ecosystems, and ourselves. The evidence suggests that a climate catastrophe has already begun.

Portents of Disaster

THE FIRST DECADE of the twenty-first century was the hottest on record.[1] The year 2012 now appears to have been the hottest ever recorded in the lower 48 states—three and a half degrees above the long-term average and a degree higher than the previous hottest-record year, 1998.[2]

Climatologists believe that before industrialization, global mean temperature had not varied as much as 1.8°F within the past 10,000 years. But in the past 100 years, the world got 1.4°F hotter[3]—66 times faster. And the Earth is not only warming astonishingly quickly by historical standards, but the sizzling pace is itself steeply accelerating.[a]

---

[a] The average warming rate over the past 50 years is nearly twice the average for the preceding century. The current decade seems to be almost a fifth of a degree F warmer, in fact, than the 1990s.

Climate change is perhaps most apparent in the increase in extreme weather and in many dramatic physical changes, such as the vanishing of northern polar ice, the melting of the world's glaciers, the disintegration of floating Antarctic ice shelves, and rapidly rising sea levels. Yet we are still in the very early and comparatively mild stages of the first human-induced climate change in history.

Given that the concentration of heat-trapping gases in the atmosphere is the highest in millions of years, the world is in a sense on borrowed time, saved from experiencing the full consequences of this atmospheric carbon only by long time lags built into the climate's system and, ironically, by some of the pollution generated by the fossil fuel combustion that's gotten us into this mess. Aerosol pollution, comprised of shiny sulfate particles, reflects some of the sun's heat spaceward, delaying for a time the full effects of the longer-lived, heat-trapping gases that are also released when fossil fuels are burned. (See chapter 5 for more discussion of aerosols.) We are thus now in a race against time to see if we can take our foot off the fossil fuel pedal soon enough to bring emissions down before the climate system moves into a permanently hotter mode for the next several thousand years.

Even if the planet's air, water, and soil were warming gradually instead of very rapidly, the short-term responses of Earth's natural systems to climate disruption would not necessarily be gradual, but more like lurching "step changes." That's because even if local temperature in a particular region might for some reason be changing slowly, plant or animal species or ecosystems at risk there could still fail quite abruptly should that slowly rising temperature eventually exceed the species' or ecosystem's tolerance limit for heat or drought.

Another example of lurching climate change is the ice shelf clinging to the coastal sea floor that suddenly breaks loose when the surrounding water warms by just a fraction of a degree past the point at which the base of the ice remains solidly frozen to the sea floor.

This principle of potentially jerky or sudden step changes in natural systems has broad relevance. In far northern forests, for example, lakes and other surface water are held above ground by perennially frozen ground known as permafrost. When the soil temperature increases above freezing, permafrost melts. That allows surface water to drain below to ground water. Whole regions can then dry out relatively quickly. This process has already begun. Thus, the plants and animals that depend on the millions of lakes and ponds preserved by frozen ground in the planet's northern latitudes will eventually be lost unless the warming in progress can be arrested. Melting permafrost also allows the organic matter frozen within it to thaw and oxidize to greenhouse gases. (See "Melting Permafrost and Frozen Methane," page 33.)

A long and sobering litany of other climate-related disasters is already occurring, and more are expected. One cannot attribute these disasters entirely to climate change; however, climate change in some cases makes these events more likely or more severe, or both.

Huge, Powerful "Superstorms"

Overall, financial losses from weather-related disasters are up sharply, setting global records. Hurricane Katrina in 2005 killed more than 1,800 people, displaced more than a million, and did more than $100 billion of damage. Hurricane Sandy in 2012 destroyed lives, brought vast coastal flooding, and knocked out power for millions. Hundreds of thousands of homes and businesses were destroyed, damaged, or disrupted. Damages in New York and New Jersey exceeded $70 billion dollars.

Tropical ocean surface temperatures cause moist, warm air to rise and help power storms. Sea surface temperatures off the coasts of New York and New Jersey shortly before Sandy struck were some 9°F warmer than normal, which contributed to Sandy's severity.

FIGURE 2-1. Much of Manhattan went dark after a Consolidated Edison power plant on the Lower East Side was flooded by Hurricane Sandy in October, 2012. Some downtown residents were without lights, heat, and power for five days. Most land-based phone service in the area was also out of commission. Photo © Spencer Platt/Getty Images/AFP.

As sea levels rise and storms intensify, many large coastal US cities are increasingly subject to flooding from high tides and storm surges. New York City was hit by an 11-foot-high storm surge in conjunction with Sandy. Even larger storms are likely in the future.

Floods

As a result of global warming, the atmosphere contains 4 percent more moisture now than it did 30 years ago. More moisture is therefore available to storms. Climate change can also produce abnormally heavy rain or snow by displacing normal storm tracks or ocean currents. Wet weather systems that would normally progress over an area stall instead, dumping huge amounts of rain or snow in one area before moving on. In September 2013, a 4,500-square-mile area across Colorado's Front Range was hit by devastating record floods. Exceptionally high (86°F) sea surface temperatures west of Mexico combined with an extratropical weather system and brought a relentless flow of humid air known as an atmospheric river to Colorado. According to Dr. Kevin Trenberth of the National Center for Atmospheric Research, the floods were created by an atypical series of climate events that might have brought a 500-year flood to Colorado even without climate change;

FIGURE 2-2. Heavy rains caused widespread flooding in numerous Colorado towns. These homes in a residential neighborhood in Longmont, Colorado, were among thousands submerged in mid-September 2013. Photo © John Wark, Associated Press.

FIGURE 2-3. A raging waterfall destroyed a bridge along Highway 34 leading to Estes Park, Colorado, as flooding forced thousands of Front Range residents to evacuate their homes in mid-September of 2013. Photo © Dennis Pierce, Associated Press/Colorado Heli-Ops.

FIGURE 2-4. A Longmont, Colorado, homeowner was comforted by a family friend in front of her possessions as they cleaned up after floodwaters ravaged her home during the 2013 mid-September floods. Photo © Chris Schneider, Associated Press.

FIGURE 2-5. A woman and her young daughter took a close look at a flood-damaged bridge in Longmont, Colorado, in mid-September 2013. Photo © Marc Piscotty, Getty Images North America.

with climate change, however, the floods turned into a once-in-a-thousand-year event that killed six people and forced thousands to evacuate. Powerful floodwaters demolished thousands of homes and farms, and many miles of roads were washed away along with bridges. The preceding images show the impacts that big floods have on communities and their residents.

Heat Waves

The European heat wave of 2003 killed 35,000 people and did $15 billion in damage to agriculture alone.[b] The 2010 Russian heat wave killed 55,000 people and produced massive crop damages. Five hundred wildfires raged over the bone-dry land around Moscow.[c] While heat waves like the European disaster were formerly expected only once in 500 years, they may become fairly common in the overheated world we're now creating.

Along with its worst drought in 130 years in 2002, India had a heat wave in May 2002 that killed more than 1,000 people. The temperature reached

---

b Sir Nicholas Stern, The Economics of Climate Change (Cambridge, UK: Cambridge University Press, 2007).

c The World Bank and the Potsdam Institute, *4° Turn Down the Heat: Why a 4° Warmer World Must Be Avoided*, A Report for the World Bank by the Potsdam Institute for Climate Impact Research and Climate Analysis (Washington, DC: The World Bank, November 2012).

nearly 124°F in one village in state Andhra Pradesh, where a local official reported that birds fell dead out of the trees from the heat.[4]

Tropical Diseases

Lethal, insect-borne and waterborne illnesses formerly restricted to the tropics are now spreading to large regions previously free of them. South Africa, for example, virtually malaria-free in the early 1970s, has nearly 60,000 cases a year.[5] Malaria has also reached highland regions of Kenya and Tanzania, where it was previously unknown.

Drought

Although individual droughts have complex causes, it is well known that higher temperatures associated with climate change increase the intensity, likelihood, and thus the frequency of hot weather extremes. So it should come as no surprise that nearly 1,700 counties in the United States were declared primary natural disaster areas in 2012. A widespread drought that

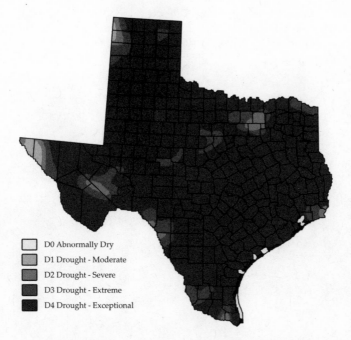

D0 Abnormally Dry
D1 Drought - Moderate
D2 Drought - Severe
D3 Drought - Extreme
D4 Drought - Exceptional

FIGURE 2-6. A historic drought of exceptional intensity afflicted almost the entire state of Texas in 2011. Parts of Texas were still experiencing drought in 2013. Courtesy of the National Drought Mitigation Center.

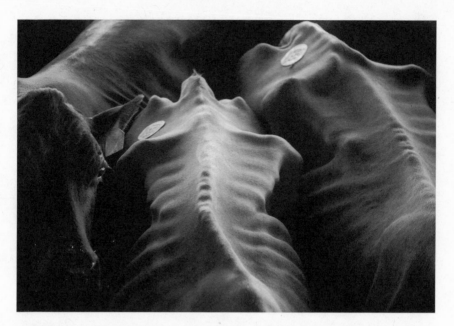

FIGURE 2-7. Emaciated cattle wait in a Gillespie Livestock Company pen to be sold at auctions during the 2011 drought. Many Texas ranchers could not find hay for their cattle or were unable to afford it. Photo © Jay Janner, *Austin American-Statesman.*

FIGURE 2-8. Two ranchers fold their hands and bow their heads during a 2011 gathering to pray for rain at a downtown park in Fredericksburg, Texas. Photo © Jay Janner, *Austin American-Statesman.*

started in 2010 as a heat wave and drought in the southern United States expanded to engulf more than 80 percent of the nation along with parts of Mexico and eastern and central Canada. The related heat wave took more than 80 Canadian and US lives. Some parts of the country, including Texas, were still in extreme drought in 2013. Severe droughts have also stricken African, Middle Eastern, and other nations in recent years.

Melting Glaciers and Ice Sheets

Vast floating ice shelves along the West Antarctic Ice Sheet have broken up in recent years, with the glaciers behind them flowing more rapidly into the sea. The West Antarctic Ice Sheet is also shrinking and becoming unstable due to loss of its sea ice shelves. These developments are not really that surprising given that the average temperature of the Antarctic Peninsula has gone up by 4.5°F since the mid-1940s, and in some seasons, by 7 to 9°F.

Scientists are observing that the Greenland Ice Cap is melting at an accelerating rate. The rapid loss of the Greenland ice—faster than climate models have predicted—is contributing to sea-level rise and is increasing coastal flooding, including parts of major cities. The melting of the Greenland Ice Cap could eventually lead to massive increases in sea levels and profound disturbances to ocean currents that transfer heat on a global basis.[6]

Glaciers around the world have also been melting rapidly, contributing to sea-level rise. About half the glacial ice in the European Alps has been lost in the past century. More than half the glaciers of Montana's Glacier National Park are gone. Extensive glacier melting has also occurred in Alaska as the images of the Muir Glacier (figures 2-9 and 2-10) illustrate.

Glaciers in the Himalayas, the world's largest mass of ice outside the polar regions, are receding faster than anywhere else in the world. They are the source of many important rivers, including the Indus and Ganges. The loss of high-altitude glaciers and diminishing snowpacks is already starting to bring more frequent droughts and threaten the water supplies of more than two billion people who live in parts of India, Nepal, China, and Pakistan and other regions of water scarcity. The East Asian monsoon has also been unreliable over the past 30 years, reducing rainfall in parts of China.

Sea-Level Rise

Sea level rises on a warming Earth because of ice and snow melt, and due to the expansion of warming seawater. Significant sea-level rise has occurred—about eight inches in the past century—a much faster rate than

FIGURE 2-9. Muir Glacier, Alaska, 1941. Photo by William O. Field, August 13, 1941. *Long-Term Change Photograph Pairs.* National Snow and Ice Data Center/World Data Center for Glaciology, Boulder. *Online Glacier Photograph Database.* US Geological Survey.

FIGURE 2-10. Muir Glacier, Alaska, 2004. Photo by Bruce F. Molnia, August 31, 2004. *Long-Term Change Photograph Pairs.* National Snow and Ice Data Center/World Data Center for Glaciology, Boulder. *Online Glacier Photograph Database.* US Geological Survey.

forecast.[7] As noted in chapter 1, as seas rise, parts of heavily populated, low-lying coastal regions and major cities around the world will be below sea level. Using Google Earth, anyone with a personal computer can now see a simulation of what these cities will look like at various expected sea levels. A recent study published in the journal *Nature Climate Change* has found that damage to coastal cities from sea-level rise will mount to $1 trillion *each year* by 2050 if cities do not make adequate preparations.[8]

Sea Ice Melting

Highly reflective summer sea ice in the Arctic is disappearing. It is being replaced by darker-colored water that absorbs vastly more of the sun's heat and acts as another very strong positive feedback that contributes to global warming and climate change. (For more explanation about climate system feedbacks, see chapter 3.)

Melting Permafrost and Frozen Methane

Permafrost in northern latitudes and frozen methane deposits in the ocean known as clathrates or hydrates together contain trillions of tons of stored carbon. The permafrost and clathrates both have slowly begun to thaw and release methane and carbon dioxide to the atmosphere. The more permafrost and clathrates melt, the hotter the earth becomes, and so the more melting occurs. At some point, the continued melting of these vast stocks of carbon would alter the climate beyond recognition and lead to a "runaway" greenhouse effect, transforming the Earth into an ice-free planet as it was millions of years ago. (Methane is 84 times more powerful a greenhouse gas than carbon dioxide on a per molecule basis over a 20-year period.) According to the geological record, when Earth was ice-free, sea level was about 250 feet higher than at the present.

Extinctions

Plant and animal extinctions are accelerating on land and in the sea. The Amazon tropical rainforest is already beginning to suffer from repeated severe droughts and increased tree mortality. Millions of acres of the forest were devastated in 2010; other severe droughts preceded it in 2007 and 2005.[d] If unchecked, this will destroy the rainforest ecosystem, which stores globally

---

[d] In the most severely affected areas, one tree in 25 was killed, according to Eli Kintisch, "Widespread Devastation in 2010 Amazon Megadrought," ScienceNOW (American Association for the Advancment of Science), December 7, 2012.

significant amounts of carbon, with disastrous consequences for global climate, wildlife, and humanity.

Harmful Ocean Changes

Vast harmful oceanic changes (temperature increases, acidification, coral
reef death, low-oxygen zones) are already occurring. Coral reef bleaching, a
sign of severe and potentially fatal stress caused by ocean warming and other
factors, is evident in oceans of the world. Half the reefs of the Indian Ocean
and around South Asia have already lost most of their living coral. Current
trends suggest that 95 percent of the reefs will be dead by 2050 if this continues.[9] Other kinds of marine life are already suffering from oxygen depletion
in warming, polluted coastal waters and from the increased acidification of
surface waters.

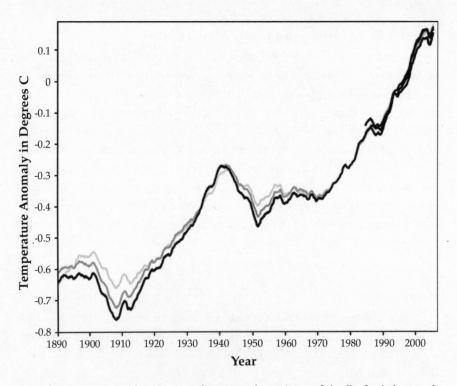

FIGURE 2-11. Using five data sets from several prestigious federally funded research
organizations, the graph shows an overall rising temperature trend for most of the last
hundred years. Data, compiled by Kelly O'Day and courtesy of Skepticalscience.com,
are from the US National Aeronautics and Space Administration's Goddard Institute for
Space Studies, the National Oceanic and Atmospheric Administration; satellite data is
from Remote Sensing Systems, under contract to NASA.

The drastic climate changes that have already occurred after only 1.4°F—less than 1°C—of warming show the climate system is very sensitive to heat-trapping gases and has already been gravely destabilized. This indicates that greatly amplified additional warming and disastrous consequences are likely should warming continue unabated. The steep, post-1970 rise in global temperature is apparent in figure 2-11.

I will pause now in describing the current effects of climate change to focus in chapter 3 on how the climate system normally operates. This explanation will subsequently be used as a basis to understand the mechanisms by which humans are now dominating climate processes and causing climate change.

CHAPTER 3

Natural Climate Change

*Climates found at present on 10–48% of the planet
are projected to disappear within a century,
and climates that contemporary organisms have never experienced
are likely to cover 12–39% of Earth.*

—Anthony D. Barnosky et al.[1]

This chapter explains how the climate naturally functions in the absence
of interference from the effluents of large-scale industrial and agricultural activities.

To better understand how humans are now affecting the planet's
climate machinery—the topic of chapter 4—it is helpful first to know
how nature operates on autopilot, without human interference. Then we can
clearly infer whether humans are but "innocent bystanders" or the perpetra-
tors of climate change. It turns out that the key to that puzzle is in the sky and
is related to subtle changes in "orbital geometry"—the shape of the Earth's
orbit around the Sun.

But it was not until the nineteenth century that scientists like the self-
taught Scot, James Crowell (1821–1890) begin to suspect the amazing fact that
these variations in the Earth's orbit could actually initiate ice ages on Earth.
Earth's orbit around the sun is almost circular but is very slightly stretched out
to form an ellipse. Over a 100,000-year cycle, the shape of the orbit oscillates
from more circular to more elliptical. As the Earth travels around the sun and
spins at an angle to its plane of rotation, the angle itself varies on a 41,000-
year cycle. Meanwhile, the tilted axis wobbles on yet another regular cycle so
that over 23,000 years, the Earth inclines in different directions with respect
to fixed stars and more importantly, the sun.[2] All these processes affect the

climate, as indeed do volcanic eruptions, large meteor impacts, and even sunspots. (See figure 3-1.)

I will now explain how the relatively slight-to-modest changes which these orbital and other variations induce in the total amount of solar energy absorbed by the Earth can trigger massive climate changes.

Clearly, the Earth's distance from the sun and its inclination toward it affect whether the Earth absorbs more or less of the sun's heat. That much certainly isn't rocket science. However, not only do the orbital and angular changes independently affect the amount of solar energy the Earth absorbs, the interactions of these cycles at times throughout the ages coincide or tend to offset each other. Thus, periodically they reinforce and intensify their independent effects. When this occurs, the enhanced effects set in motion conditions on Earth that gradually amplify the initial reduction in the energy Earth absorbs, and take the planet into an ice age. Similarly, when celestial geometry dictates that orbital and axial cycles combine to increase the peak energy absorbed in the Northern Hemisphere, the ice sheet created during the ice age warms, and thaws.

The warming transpires far more quickly than the freezing process, however. The freezing takes tens of thousands of years, whereas ice sheets have collapsed in a few thousand.[3] The enormous mass of ice sheets causes the land below to subside by as much as a half mile or more. At the lower elevation, temperatures are higher and speed up the melting.

The details of how the Earth's major climate cycles unfold are complicated by the many ways in which the climate system, with all its biology, chemistry, and geology, and the Earth's asymmetrical distribution of continental land masses and oceans respond in myriad ways to the changes in incoming solar heat. But in brief, the Earth's normal long-cycle orbital changes serve to initiate relatively small changes in the amount of solar energy absorbed by the Earth. Over long periods of time, these slight variations in absorbed energy are then powerfully amplified or dampened by terrestrial processes so as to trigger massive climate changes on Earth.

In essence, when less solar heat is absorbed, the area of the Earth covered by snow and ice expands. The whitened landscape then increases the reflection of solar energy to space. This intensifies the cooling begun by the orbital changes.

Planetary cooling or heating triggered by orbital changes also affects the Earth's natural cycling of heat-trapping gases, such as water vapor, carbon dioxide, and methane. These affect the physiology of plants and microorganisms that in turn influence the composition of the atmosphere.[4] The resulting atmospheric changes make the atmosphere more (or less) opaque to the solar heat that Earth constantly receives and then re-radiates to its own atmosphere and to outer space.

Thus, for example, as the Earth's temperature falls, less water evaporates from the land and ocean, so the atmosphere is drier. With less heat-trapping water vapor aloft, the atmosphere is less able to capture heat. In addition, because the planet is cooler, more carbon dioxide dissolves in the ocean, so less remains airborne to heat the Earth. The atmosphere then is more transparent to heat radiated from the Earth, and so more can escape to space, further cooling the Earth.

Climate as Distinct From Weather

Climate is another word for average long-term weather, an ensemble of conditions marked by temperature and moisture, as well as atmospheric motion and transparency. Weather, by contrast, is a relatively short-term phenomenon, even though individual episodes can last for days, weeks, or even months. People sometimes lose sight of these distinctions between climate and weather, however. They mistakenly allow their judgment about climate change—which can be reliably deduced only from the careful analysis of long-term weather trends—to be clouded by their perceptions of weather. A particularly cold or snowy winter is often enough to convince them that the climate is cooling rather than warming. A sudden heat wave or violent storm is then taken as certain evidence of the opposite conclusion.

Weather is notoriously fickle, of course. Sometimes placid, sometimes turbulent, its rapid oscillations can be like noise on an audio channel that makes the main signal—climate change—difficult to hear. Filtering out the noise of weather and random or chaotic climate fluctuations is difficult in the short term. Climate also has normal variation cycles, sometimes getting warmer, sometimes cooler. Moreover, it goes through long-term periods of greater and lesser relative stability. This makes it hard to discern underlying trends through casual observation.

Long-term climate trends are also somewhat obscured by large seasonal climate changes in the same geographic location, where temperatures often differ by tens of degrees. Day and night also bring large temperature swings. All these complications make it very hard for the untrained observer to detect small-to-moderate underlying trends. Without systematic scientific analysis, it is challenging to make accurate observations about global climate, let alone forecast it. But by focusing on long-term trends and by synthesizing millions of temperature observations taken on land, sea, and in the atmosphere, scientists have developed consistent and reliable measures to track the evolution of global average temperature over time. That, then, gives us a reliable indicator of climate change, such as the data in figure 2-11, page 34.

Lessons from Ancient Climates

At first it may seem strange that a few degrees of change in average global temperature could have much significance for the planet. These seemingly small changes do matter enormously, however. First, they are enduring, and they increase the probability of extreme conditions, such as heat waves, floods, and intense storms. Second, average conditions are not uniform over the planet. Thus, as mentioned earlier, average warming is magnified in the interior regions of continents and especially in high latitudes near the polar regions, creating large stresses on ecosystems there. So a relatively small change may be a great deal more significant when doubled or tripled or even raised by half.

Finally, relatively small average temperature changes are themselves highly significant, because the climate system itself is sensitive to them. Physical evidence from past geologic eras shows how powerfully the Earth has responded over time to small shifts in the amount of energy received from the sun that are ultimately reflected in global average temperature. This knowledge has been derived from data embedded in historical climate records stretching back hundreds of thousands of years—captured in ancient sediments from the ocean floor and from ice cores extracted from the Arctic and Antarctic, as well as from glaciers. Sophisticated analytical methods make it possible for scientists to read these records, as clearly as you're reading the pages of this book.

The Sun's Role

The Earth's reaction to long-term variations in absorbed solar radiation is in turn modified by responses from the land, sea, and atmosphere. Climate is thus determined by everything on Earth—plus the sun and the orbital dynamics that affect how much of its energy we absorb. The sun, however, establishes overarching conditions for our climate and sets climate changes in motion.

Vast quantities of solar energy radiate continuously toward the Earth. As the bright side of the Earth receives both visible and invisible solar radiation, the Earth absorbs some of that energy and both re-radiates and reflects some of it back into space. The portion that the Earth normally retains through absorption by the land, sea, and atmosphere keeps the planet's temperature far above freezing, powers the winds, and produces ocean waves. It also provides the energy that lifts water into the air by evaporation, which eventually results in precipitation.

How can a body 93 million miles away have such profound influence upon the Earth and control our climate?

The sun accomplishes this by being almost unimaginably large and rich in energy. With 99.8 percent of all the mass in our solar system, the sun has plenty to spare. Every second, it crushes 700 million tons of hydrogen in its core into helium and energy. The amount of energy it produces is also far, far beyond ordinary comprehension. Phrases like "billions of watts" do not begin to describe it.

As some of this incomprehensibly vast flood of energy first reaches Earth's outer atmosphere, it encounters an obstacle course of gas molecules, dust, and clouds. Most of the sun's incoming radiation is absorbed or scattered back into outer space or toward Earth by the atmosphere and clouds. However, about a quarter of radiation misses the clouds and dust. This is the direct beam radiation we perceive from the Earth as bright sunlight.

Of the energy that reaches the Earth's surface, some is reflected, and about half is absorbed by surface water, land, and vegetation. As the energy is absorbed, it heats these surface features and is transformed into infrared radiation (heat), which radiates skyward.

The infrared radiation must again pass through a gauntlet of gas molecules, dust, and other tiny particles in order to leave the atmosphere. Among the gases that it encounters is carbon dioxide, the veritable protagonist of this book. Just as clear glass transmits visible light while a black piece of metal absorbs it and heats up, so, too, atmospheric gases vary in how transparent or opaque they are to radiation.

Each gas absorbs, transmits, or reflects radiation differently, according to the gas' characteristics and the wavelength of the radiation hitting it. Although they make up a very small portion of the atmosphere (well under 1 percent of its volume), carbon dioxide and other heat-trapping gases are powerful absorbers of infrared radiation, in contrast to visible light, which they largely transmit. So, contrary to intuition, even tiny percentages of these gases in the atmosphere have very powerful effects on its ability to retain heat. As heat-trapping gases intercept outgoing heat, the energy they absorb then warms the atmosphere.

Some of that energy is radiated Earthward again, so that it ricochets back and forth between the Earth and the atmosphere.[a] The longer the heat remains trapped in the Earth's atmosphere, the warmer it makes the system.

Climate Cycles

The temperature of the Earth is controlled by the balance of the radiation coming from the sun and the radiation sent back to outer space. When the incoming and outgoing radiation are equal, the overall heat loss or gain is zero. In this condition of thermal equilibrium, the Earth emits as much radiation as it receives, so its temperature stays constant. But incoming and outgoing radiation are not always in exact balance, so the Earth's climate is not constant. The degree and sign of the imbalance (whether positive or negative) indicates Earth's future temperature and climate.[5] Prolonged periods of substantial cooling result in ice ages, when parts of the Earth are covered by layers of ice up to five miles thick. During prolonged warming periods, much of the planet's icy cover melts. Ice sheets retreat toward the poles. Vegetation and animal life expand northward and southward.

The Earth is currently in a warm period known as the Holocene Epoch following the last glaciation, which ended about 10,000 years ago. (We are actually in an interglacial stage of the Quaternary Ice Age, one of the Earth's five known major glacial periods, each lasting millions of years.)[6] In the natural course of events—that is, without the powerful influences of humans on the climate—glaciers would advance again thousands of years from now. However, human influence is now so strong that a return to ice age conditions is impossible for the foreseeable future.[7] Indeed, the Earth's rapid warming in what would naturally be a *cooling* period is powerful evidence that humans are responsible for the climate change being observed.[8,b]

---

[a] A fundamental principle of physics known as Planck's Law tells us that the amount of radiation emitted by a perfectly absorbing object or mass known as a black body is proportional to its temperature. A portion of the energy blocked and absorbed by a heat-absorbing gas in the atmosphere is ultimately reradiated to space. Because the gases are aloft, primarily in the troposphere (lower atmosphere), they are at a lower temperature than the Earth. By Planck's Law, they are less efficient energy radiators. As a consequence, the Earth-atmosphere system begins to accumulate more energy than it radiates. So to restore its thermal equilibrium between the incoming and outgoing radiation, the Earth and atmosphere must warm up to radiate more energy back to space. In this manner, the Earth's temperature increases, causing climate change.

[b] For additional powerful evidence that the observed climate changes of the late twentieth and early twenty-first century are of human origin, see the definitive work of Benjamin D. Santer, "Human and Natural Influences on the Changing Thermal Structure of the Atmosphere," *Proceedings of the National Academy of Sciences* (early edition), Released from embargo on September 16, 2013, www.pnas.org/cgi/doi/10.1073/pnas.1305332110.

The Milankovíc Cycles

As a child in grade school, I had a simple image in my mind of the Earth traveling around the sun on a fixed path. Reality, however, is far more complex and interesting. Not only does the shape of the Earth's orbit change, but so does its position in space relative to the sun. Even the plane of the orbit itself changes slowly over time. The gravitational forces exerted by Jupiter and Saturn are responsible for tugging on the Earth's orbit and changing its shape. The time when the Earth is closest to the sun, the perihelion, also varies cyclically. It is slightly earlier each year, altering the Earth's seasonal changes over a 20,000-year cycle.[c]

The seasons themselves are caused by the tilt of the Earth. As the tilt angle—which varies about two and a half degrees over a 41,000 year cycle—gets larger, the seasons become more intense: summers get hotter and winters get colder.[9] The shape of the Earth's orbit not only varies on the well-known 100,000-year cycle, but also on a "beat cycle" of 400,000 years.[10]

The Earth's tilt, wobble, orbital shape changes, and the changes in timing of the perihelion are collectively known as the Milankovíc cycles for their discoverer, the painstaking and brilliant Serbian mathematician, geophysicist, and engineer Milutin Milankovíc (1879-1958).[d,11]

The combined effect of the Milankovíc cycles slightly varies the amount of radiation reaching the Earth over geologic time and—most significantly—alters the distribution of the radiation on the Earth's surface. The northern hemisphere, with more of Earth's land mass, responds more quickly than the southern hemisphere, which has more ocean. The differential hemispheric distribution of land and water influences the manner in which the Milankovíc cycles intensify characteristics of summer or winter in each hemisphere.

When the Northern Hemisphere is tilted more steeply away from the sun during the winter and the Earth is simultaneously farther from the sun due to orbital changes, winters are at their coldest in the north and warmest in the south. This makes for rainier summers, because the Southern Hemisphere oceans receive more solar heat, causing increased evaporation. As the Earth returns to a more vertical alignment, this shift, along with other Milankovíc cycles changes, produces relatively cool summers in the Northern Hemisphere and warmer winters. Winter snowfall then increases as evaporation

[c] US Naval Observatory. http://aa.usno.navy.mil/faq/docs/seasons_orbit.php. Consulted January 23, 2012.

[d] Astronomical Society. http://www.astrosociety.org/education/publications/tnl/45/globe4.html. Consulted January 23, 2012.

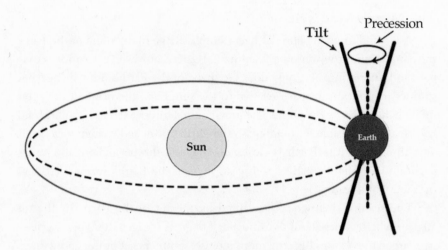

FIGURE 3-1. Ice ages are initiated on Earth by changes in orbital geometry .The diagram illustrates the cyclical changes in the shape of the Earth's orbit and in the tilt of the Earth's axis (T) as it spins on its elliptical path (E) with a slight wobble, its precession (P), which alters the direction in which the axis points.The periodicity of these variations in orbital geometry and in the timing of the perihelion, when the Earth is closest to the Sun, and aphelion, when it is farthest away, constitute the Milanković cycles.

increases. But cooler summer weather permits the winter snow and ice to last longer into the spring and thus to accumulate over larger areas.The increased snow and ice cover increases the reflection of sunlight. This, in turn, reduces the Earth's absorption of heat, further cooling the Earth—an example of what's known as a negative climate feedback. Gradually, summers get cool enough for snow to persist to the following winter over a larger and larger area.[12] As successive layers of snow fall and compress the layers below, snow crystals are transformed into ice. Over millennia the glaciated areas spread and deepen, eventually culminating in ice ages. At the peak of the last glacial period, the ice was up to two and a half miles thick.

Conversely, when the Northern Hemisphere is more strongly tilted toward the sun in summer, ice melts earlier and freezes later in the year, and the total land and sea area covered by ice declines, especially when a steep tilt occurs while the Earth's orbit is closest to the sun during the summer. With the resulting accelerated melting of ice, more and more sea ice is replaced by darker-colored seawater, and bright land ice is replaced by darker-colored soil and vegetation, both of which absorb far more of the sun's heat than does ice. This decrease in Earth's surface reflectivity (its "albedo") is known as a positive feedback effect that increases the Earth's absorption of heat and tends to warm the Earth's climate.

All else being equal—that is, when no massive volcanic eruptions release cooling aerosol particles along with carbon dioxide, and when no movements of the Earth's crust cause a release of large amounts of heat-trapping gases[e]— the overall net effect of the Milanković cycles will then suffice to control the Earth's heat balance by changing the amount of sunlight being absorbed by the Earth.

Earth's Gas "Thermostat"

Milanković today is justly lauded for demonstrating "the interrelatedness of celestial mechanics and Earth Sciences."[13] So how exactly is the variation of climate on Earth coupled to the stuff happening in the sky? Well, we just explained how the Milanković cycles bring on cooling periods. Warming, however, is a different story.

Although slight at first, an initial Milanković warming is sufficient to set in motion natural processes that amplify the warming by raising the concentrations of heat-trapping atmospheric gases here on Earth. Then, in a positive feedback process, those elevated gas concentrations cause a further increase in the Earth's temperature, which results in even higher concentrations of heat-trapping gases. Thus the Earth enters a distinctly warm period, like the one of the past 10,000 years. For these reasons, one can think of the Milanković cycles as triggering climate change. By contrast, the heat-trapping gases in the atmosphere can be thought of like a thermostat that establishes the Earth's temperature once the heat is turned on.

One of the intriguing things about the Earth's climate is that many things are going on at once that affect the concentration of carbon dioxide and other gases in the atmosphere, but all are happening on multiple time scales. Over short time periods of days and seasons, for example, plants remove carbon dioxide through photosynthesis. On the other extreme, some oceanic processes and rock weathering, which alters atmospheric carbon dioxide levels, occur on millennial time scales.[14]

---

[e] Enormous volcanic eruptions have occurred in the Earth's past and have dumped vast volumes of carbon dioxide into the atmosphere, causing rapid global warming. At other times, the friction caused by continental plates sliding over methane clathrate–rich areas of the seafloor have warmed those frozen methane deposits and released vast quantities of methane to the atmosphere, which also served to warm the Earth.

Water Vapor

Water vapor is actually the most powerful of the common heat-trapping gases, because it contains hydroxyl ions (OH⁻) that strongly absorb heat.[15] Unlike carbon dioxide and the other important heat-trapping gases, however, water vapor only stays in the atmosphere a very short time—just eight to ten days.[16] Natural processes normally keep water vapor in something close to equilibrium. Evaporation from land and water surfaces increases the quantity of water vapor in the atmosphere, whereas condensation decreases it.

Human activity does not directly affect the amount of water vapor in the atmosphere to any appreciable extent globally, although it can increase it locally.[17] But human action has a strong *indirect* effect on water vapor concentration through an intermediary: carbon dioxide. Warm air holds more water vapor than cool air, so water vapor concentration increases as temperature rises. Thus, as carbon dioxide and other heat-trapping gases warm the planet by trapping more of the sun's heat, the Earth's temperature rises, and so the amount of water vapor in the atmosphere also increases, further intensifying the inaccurately named "greenhouse effect" (see box, "Greenhouse Gases and the Greenhouse Effect," page 56).

Notably, the effect of carbon dioxide on global temperature and water vapor is the same whether it originates naturally or whether it arrives in the atmosphere because of human activities, as discussed in the next chapter. Despite the powerful effects of water vapor and other atmospheric gases, proof that the *trigger* for Earth's natural climate change is indeed the orbital and axial changes of the Milanković cycles exists in the Earth's long-term natural climate records.

Antarctic ice core data extending back 800,000 years reveal that past temperature cycles between ice ages and warm periods were not initiated by either water vapor or by changes in atmospheric carbon dioxide concentration.[17] Instead, after each major orbitally induced warming began, carbon dioxide concentrations rose, but only after a time lag of about 800 years. If indeed carbon dioxide and other climate-destabilizing gas levels initially rise only in a delayed response to shifts in the Earth's orbit and axial tilt, does this mean that we shouldn't be concerned about adding carbon dioxide to the air?

That would be like saying that because a spark is needed to fire gunpowder in a cartridge, gunpowder plays no role in powering the bullet. For while carbon dioxide concentrations don't rise of their own accord to provoke Earth's great climate cycles, once they climb in a natural response to warming caused by orbital changes, they soon interfere with the departure of heat from the Earth and thereby raise the Earth's temperature in a positive feedback process. The 800-year delay in the increase in atmospheric carbon dioxide—and hence the climate's delayed response—is caused by a physical process known as thermal inertia. The oceans provide a prime example.

The Oceans and the Climate

Thermal inertia is a measure of a substance's ability to absorb and release heat. The larger and denser a mass, the greater its ability to absorb heat and the greater its thermal inertia.[f] The thermal inertia of the world's oceans is very great because of their enormity and because water has a relatively high specific heat, meaning it requires a lot of energy to raise its temperature. These factors ensure that the ocean warms very slowly. Moreover, because the ocean's surface waters are much warmer than the deep ocean—and because heat naturally rises rather than sinks—the transfer of heat from the warmer surface to the ocean depths is further delayed. Thus there is a long lag of up to a thousand years before heat added to the ocean's surface is well mixed into the ocean as a whole.

For these reasons, the ocean warms very gradually relatively to the atmosphere. Eventually, however, it does warm up. As it does, it absorbs less and less carbon dioxide from the air. (The solubility of carbon dioxide in water declines as temperature rises.) At a certain point, the ocean may even, on balance, begin releasing carbon dioxide *to* the atmosphere rather than removing it *from* the atmosphere. The additional carbon dioxide then further warms the climate. The slowly warming ocean eventually warms the climate in other ways, too.

During El Niño phases and storms, some of the thermal energy in the ocean surface is released to the atmosphere. In addition, over long periods of time, a warming ocean eventually melts frozen methane in the seafloor. Once in the atmosphere it, too, warms the climate. The warming ocean also melts more sea ice. This increases the absorption of solar energy by the sea surface and adds to global warming.

Warmer ocean water will also melt the frozen subsurface anchorages of ice shelves. That will contribute to the breakup of ice shelves. In turn, the loss of these "ice dams" will allow glaciers formerly held behind the ice shelves to flow more quickly into the sea. As the extent of this land ice is thus reduced, the newly ice-free land will absorb more solar energy, and that, too, will further warm the climate.

---

f Thermal inertia is defined in physics as the square root of the product of thermal conductivity, density, and heat capacity.

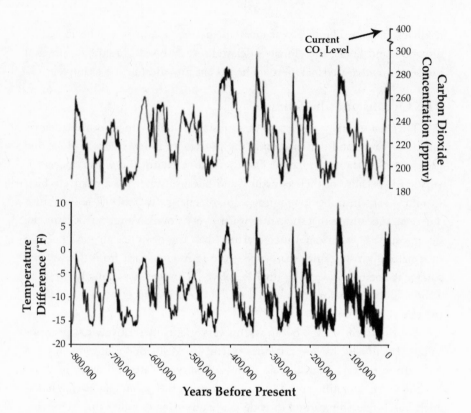

FIGURE 3-2. Estimates of the Earth's changing carbon dioxide concentration (top) and Antarctic temperature (bottom), based on a meticulous analysis of ice core data extending back 800,000 years. The graph highlights the close correlation between carbon dioxide concentrations and global temperature, offset in time by a lag of about 800 years during which the Earth's carbon cycle responds to the warming signal sent by the Milanković cycle changes described in the text. Source: Courtesy of US EPA.

With rising air and ocean temperatures, the rate at which water evaporates from the Earth's land surface and oceans will increase slightly. That will slightly raise the concentration of water vapor in the atmosphere by a couple of percent. This, too, contributes to global warming in yet another example of a positive feedback cycle.[g] (See figure 3-2.)

---

[g] The solubility of a gas in liquid diminishes with temperature, all else being equal, so as seawater warms, more carbon dioxide comes out of solution passing across the ocean surface boundary back into the atmosphere. The long turnover time of the ocean—a measure of internal circulation, mixing, and, hence, heat assimilation—limits the speed of this process and therefore delays the response of atmospheric carbon dioxide concentration to natural warming of the Earth-ocean system. This is particularly relevant for long-term climate cycles arising from changes in incoming solar radiation due to changes in the Earth's orbit, tilt, and wobble. (See footnotes a and b.)

Thus, the more the Earth and oceans warms, the more stored carbon dioxide and methane become available to the atmosphere from the oceans and the soil, and the warmer the Earth gets in a continual positive feedback cycle that magnifies the effect of the Earth's initial orbital changes. Although 800 years seems a long time between increases in temperature during Earth's ancient climate history and ensuing increases in airborne carbon dioxide and methane concentrations, these 800-year time lags are very small in relation to the length of ice ages and therefore do not cast doubt on the critical role of carbon dioxide and other heat-trapping gases in bringing the Earth out of ice ages and in "driving" or "forcing" global heating in general.

Now that we have described the basic elements of natural climate change, we are ready for a deeper understanding of the processes by which human activities have dangerously destabilized the climate, which is the subject of chapter 4.

CHAPTER 4

Unnatural Climate Change

*Today we're seeing that climate change is about more than
a few unseasonably mild winters or hot summers.
It's about the chain of natural catastrophes and devastating
weather patterns that global warming is beginning
to set off around the world . . . the frequency and intensity
of which are breaking records thousands of years old.*

—Remarks by then-Senator Barack Obama[1]

During the twentierth century, humans wrested control over the Earth's climate machinery, overpowering the natural processes described in chapter 3. The current chapter explains how massive changes in the Earth's atmosphere caused by human activities have altered the world's climate.

HAVING SUMMARIZED THE NATURAL WORKINGS of Earth's climate cycles, I will now focus on how human actions influence climate through climate "forcing agents." Forcing agents are anything that perturbs normal climate cycles. These include heat-trapping gases, such as water vapor, carbon dioxide, methane, nitrous oxide, ozone, or other chemicals.[a] All these are positive forcing agents that tend to warm the Earth. Another important but nongaseous positive forcing agent is black carbon (soot particles).

By contrast, negative forcing agents tend to make the Earth cooler. The most prominent of these are shiny sulfate aerosol particles. Produced in fossil-fuel burning and in volcanic eruptions, they cause cooling because they

---

[a] These include a class of chemicals known as halocarbons, which include chlorofluorocarbons (often used as refrigerants), and sulfur hexafluoride, a gas 22,800 times more powerful than carbon dioxide in trapping heat but, fortunately, present in the atmosphere only in trace amounts.

reflect sunlight back to space or provide nuclei for the formation of cloud droplets, which in turn reflect sunlight.

The per-molecule warming or cooling power of each forcing agent can readily be determined from principles of physics and chemistry or in an empirical laboratory test. Figuring out what effect they actually have on the climate in the real world is far more complex, however. The quantities of forcing agents are in flux. Multiple positive and negative feedback effects (some natural, some caused by humans) occur simultaneously along with physical and chemical atmospheric processes that alter both the quantities and effects of forcing agents.

Much of the reliable information today about how sensitive climate is to forcing agents is deduced from the planet's climate history—its paleoclimatic record. The record is composed of information left in ancient ice, fossils, and sediments. These natural temperature records are interpreted with the benefit of current scientific knowledge of each forcing agent's physical and chemical properties.

It is convenient to express the strength of a forcing agent in units of power per area, in this case, watts per square meter, averaged over the entire surface of the Earth. The sum of all the positive and negative forcing agents in operation at any time then indicates how much the Earth's temperature is going to increase or decrease.

For example, Table 4-1 shows the strength of the most important forcing agents, taken from data presented by distinguished climatologist Dr. James Hansen, one of the earliest and most prominent climate researchers to sound the alarm about the risks of global warming.

Table 4-1 does not include water vapor because, for reasons explained earlier, it is under the control of the direct climate forcing agents. Thus, it can be accounted for by assessing the impacts of the direct forcing agents that affect it. Because the form that water vapor takes when it condenses on

TABLE 4-1 (OPPOSITE). Major human and natural climate "forcing agents" that disturb Earth's energy balance are shown with the strength of their heating or cooling effects on climate as indicated by their "global warming potentials" in watts/m^2. (Positive signs indicate heating, and negative signs indicate cooling.) The forcing agents are summed to show the net magnitude of human-induced climate forcing from the addition (to date) of about 120 ppm carbon dioxide and other heat-trapping gases to the atmosphere from preindustrial to modern times, causing global average temperature to warm 1.4°F. Scientific uncertainties about the exact effects of aerosols and cloud changes are indicated by the large ranges associated with these values. Data adapted, with permission, from James Hansen, *Storms of My Grandchildren* (New York, NY: Bloomsbury USA, 2009).

TABLE 4-1.
Primary "Forcing Factors" Affecting Earth's Climate and Temperature Since the Start of the Industrial Era (watts/m²)

Positive Climate Forcing Agents		Negative Forcing Agents	
CO_2	1.5	Negative reflective aerosols	(–1.5 to –0.5)
N_2O, CFC, CH_4	1.25	Induced cloud changes	(–1.5 to –0.5)
O_3	0.3	Human land cover changes	(–0.1)
Soot	0.5	**Total all negative aerosols**	(–1 to –3)
Total	3.55	**Total (mid-range estimate)**	~ –2
Net estimated human-induced positive forcing (column 1)	**3.55**		

Scale of Natural Forcing Processes (for comparison with human forcing agents):

Solar cycles (total forcing from minimum to maximum)	0.2 (–0.1 to 0.1)[*]
Volcanoes (episodic effects)	~0.10[‡] (–0.05 to 0.25)
Peak effect of solar cycles and volcanoes combined	~0.4
Changes in Earth's surface reflectivity due to ice sheet changes in going from ice ages to warm periods	3.5

Grand Total Estimated Net Human Forcings   **1.85[§]**
(All positive minus all negative human forcings)

Magnitude of Natural Forcing Processes During Transitions To and From Ice Ages

Effects of changes in all heat-trapping gases combined from ice ages to warm periods 3

Note: If actual net forcing is actually close to 2W and produced 1.4°F of warming since 1750, then the short time-scale climate response equals ~ 0.7°F/W. For comparison, the Intergovernmental Panel on Climate Change's 2007 mean net total anthropogenic radiative forcing estimate was 1.6W, with a large uncertainty range: (0.6–2.4W). However, the panel increased its estimate of net human forcing by 43 percent in its 2013 report to 2.29W (1.13–3.33W) based on mean forcing in 2011 relative to 1750.

[*] 0.1W of the 0.3W is from increased forcing due to induced increased O_3 production; when solar radiation increases, ultraviolet radiation increases, which increases the concentration of ozone in the upper atmosphere.

[‡] Although volcanic eruptions produce global cooling by discharging sulfur dioxide to the atmosphere where it produces a reflective haze of sulfuric acid particles, volcanoes are treated as a minor positive forcing agent in this table because, during the period since industrialization, volcanic activity has been markedly less than in the eighteenth century, per Hansen in *Storms of My Grandchildren*, as cited (see endnotes).

[§] The 1.85 W estimated total forcing for the period from the start of industrialization to the present is the human-induced forcing of 3.55W minus the estimated net aerosols (-2W) plus the 0.3W of combined solar cycle (0.2) and diminished volcanic cooling of 0.1W (relative to the eighteenth century). This yields 1.85W (3.55 – 2 + 0.3), which is on the order of the 2W figure cited in the note to Table 4-1.

tiny particles of dust to form clouds acts on the climate in complex ways, the effects of clouds are included in climate models.

Clouds' effects, in terms of both reflecting and absorbing radiation, depend on their composition, type, and altitude. Broadly speaking, they tend to shield and cool the Earth by day and to warm it at night by intercepting outgoing infrared (heat) radiation—much like an insulating blanket.[b]

Aerosols, which are simply tiny particles that have become airborne, are explicitly included among the direct forcing agents of Table 4-1. Volcanic eruptions, for example, blast sulfur aerosols into the stratosphere.[c] This causes the stratosphere to become more reflective to incoming solar radiation, and so the Earth cools markedly until these particles drop out of the atmosphere over a few years. Volcanic eruptions also add carbon dioxide to the atmosphere, but the amount is very small (less than one percent) relative to the amount released by human activity.

Particles of soot, produced by burning coal and oil, tend to warm the Earth, because the black particles, when deposited on snow or ice, make the ground less reflective and warm it locally by increasing the absorption of heat. They thereby accelerate the melting of glaciers and ice caps.

Industrial Activity

Obviously humans have no influence over how much energy the sun radiates, the tilt of the Earth, the wobble of its spin, its orbit, or the eruption of volcanoes. But industrial activities, as well as certain human land use activities, collectively do affect the composition of the atmosphere. And that, in turn, through the feedback processes described, exerts a powerful influence upon our climate.

Ninety-nine percent of the Earth's atmosphere is within eighteen miles of the Earth's surface. Seen from space, this near-Earth atmosphere is an astonishingly shallow film of gas veiling the Earth. It is similar in its relative thickness to the Earth as an apple peel is to an apple. This sparse gaseous coating consists almost entirely of nitrogen (78.084 percent) and oxygen (20.946 percent). These two dominant gases thus make up ninety-nine percent of the whole atmosphere by volume.

---

[b] Complexities like the diverse effects of clouds and aerosols have provided a field day for climate science deniers, who, by selectively overemphasizing or deemphasizing these opposing climate forces out of context, discover "discrepancies" in climate science.

[c] The atmosphere has five main layers. The one closest to Earth is the troposphere. Above that is the stratosphere which extends from 6 to 32 miles above the Earth.

Mixed into the atmosphere, however, are small but critically important quantities of carbon dioxide (0.0400 percent), methane (0.0001745 percent), nitrous oxide (0.00003 percent), ozone (0 to 7 x 10^{-6} percent), as well as tiny amounts of inert and other trace gases and water vapor. As suggested earlier, it may at first seem astonishing that changing the amount of a gas like carbon dioxide in the atmosphere by just a hundredth of a percent can profoundly impact the Earth's climate. But even at this seemingly low concentration, carbon dioxide does markedly decrease the transmission of heat through the atmosphere. The other heat-absorbing gases operate at even lower concentration ranges: in the ten-thousandths of a percent for methane and the hundred-thousandths of a percent for nitrous oxide.

Nonetheless, because the atmosphere is so vast, it takes enormous amounts of carbon-rich and nitrogen-rich gases to raise their atmospheric concentration even by the small amounts necessary to affect the atmosphere. However, we are indeed releasing these gases in huge and increasing quantities. Humans currently add 33.5 billion metric tons of carbon dioxide to the atmosphere every year by burning coal, oil, and natural gas, and by inadvertent releases of carbon dioxide when using carbon-based fuels as chemical feedstocks.[2] This enormous quantity has itself increased by close to 120 percent between 1971 and the present.[3] Unintended releases of methane also occur from gas wells, gas pipelines, coal mines, and landfills.

In addition, burning carbon fuels in engines discharges vast amounts of oxidized nitrogen to the atmosphere. Nitrogen fertilizer use puts out still more nitrous oxide.[4] Finally, in addition to the carbon dioxide from fossil fuel burning, another 4.3 billion metric tons of carbon dioxide are discharged to the air annually from agriculture, logging, forest burning, and other land use changes.[5]

As all these gases absorb heat and slow its escape from the atmosphere, the air gets warmer and, in turn, releases extra heat to warm the oceans and land masses. The oceans, which cover 70 percent of the planet's surface, are also warmed by direct solar radiation. Over the past 50 years or so, they have absorbed about 90 percent of the total excess heat the entire planet has gained. Though the oceans do warm slowly relative to the air—because of their huge mass and for reasons of basic physics explained in chapter 3[d],—the ocean surface layer eventually warms the deeper ocean and reaches an equilibrium with it.[e] As the warming progresses, the ocean eventually releases both carbon

---

[d] Water has a higher specific heat than air.

[e] In equilibrium, the rate at which heat is absorbed from the atmosphere will equal the rate at which heat is released to it.

dioxide and methane to the atmosphere and initiates various positive feed-backs described in chapter 3. The warming ocean thus eventually adds a total of about 1 to 2°F to the average temperature of the warming climate.[6]

The warming caused by heat-trapping gases is often euphemistically referred to as, "the greenhouse effect," and the heat-trapping gases that produce it are commonly known as "greenhouse gases." But the term *greenhouse*, with its warm and faintly pleasant connotations, is a misleading characterization of the carbon dioxide, methane, nitrous oxide, halocarbons, and other gases that destabilize the climate when released in large amounts compared to natural sources.

GREENHOUSE GASES AND THE GREENHOUSE EFFECT

A greenhouse readily admits visible light and some ultraviolet light. Once inside, some of this solar energy is reflected outward; the rest is absorbed by soil, plants, and air molecules and then is re-radiated as heat (infrared radiation), warming everything in the greenhouse.

Because glass allows visible light and short-wave infrared energy to pass but doesn't transmit the longer infrared radiation, much of the sun's incoming energy remains trapped inside the greenhouse, where it keeps plants warm, promotes their growth and health, and provides a barrier against frost damage by preventing convection—the mass movement of warm air out of the greenhouse.

Unlike the dynamic of a greenhouse, however, excess heat-trapping gases do not warm the Earth by preventing convection. Nor do they make it healthier. They simply heat the Earth by continuously absorbing and re-radiating solar energy that otherwise would have escaped to space. The higher the atmospheric concentration of heat-trapping gases, the hotter the planet becomes.

Global Warming Potential

As explained in chapter 3, the atmospheric concentration of water vapor is largely determined by the Earth's temperature, since heat powers evaporation. So, the quantity of water vapor is largely a function of the atmospheric concentration of the primary heat-trapping gases—carbon dioxide, methane, nitrous oxide, ozone, and the synthetic halocarbons. Their effect on the Earth's temperature directly determines the amount of energy available to drive the hydrological cycle, a fancy way of saying the evaporation and condensation of water vapor and its circulation throughout the planet. See chapter 3, pages 46–48 for further discussion.)

The most abundant of the heat-trapping gases in the atmosphere apart from water vapor is carbon dioxide, which (as indicated previously) is produced any time a fuel containing carbon burns, or when any living thing breathes. The total heating effect of any heat-trapping gas on the atmosphere is determined by the number of molecules of the gas added to the atmosphere and the amount of warming each molecule causes, which depends on the type of the gas. The cumulative warming depends not only on the volume of gas and the energy absorbed per molecule, but the time the gas stays in the atmosphere (its "residence time"), which also varies from gas to gas.

The overall effect of an atmospheric gas on climate is known as its "global warming potential," which is based on its atmospheric volume, per molecule heating potency, and residence time. Carbon dioxide's global warming potential makes the most important of the primary global warming gases by far, even though it makes up only a tiny fraction of the atmosphere and, molecule for molecule, is not the most powerful heat-trapping gas.

Other Heat-Trapping Agents

As mentioned earlier, the other leading heat-trapping gases are methane (the main component of natural gas), nitrous oxide (a component of auto exhaust and a byproduct of burning coal), halocarbons (which include some refrigerants), and ozone, a highly reactive gas consisting of three oxygen atoms rather than the two found in an ordinary molecule of oxygen.[f]

In addition to the burning of fossil fuels, human activities that add heat-trapping gases to the air include livestock raising, tilling the soil, fertilizing crops, making cement, producing certain industrial chemicals, dumping waste in landfills, and logging and burning forests.

Although carbon dioxide, methane, and nitrous oxide—the most common heat-trapping gases in the atmosphere—are also naturally present there, human activities greatly add to their concentrations. By contrast, the powerful heat-trapping halocarbon gases were unknown until they were synthesized for industrial use.

---

[f] Ozone is created in the upper atmosphere by cosmic radiation and at ground level by chemical reactions that produce smog and by some air purifiers. Ozone in the lower atmosphere is a corrosive pollutant harmful to human lungs, plants, and buildings, whereas in the stratosphere, ozone protects the Earth by absorbing harmful ultraviolet radiation.

The CO₂-Temperature Link

As shown on Table 4-1, carbon dioxide is more responsible for changing our climate than any other gas humans produce. That's in part because carbon dioxide has a very long atmospheric "residence time." A pulse of carbon dioxide doesn't just enter the atmosphere and then leave forever at a specific date. Carbon dioxide molecules make many roundtrips, recycling to and from the atmosphere through the complex dynamic global network of carbon sources and sinks known as the Earth's carbon cycle. Thus, a substantial fraction of the added carbon dioxide is not removed quickly by the ocean or by other carbon "sinks," such as forests, where carbon on balance is absorbed. More about this shortly.

Currently, the concentrations of heat-trapping gases in the atmosphere are at their highest levels in about two million years[11] (figures 4-1 and 4-2) and possibly far longer. As previously discussed (see figure 3-2), the Earth's temperature and the atmospheric concentration of carbon dioxide have risen

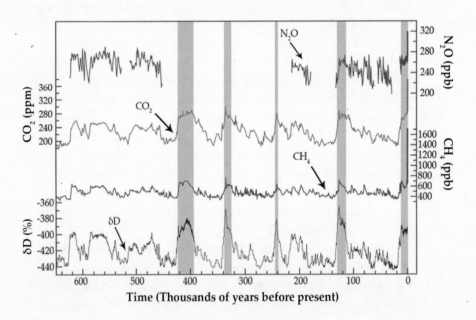

FIGURE 4-1. The graph shows the coordinated increases and decreases of the greenhouse gases carbon dioxide (CO_2), methane (CH_4), and nitrous oxide (N_2O) over the past 650,000 years. All three are highly elevated compared to natural concentrations. The graph also shows changes in the deuterium concentration of the atmosphere. Deuterium (δD) is a form of hydrogen with an extra neutron and its rise is correlated with increases in temperature. Source: Intergovernmental Panel on Climate Change.[7]

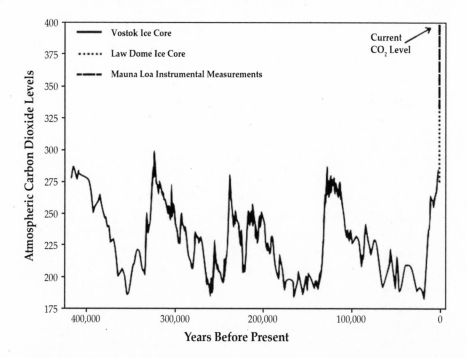

FIGURE 4-2. Variation of atmospheric carbon dioxide concentration in parts per million (ppm) over the past 420,000 years showing that the current level of 400 ppm far exceeds natural carbon dioxide concentrations during previous warm periods and ice ages. Data like this constitute powerful evidence that the current high carbon concentrations are unlikely to have any natural explanation other than the recent large and rapidly growing releases of carbon dioxide from human industrial activities and land use changes. Source for Vostok data: J.R. Petit et al.[8]; for Mauna Loa data: NOAA-ESRL[9]; and for Law Dome data: NOAA-NCDC.[10]

and fallen together for at least 800,000 years—as far back in time as our ice core data go.[12]

This correlation is extraordinary evidence for an intimate connection between carbon dioxide and temperature. However, only in the past 250 years, since the start of the Industrial Revolution, have human actions altered atmospheric carbon dioxide levels sufficiently on a global scale to drive up temperature.

The Rapid Increase of Heat-Trapping Gases

Right before the Industrial Revolution, the Earth's atmosphere had about 280 parts per million (ppm) carbon dioxide. Human activity has added about 120 ppm, most of it in the past century and a half, raising the concentration to 400 ppm. We're now adding carbon dioxide to the atmosphere at 2 to 3 ppm per year—about 1,000 times the natural rate during past normal warming periods. Over the previous 8,000 years when industrial activity was essentially absent, atmospheric carbon dioxide concentration increased by only an average of a quarter of a ppm per century due to natural forces.[13]

Yet despite the dramatic increase in heat-trapping gases, we have only experienced a 1.4°F rise in average global temperature over the past century. However, now that the Earth's average temperature has begun increasing quickly, it could increase by up to 10°F or more by 2100 if heat-trapping gas releases continue rising steeply. In the (hopefully unlikely) event, however, that concentrations reach extremely high levels equivalent to 1,000 ppm of carbon dioxide before leveling off, the Intergovernmental Panel on Climate Change (IPCC) estimated that the resulting global temperature would likely rise between 6.7 to 15°F (3.7 to 8.3°C).[14,g]

The effect of atmospheric carbon dioxide on temperature can be seen very dramatically on Venus. Venus is closer to the sun than the Earth is, so Venus should be somewhat warmer based on its proximity and because it has clouds of sulfuric acid that absorb infrared radiation. But the temperature of Venus is not "somewhat warmer." It is a furnace-like 900°F, which is due primarily to the 95 percent carbon dioxide content of its atmosphere, about 2,500 times the concentration in the Earth's atmosphere.

Carbon Dioxide's "Staying Power"

Undisturbed by humans, the carbon in a coal seam or a petroleum deposit probably would not find its way into the atmosphere for many thousands of years or more. Burning fossil fuels, however, injects carbon immediately into the air. The carbon cycle then takes a long time to remove and immobilize it again within the Earth or the ocean floor.

Unlike some other heat-trapping gases, carbon dioxide is not destroyed in the atmosphere. Natural processes do gradually extract it from the air, but that's not the end of its life cycle. Once removed, it is redistributed to other carbon "sinks," which can be thought of as reservoirs through which carbon flows. Thus, carbon may be dissolved in the ocean, buried in the sediments of the seabed, marsh, or swamp, or incorporated in soil and rocks or in the tissues of living or dead plants and animals. These reservoirs, however, also release

---

g The highest projected concentration of carbon dioxide reported on by the IPCC was 1,200 ppm. That would be roughly four and a half times preindustrial levels—a shock to the climate system's heart. The IPCC's best estimate was that this would cause an 11°F (6.3°C) temperature increase by 2100.

carbon to the air and to other carbon sinks. When a molecule of carbon dioxide is incorporated into water, soil, rock, or vegetation, the carbon may be securely stored for a long period, or quickly released back to the atmosphere.

For example, a plant takes carbon dioxide from the air during photosynthesis and, combining it with water, turns it into carbohydrate and oxygen. The plant might then be eaten by an animal and its carbon content exhaled by the animal as carbon dioxide. Meanwhile, the carbon in the plant's root might decay and be incorporated into the soil. The soil could then be eroded and washed into a riverbed or compressed into rock, uplifted over millions of years, and then slowly weathered and dissolved in rainfall, perhaps to become runoff that finds its way to the ocean. But the carbon is not destroyed. It might then nourish marine plankton, become incorporated in the marine food chain, and eventually join with a mass of dead or dying debris on the ocean floor to be gradually buried into bottom sediments. Even there, its journey may not be done, as ocean sediments are slowly forced under continents. It's quite an extraordinary feat that scientists, through painstaking study of the natural world, have been able to measure the stock and flow and residence times of carbon through such a complex cycle.

As it travels the many pathways of the carbon cycle, some of the carbon dioxide we put into the air ultimately spends thousands of years circulating from the atmosphere to the other carbon sinks and back to the atmosphere before drifting down to its very long-term fate in the sediments of the ocean floor. Complete removal from the atmosphere of any particular pulse of carbon dioxide is thus a very, very slow process. On the average, half of any such influx—for example, from burning fossil fuels or biomass, or from forest fires or deforestation, or from agricultural processes[h]—will endure in the atmosphere for hundreds to thousands of years and will affect the climate for tens of thousands of years.[15] As a result, *any* additional release of carbon dioxide to the atmosphere in excess of the carbon cycle's natural rate of carbon removal will cause accumulation of atmospheric carbon. This additional carbon dioxide then heats the planet, damaging the climate. The longevity of carbon in the atmosphere places a great responsibility on us to be prudent in energy use, since our actions today have such far-reaching, long-range consequences for future generations and for the environment. As the scientists of the IPCC explained in 2007: about half a carbon dioxide pulse to the atmosphere is removed over a time scale of 30 years; a further 30 percent is removed within

---

[h] Such as tilling the soil or flooding rice paddies.

a few centuries; and the remaining 20 percent will typically stay in the atmosphere for many thousands of years.[16]

Given that complete removal of carbon dioxide is so slow and that the current concentrations of heat-trapping gases in the atmosphere are *already* destabilizing our climate, the burning of *any* additional fossil fuels clearly is harming the Earth in ways that may be irreparable. Even burning biomass fuels—like wood and agricultural waste—and releasing their carbon to the overloaded atmosphere has a detrimental effect.

How Hot Will It Get?

Based on the record of climate in the geologic past in ancient sediments, fossils, and ice cores, scientists know that thousands of years ago, when only about 100 ppm carbon dioxide was added to the atmosphere, the planet warmed 9°F in response and came out of the last ice age. Modern civilization has now added 120 ppm—but only a very short period of time has elapsed for the full spectrum of effects to be expressed. Moreover, the Earth's current short-term response to the surfeit of carbon has been dampened due to a phenomenon known as "global dimming." This is a reduction in the intensity of incoming solar radiation due to the presence in the atmosphere of shiny sulfate aerosol particles—byproducts of the same fossil fuel burning that is initiating global climate change.[i]

As described in chapter 3, when the Earth heats up during the warming phase of the natural climate cycle, the higher latitudes (both north and south of the equator) get disproportionately warmer than the global average. Thus, the 1.4°F average heating in the past century is widely experienced as an increase in temperature of about 3°F at higher latitudes; temperatures have risen even more than that in some high latitude regions. Parts of Alaska and Canada, for example, have already experienced increases of 7.2°F in winter nighttime temperatures.[17] These large increases have already profoundly disturbed native ecosystems, their wildlife, and the people who depend on them. Many northern forests of spruce and aspen are now infested with outbreaks of spruce bark beetle and aspen leaf miner (figure 4-3). Parts of Colorado and Wyoming have experienced severe infestations of pine bark beetles.

---

[i] Aerosols, which appear in the atmosphere as haze, are produced both by natural phenomena, such as volcanoes and dust storms, and by humans in burning fossil fuels. In general they only remain airborne for days or weeks, but since they exert a cooling effect on climate, they mask some of the effects of the dangerous, long-lasting heat trappers that often are discharged with them.

FIGURE 4-3. Aspen leaves damaged by aspen leaf miner. Copper Center, Alaska. Photo © by Benson Lee.

Forecasting Future Temperatures

If we know the future atmospheric concentration of carbon dioxide and other heat-trapping gases and aerosols, we can forecast future average global temperatures. Unfortunately, no one knows what future concentrations will be because no one knows exactly what future energy use patterns will be because they depend on unknown future policy decisions. Hence, we don't know exactly how much carbon the world will add to the atmosphere in the years ahead. Scientists can only make forecasts based on plausible future energy use and emissions scenarios.

The total amount of carbon that will be released depends on factors such as trends in world population, global income (and thus the amount of energy and resources used), and on energy technology choices. Scientists therefore use a range of scenario-based forecasts rather than a single forecast. These forecasts, in addition to measuring the effects of carbon dioxide, also estimate the effects of increases in methane and other heat-trapping gases. All these gases are accounted for in a unit known as the carbon dioxide equivalence unit (CO_{2eq}). The unit measures the combined global warming effect of all heat-trapping gases expressed as an equivalent amount of carbon dioxide. Thus, whereas the Earth's atmosphere is currently at 400 ppm of carbon dioxide, it is at about 480 ppm in CO_{2eq} when all major heat-trapping gases are also accounted for. The last time in the geological record when the Earth had such high levels of carbon dioxide was at least 2 million years ago and may have been as long as 15 to 20 million years ago.[18]

At that point, the Earth's temperature was 5 to 10°F hotter than today. Sea level was 75 to 120 feet higher, and the Arctic had no permanent ice cover. Small amounts of ice remained in the Antarctic and Greenland. Thus, carbon dioxide levels similar to today's were associated with an enormously elevated ocean rise and a climate profoundly different from our own. Moreover, when the Earth was but 1.8°F hotter than today during a more recent warm period known as the Eemian Period, sea level was 13 to 20 feet higher.)[19] So sea level over time is *very sensitive* to global temperature—another powerful reason to protect today's climate.

Regrettably, climate and energy experts believe that within only a generation or so, we will reach 550 ppm of carbon dioxide equivalent gases in the atmosphere—about twice preindustrial levels—unless profound changes are made in human energy use and behavior to reduce carbon emissions.[20]

As noted earlier, the Earth is now already more than 1.4°F above preindustrial global average temperatures, and about 90 percent of the excess heat the Earth has gained in the past half century has been taken up by the oceans, mostly in the surface layer (to a depth of about 2,300 feet). The other 10 percent of the excess heat captured by the Earth has warmed the atmosphere and the land surface and gone into melting ice in the Arctic sea, on Greenland, and in glaciers. What does this imply about additional warming? It means that even if *no* additional heat-trapping gases were added by society to the atmosphere, the world would continue warming at least by another 1 to 2°F because of climate forcing by the heat-trapping gases already in the atmosphere, and the additional carbon dioxide and methane to be released from the warmer ocean.[21,j]

It takes the atmosphere-ocean-land system time to adjust to the relatively rapid addition of heat-trapping gases to the atmosphere. Time lags operate on multiple scales within the system while the climate adjusts to new conditions. For example, as the Earth's surface warms, it gradually heats the atmosphere by convection, but the ocean heats much more slowly. Gradually, the whole system warms and radiates more heat. Ultimately, over a very long period of time, the flow of energy radiated outward to space thus rises

---

j Warming water will be able to hold less carbon dioxide because the solubility of carbon dioxide in water diminishes as temperature rises.

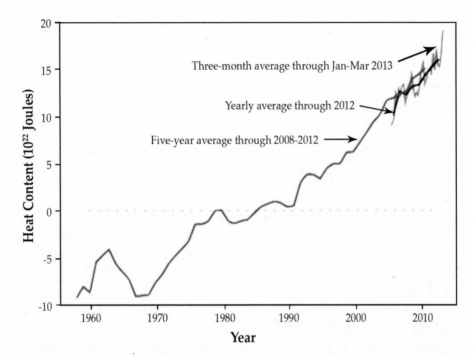

FIGURE 4-4. This graph shows how much extra heat has been stored in the ocean since the mid-1980s. Source: NOAA. "Global Ocean Heat and Salt Content," National Oceanographic Data Center, National Oceanic and Atmospheric Administration.[22]

to roughly balance the increased solar energy absorbed from space. This then reestablishes the Earth's thermal equilibrium with energy reaching the Earth from space, but at a higher temperature than before human activity altered the atmosphere.[k]

The Intergovernmental Panel on Climate Change (IPCC) anticipates that if we continue increasing heat-trapping gas emissions at the current breakneck rate through the end of this century, the concentration of CO_{2eq} will reach 650 to 1,200 ppm—more than four times the preindustrial level. By 2100, as much as 200 ppm of this buildup could be coming from methane and carbon dioxide released from natural sources as a result of uncontrolled positive feedback.

We don't know exactly how sensitive climate is to heat-trapping gases, because the precise effects of various aerosols that cause global cooling and .

---

[k] The planet is currently not in thermal equilibrium, as it is now retaining more heat from the sun than it is radiating back to space. At higher temperatures, radiative efficiency is higher. Thus, to attain a new thermal equilibrium given all the additional heat-trapping gas added to the atmosphere by human activity, the Earth's temperature must rise so that the influx and efflux of heat can equilibrate.

therefore conceal the effects of heat-trapping gases still have not been precisely quantified. If these effects prove to be on the high end of the uncertainty range, then these powerful aerosols could, in effect, be hiding a great deal of the climate's inherent response to the heat-trapping gases we are putting aloft.[23]

Eventually, however, as combustion technology improves and environmental concern increases, humans will very likely reduce pollutants from industry and transportation, and that will reduce the concentration of aerosols and their cooling effects. Based on the capability of human-released, heat-trapping gases to produce an estimated three watts per square meter of positive climate forcing (i.e., heating) over the Earth's surface, world renowned climatologist Dr. James Hansen stated:

If the net [human-made] forcing is 2 watts, aerosols have been masking about one-third of the greenhouse forcing. So if humanity makes a big effort to clean up particulate air pollution (say, reducing human-made aerosols by half), the net forcing will increase by only a quarter, from 2 to 2.5 watts. The additional global warming would not be welcome, but it might not be earthshaking. On the other hand, if the net forcing is only 1 watt, that is, if aerosol forcing is -2 watts, that means aerosols have been masking most of the greenhouse warming. In that case, if humanity reduces particulate pollution by even half, the net climate forcing would double. That increased forcing, combined with a continued greenhouse gas increase, might push the planet beyond tipping points with disastrous consequences.[24]

Apart from the issue of climate sensitivity[1] and the aerosol uncertainties that cloud it, scientists have forecast that without corrective action to reduce emissions, carbon dioxide levels might triple or even quadruple by the year 2100. Global average temperature could then soar by double digits, as noted earlier. What would this mean for life on Earth? Let's take the answer step by step.

With a doubling of atmospheric carbon dioxide concentrations, scientists project that the world's average temperature might increase anywhere from 2.2 to 16.4°F, with the most likely projection being a rise of 4.3 to 9.5°F.[25,26] How bad would that be?

According to some climate models, even a warming of 3.6°F (2°C)— which we are already expected to overshoot[27]—would dry out the Amazon rainforest, now an important source of global oxygen, biodiversity, and climate

---

[1] Climate sensitivity is a measure of how much Earth's average temperature will rise in response to a doubling of atmospheric carbon dioxide.

stability. But the consequences of doubling carbon dioxide could drive world temperatures up by *multiples* of 3.6°F (2°C), a totally unacceptable future. Dr. Hansen has pointed out, for example, that the near-term consequences of raising carbon dioxide by a little more than double would impose an economy-shocking, semipermanent drought over "the Western United States and the semiarid region from North Dakota to Texas."[28] Later in this book, I explain in depth how devastating it would be for the planet, the economy, human health, and national security if the travesty of a doubling or tripling of carbon dioxide were allowed to occur.

The Road to Ruin

It should now be clear that much more global heating—and far worse environmental consequences than in the past half century—are yet to come within the lifespan of people born today if we continue adding ever more carbon, methane, and nitrogen oxide to the air. Continuing to raise their atmospheric concentrations is not compatible with a stable climate.

Figure 4-6 shows how quickly emissions are rising. The first step toward reducing carbon emissions is to stop increasing them. Once capped, emissions can then be abated. The ultimate goal is not only to eliminate emissions but also to find safe bioengineering methods to enhance the natural removal of carbon dioxide from the atmosphere to accelerate the return of safer concentrations more consistent with long-term climate stability. Scientists today recommend that the concentration be brought down at least 50 ppm (12.5 percent) to 350 ppm and preferably lower to provide the Earth a margin of climate safety.

Even that target—125 percent of normal—is unlikely to be safe in the long run once we clean up the cooling haze of reflective industrial pollution. Therefore, striving for 350 ppm will take us in the right direction, but 350 ppm should not be regarded as the final goal.

Limits to Our Knowledge and the Case for Caution

In the preindustrial era, when the Earth experienced gradual episodes of natural warming over millennia, natural ecosystems had an opportunity to adjust by moving across the landscape or adapting genetically. The Earth's natural resource systems, however, had no industrial pollution to contend with, no barriers of asphalt and concrete, nor the other harsh impacts that agriculture, commercial fishing, industrial forestry, and urbanization inflict today. By continuing with our present climate policies, we may be tacitly overestimating how much stress natural resource systems can withstand before collapsing. Similarly, we may be overestimating the capacity of social systems to cope with increasingly severe climate-related hardships and disasters (see chapter 8).

The Earth's climate is not completely predictable. It depends on some processes that have well-defined outcomes, some that depend on chance, and some that are chaotic. Much is still unknown. But we do know these facts with certainty: humans have changed the composition of the atmosphere and, as a result, the climate is warming rapidly at an accelerating rate. Thus, ice is melting, seas are rising, and storms, forest fires, floods, and droughts are on the rise.

We do not yet know enough about the behavior of the Antarctic or Greenland Ice Sheets in a rapidly warming world to predict exactly how they will respond. These two great ice sheets together constitute the world's largest freshwater supply. Melting of even a small fraction of their mass could cause a very large and fast rise in sea level. Both are losing mass.[30]

The West Antarctic Ice Sheet alone holds 10 percent of the world's freshwater and, until fairly recently, it was thought that its melting would raise sea

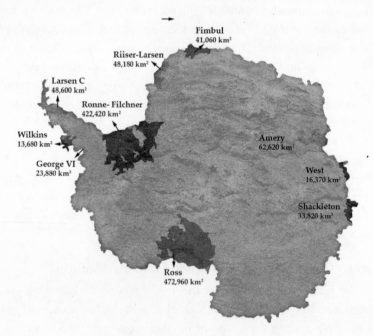

FIGURE 4-6. Antarctica's major ice shelf areas. The darkened area of the Ross Ice Shelf had already broken off in March 2000 to create the world's then-largest iceberg, a floating block of ice the size of Belgium, then. Six years later, 160 square miles of ice broke away from the Wilkins Ice Shelf. Courtesy of Ted Scambos, National Snow and Ice Data Center, University of Colorado, Boulder.

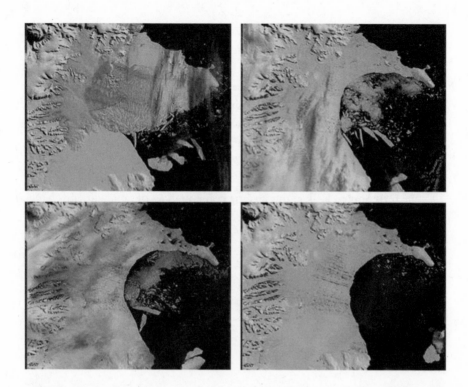

FIGURE 4-7. NASA satellite images document the collapse of the Larsen-B ice shelf between January 31 and March 5, 2002. Courtesy of NASA/Goddard Space Flight Center, Scientific Visualization Studio.

level by 16.4 to 19.7 feet worldwide. New research, however, has shown that loss of the ice sheet might raise sea level by as much as 23 feet, which would put coastal cities like Washington, DC, under water.[31] As we have indicated earlier, if the Greenland Ice Sheet were to melt completely, a process that would take centuries, this would raise sea level about 23 feet.[32] If all of the Antarctic Ice Sheet (see figure 4-6) were to melt someday, that would raise sea level by more than 200 feet).[33] This possibility has been dismissed by some scientists who argue that it would require an increase in global temperature of 36°F,[34] but serious flaws have been found in some of the modeling assumptions used for those conclusions.[35]

With regard to the massive IPCC *Climate Change 2007* report, Dr. Hansen has pointed out that "Disappearing ice shelves, ice stream dynamics, and iceberg melting were not included in global climate models used for IPCC studies (see figure 4-7). This failure to take into account the increased discharge of icebergs to the ocean, where they melt much more rapidly than they would if they had remained as an ice block on land, probably explains the models' inability to predict realistic sea level change."[36]

Because of all the unknowns and the enormous consequences of miscalculation, common sense dictates proceeding with utmost caution, not seeing how close to the brink of climate disaster we can go before turning back. This means doing everything we can to minimize the release of heat-trapping gases and to maximize their recapture. (My forthcoming book *Climate Solutions: Turning Climate Crisis Into Jobs, Prosperity, and a Sustainable Future* offers some ideas on how to do this.)

We turn now to look at the specific consequences of rapid climate change for the US[m] We will then explain the staggering financial consequences of major climate change.

---

[m] John J. Berger, *Climate Solutions: Turning Climate Crisis Into Jobs, Prosperity, and a Sustainable Future* (Berkeley, CA: Northbrae Books, 2013).

CHAPTER 5

The United States in Peril

Surely now, then, is the time to recognize that we cannot have capitalism without nature's capital—we cannot sustain our human economy without sustaining nature's economy.

—HRH CHARLES, PRINCE OF WALES

Climate change will have astounding economic, financial, and ecological impacts in the United States. Multitrillion dollar losses to the economy are on the horizon as rising seas conquer major coastal cities and droughts, floods, heat waves, and crop failures afflict the heartland. As opposed to the future possible global impacts of climate change presented in chapter 1, this chapter focuses more closely on the current and likely impacts of climate change on the United States.

An Interagency Alarm Bell

IN 2009, THE US GOVERNMENT's interagency Global Change Research Program, a prestigious scientific effort, produced a report by 28 scientific experts and reviewers titled *Global Climate Change Impacts in the United States* (GCCIUS).[1] Reports like GCCIUS are usually respectfully noted in the press, then forgotten. But as you are about to see, this report had astounding findings about the probable impacts of climate change in the United States that are worth remembering. Like other analyses cited previously in this book, the report makes clear that even if humans were to stop increasing carbon emissions today and instead stabilized them at the current level, the planet would nonetheless very likely heat up by more than 3.6°F above the preindustrial average global temperature,[a] an amount of heating would constitute "dangerous human interference with the climate," to use the language, of the 1992 UN Framework Convention on Climate Change.

The report contains many sobering forecasts about the impacts of higher temperatures; for example, that the climate is changing at a rate incompatible with the survival of a many wild species[2] that cannot move quickly enough to more suitable habitats. It is now common knowledge that the iconic polar bear is expected to be extinct in Alaska within the next 75 years[3] but not so well known that the United States is also likely to lose all its spruce-fir forests.[4] A broader range of dangerous impacts, however, will affect not just wildlife and scenery, but our food and water supplies and vital natural resource industries, such as fishing and agriculture.

Despite its implicit stern warnings, the GCCIUS report paradoxically remains a conservative document for it often projects the effects of climate change for only 50 to 60 years and rarely for more than the next 100 to 150 years, even though the effects of climate change, particularly sea-level rise, will continue for millennia during which they will be compounded through positive climate feedbacks. In its totality, the report shows that climate change is initiating so many processes whose consequences we know so little about that it is fair to infer that we are recklessly experimenting with the natural world—also known as our life-support system. With considerable restraint, the authors remark, "Human-induced climate change is projected to be larger and more rapid than any experienced by modern society so there are limits to what can be learned from the past."[5] That's a polite way of saying, "We don't quite know what's going to happen." Yet the report was able to make some reasonably confident forecasts.

North American Temperature Forecasts

While the Earth has on average warmed 1.4°F over the past century as noted, the average temperature increase in North America has been about a third greater than the global average—2°F. The report concludes that under a scenario assuming that the world continues on its current high-emissions trajectory, the United States should expect a massive average temperature increase of 7 to 11°F by the end of the century. The GCCIUS report warns,

---

[a] Were the atmospheric concentration of carbon dioxide stabilized at 400 ppm, the only hope for avoiding an eventual 3.6°F (2°C) temperature increase would be if the emission of other global warming agents like methane and black carbon soot were very sharply reduced to compensate for the high carbon dioxide concentration.

however, that we may be facing "even larger changes in climate than current scenarios and models project."

Offshore, US coastal waters could get 4 to 8°F hotter this century, both summer and winter.[6] Changes of 4 to 8°F are very large for ocean temperatures, as anyone who has ever kept an aquarium will understand. Aquarium fish comfortable at 76°F will not necessarily even survive at 84°F. Some coastal waters have already gotten 2°F hotter in just a third of a century.[7] Certain Pacific Northwest coastal waters have warmed as much as 3.2°F since the 1960s.[8]

US Economic Impacts[b]

As US temperatures continue climbing, air and water quality both will tend to deteriorate,[c] and the quality of life in many places will be impaired by heat waves and by both heat-related and tropical illnesses. With the heat, seasonal water shortages in many parts of the United States will become more common and severe. Drought areas also will expand, impairing agriculture, forestry, and other water-dependent industries, including fishing and inland waterborne freight transport. Many power plants and factories will encounter water shortages.

Agriculture. Although crops like melons and tomatoes do well in heat, most grains and vegetables do not, especially when the heat is accompanied by water shortages. Climate change will force crops to make major physiological adjustments or die. For example, crops will have to deal with more heat, changes in the length and the timing of seasons, changes in water availability, more extreme weather events, and higher atmospheric carbon dioxide levels. This set of challenges will be further complicated by new diseases, insect pests, and invasive weeds.[d] Weeds and invasive species in general will tend to outperform native species in a hotter world, and rising carbon dioxide will increase pollen production and prolong the allergy season in the United States, making hayfever sufferers miserable longer.

---

[b] Global economic impacts are discussed in chapter 7. And though human health effects clearly have important economic consequences, human health is broadly treated in chapter 8.

[c] See page 75 for a discussion of water quality impacts; the reasons why air quality declines with increased temperature are spelled out elsewhere in the book, for example, on page 158.

[d] Initially some plants will respond to higher carbon dioxide with faster growth, but the effect tends to saturate as carbon dioxide concentrations continue rising. Faster growth at higher temperatures can result in lower nutrient content per volume of plant material.

Livestock like beef cattle will be harmed by increases in temperature and carbon dioxide. These conditions lower the nutritional content of grass and sometimes make it less digestible,[e] causing rangeland and pasture forage quality to decline in the semiarid western states. These will therefore support fewer cattle. Meanwhile, high summer temperatures will increase livestock stress, parasites, and disease. That, in turn, will depress pork, lamb, beef, and milk production.[9]

Water. Power stations that consume fuel to produce electricity typically require large volumes of water for cooling, although some use dry cooling towers. The nation's water-cooled power stations withdraw almost as much freshwater for cooling as farmers withdraw for irrigation.[10] But with each percent drop in stream flow in the hotter, drier Colorado River basin, for example, power production there will fall by 6 to 9 percent.[11]

Higher temperatures will not only lower the power output of fossil fuel and nuclear power plants at times when electrical demand for air conditioning will be high, but will also reduce hydropower production. Even when the *volume* of cooling water is adequate for power plant operation, its temperature will be higher in a hotter world. This will lower the power plants' thermodynamic efficiency and power output.

According to GCCIUS, "Future water constraints on electricity production in thermal power plants are projected for Arizona, Utah, Texas, Louisiana, Georgia, Alabama, Florida, California, Oregon, and Washington state by 2025."[12] Water shortages will also block construction of some new nuclear and fossil fuel plants and reduce operations at some existing ones.

Heat, Water, Fish, and Storms. One of the effects of higher temperatures in lakes is a more strongly pronounced vertical layering of warm and cool water. This reduces or delays the lake's seasonal overturn that brings oxygen to the bottom, so less oxygen is available at depth for fish. Also, as the mercury climbs, evaporation from reservoirs increases, reducing the amount of water available for hydropower production, as well as for recreation, irrigation, and fishing downstream from dams. In addition, extreme heat above 90°F (32.2°C) also melts asphalt highways, warps rail lines, and reduces aircraft cargo capacity.[f]

---

[e] Carbon dioxide stimulates the plant to grow, but the nutrient content per volume is reduced.

[f] Because hot air expands and is less dense than cool air, aircraft engines have to do more work to keep planes airborne. As temperatures get really hot, some planes are unable to fly.

Water scarcity and high temperatures will lower water quality by concentrating pollutants in a smaller volume of water, and by contributing to algal blooms in lakes, rivers, and coastal areas. As the climate changes, a larger proportion of precipitation will be falling in heavy downpours with pelting rains resulting in increased crop damage. Costly mudslides, landslides, road failures, and huge floods—like the June 2008 Midwestern flood—will happen more often.

High-priced stormwater collection infrastructure will be needed to deal with higher stormwater flows. More violent storms also produce expensive coastal consequences. These include damage to ships in harbors and offshore drilling rigs, rerouting of ship, barge, and air traffic, delays of exports and other shipments, interruption of fuel delivery, and power outages.

Sea-Level Rise

These economic impacts in the interior United States may well be overshadowed, however, by the destructive economic impacts that sea-level rise will have on coastal areas due to the forced relocation of people, facilities, and equipment, and the submergence of valuable land. Rising seas mean additional costs to build flood barriers, elevate structures, and reconstruct coastal facilities farther inland, as well as to armor coastlines against erosion. (Coastal structures include ports, coastal airports, and freight yards, as well as thousands of miles of major highways, bridges, pipelines, and transmission lines.)

As the northeastern United States learned in 2012 during Superstorm Sandy, underground tunnels for trains and vehicles will also need to be closed or protected. Insurance costs will increase for coastal and floodplain property at risk from sea-level rise, flooding, and the more severe hurricanes, storms, and sea surges of a warmer world.

The report projects that sea level off the US coast could rise 3 to 4 feet. That would "place into jeopardy existing homes, businesses, and infrastructure, including roads, ports, and water and sewage systems." Portions of many cities would be subject to inundation by ocean water during storm surges or even during regular high tides."[13] Because "most [US] shorelines are subsiding [sinking] to various degrees—from a few inches to over 2 feet per century,"[14] a sea-level rise of 3 feet would also inundate about two-thirds of all the coastal wetlands and marshes in the contiguous United States. This would be a huge loss: wetlands are not only valuable for wildlife, recreation, fishing, and for protecting the coast from storm surges and hurricanes, but also for removing atmospheric carbon.

Coastal losses. Trillions of dollars have been invested in the coastal zones of the United States and at least a trillion dollars in economic activity are created there and in ocean activities each year.[15] The rising waters would not only cause the permanent loss of thousands of square miles of US land, but would also damage homes, businesses, and infrastructure. Marinas, restaurants, hotels, fishing fleets, and charter services would all suffer severely and many would not survive. Islands like the state of Hawaii face similar but even graver threats. They depend heavily on ports, coastal tourism, fishing, and coral reefs. They are also especially vulnerable to heavy rains and hurricanes.

Perhaps because many people still think of sea-level rise as a long-term problem, they are not yet perturbed about it. Eventually, however, as the coastal economy, real estate, and coastal ecosystems are progressively degraded, residents of these areas *will* become concerned. (See chapter 11 for more about sea-level rise.)

Dead Zones off US Coasts

The report's authors project that a rise in spring freshwater runoff along the East and Gulf Coasts, due to increased precipitation, will be high in dissolved nitrogen and will therefore promote algal blooms in the near-shore waters and subsequent dead zones littered with decaying algae on the ocean floor.

The relatively low-density layer of new freshwater runoff tends to flow above the higher-density saltwater along the coasts. The trapped freshwater "lid" is then heated by the sun and warmer air, creating an upper layer of warmer water in coastal areas. This reduces water mixing by wind and waves, and so further reduces the supply of oxygen to the deeper bottom waters, killing marine life. Dead zones are a serious problem off parts of the Eastern Seaboard, the Gulf Coast, and the Pacific Northwest but are seldom seen by an the general public.

Acidification is another serious oceanic impact that is already afflicting US coastal waters and the entire ocean. As carbon dioxide continues accumulating in the atmosphere, about a quarter to a third of it ends up in the ocean where it leads to the formation of carbonic acid. (See chapter 9 for the global implications.)

Alaskan Impacts

During the longer intervals between heavier rains, forests in Alaska and the West will parch and be much more susceptible to large, uncontrollable wildfires.[g] Alaska is already experiencing hotter, drier weather and increased forest fires.[16]

Huge forest losses have also occurred in the white spruce forests of Alaska as spruce bark beetles have spread. Warmer weather allows them to survive through the winter and mature in half the normal time. Much of Alaska's Kenai Peninsula in the Chugach National Forest has been devastated. The beetle infestations have been aggravated by drought, which dries the forest floor and reduces the flow of sap that healthy trees normally use to repel beetles. Forests sickened by beetle invasion are then defenseless against wildfires promoted by hotter, drier summer conditions.[17]

Areas burned by wildfire in Alaska and Canada have tripled since the mid-twentieth century and may quadruple in Alaska by the end of this century.[18] Other forests in Alaska are also being destroyed by thawing permafrost, which heaves shallow-rooted trees upward, destroying their root systems and creating what are sometimes called "drunken forests." (Forests are also drying and succumbing to beetle infestations in the Pacific Northwest and Canada, a disaster of epic proportions for large areas of North American forests and the people, plants, and animals that depend on them.[h])

Many lakes are shrinking in Alaska as the state's average temperature has increased by 3.4°F over the past 50 years, reducing surface water supplies. By the end of the century, temperatures in Alaska might rise by as much as 13°F if the world follows a high-carbon-emission scenario. Forest damage and reduction of water supplies mean less habitat for native Alaskan fish, mammals, and

---

g Since the GCCIUS report was written, new research published in the journal *Nature Geoscience* has found that wildfires in Alaska have become much more frequent and severe. (Merritt R. Turetsky et al., "Recent acceleration of biomass burning and carbon losses in Alaskan forests and peatlands," *Nature Geoscience,* vol. 4, no. 1, pp. 27–31, 2011.) As the ground warms up earlier in the spring and stays warm and dry later in the summer, fires are able to burn more deeply into the soil where centuries of carbon have been stored. Thus, in a single decade (from 2000 to 2009), the forests studied released twice as much carbon as in the previous five decades. (Climate Central Blog, "Climate in Context: Alaska Wildfires Release Decades of Carbon," http://www.climatecentral.org/blog/ climate-in-context-alaska-wildfires-release-decades-of-carbon.) This has important implications for the world's climate since half the world's soil carbon is stored in its northern forests, and these are likely changing from carbon sinks to carbon sources. As northern temperatures rise, more carbon will emerge from storage in the soil and, in turn, will further increase global temperature in a dangerous positive feedback cycle.

h Forty percent of the commercial pine trees of British Columbia, Canada, have already been killed by the mountain pine beetle, and almost an equal amount are likely to die in the next decade. Dry summer conditions are also limiting forest growth and altering species composition, possibly driving some species to extinction.

birds, all critical to the well-being of traditional indigenous peoples engaged in subsistence hunting and gathering.

Thawing is particularly severe in interior Alaska, where the permafrost temperature is just below freezing. This sometimes allows perched lakes to drain, making soils soggy and muddy. In developed areas, soil sinking associated with thawing permafrost undermines roads, runways, buildings highways, airports, and pipelines as well as water and sewer systems, eventually adding billions to infrastructure and private property maintenance costs to stabilize structures.

Thawing permafrost combined with increased storm activity and a decline in sea ice that once protected shores is causing Alaska's northeast coastlines to erode twice as fast as in the mid-twentieth century. Low-lying coastal communities are particularly vulnerable, some losing dozens of feet of land a year. For example, in 2003, the town of Newtok, Alaska, lost 110 feet of shoreline in some places.[19] Many native communities that live on the coast or depend on hunting, fishing, and gathering are being quietly devastated.[20]

The Pacific Northwest

Wholesale ecological changes are already underway in the Pacific Northwest, affecting precipitation, stream flow, snowpack, fisheries, forests, wildlife, and, of course, sea level. Due to warmer winter temperatures, the area's winter snowpack is melting earlier, causing peak stream flow to occur earlier in the year. That leads to major reductions in water availability during the hot, dry summer months when water is most needed for irrigation, hydropower, wildlife, and fisheries.

Salmon spawning streams will be scoured by higher winter stream flows because of the earlier, more concentrated snowmelt. This will prematurely flush salmon eggs or fry from spawning beds. During the summer and fall, stream flows will be lower and warmer, so salmon and other coldwater fish will be stressed and more susceptible to diseases and parasites, both of which thrive in warmer conditions. Because of harmful land management and other environmental conditions, especially deforestation, "Most wild Pacific salmon populations are extinct or imperiled in 56 percent of their historical range in the Northwest and California, and populations are down more than 90 percent in the Columbia River system," according to GCCIUS.[20] The report projects that rising temperatures will eliminate salmon and other coldwater fish from a third of their habitat in the Pacific Northwest by the end of the century. Some 40 percent of the salmon in the Pacific Northwest could be gone by 2050.[21] Needless to say, these are devastating impacts.

The Southwest

The impacts to the southwestern United States will be shocking to the unprepared. The Southwest has a long history of droughts and water scarcity. As the climate heats up, natural ecosystems as well as agriculture and managed landscapes will go thirsty. A return to long-term "megadrought" conditions is possible.[i] That would stifle and ultimately strangle the region's economy, bringing enormous losses in biodiversity and agricultural productivity.

Initially, as in the Pacific Northwest and in Alaska, some reductions in winter snowpack will occur,[j] but in addition, the Southwest will have to endure reductions in spring rain and snowfall as well as rising temperatures that could surge 8 to 10°F by the end of the century. This will happen in a region that is already overpumping its groundwater and experiencing forest damage.

As in the Northwest and Alaska, drought stress will make forests far more prone to insect attacks as runoff decreases and soils dry out, while climate change will expand the range of bark beetles and other pests. These factors will combine to make forests more susceptible to wildfires that, in turn, will be more frequent, severe, and difficult to extinguish. This is already evident in more than 4,600 square miles of piñon-juniper forest in the Four Corners area of Arizona, New Mexico, Colorado, and Utah. As much as 60 to 90 percent of high-elevation forests of California will also be in decline by the end of the century.[22]

It might at first seem that classic Southwest desert landscapes would be unscathed by a little additional heat, but no. Heat-tolerant, flammable invasive grasses from Africa (buffalo grass and red brome) are invading. When they burn, the blaze often kills the saguaro cactus and Joshua trees, symbols of those deserts.

California

Climate change will hit California hard. This biologically diverse state is divided by many natural and manmade barriers to the plant and animal migrations that become necessary when climate change makes habitats inhospitable. In part for this reason, GCCIUS warns that "two-thirds of the state's more than 5,500 native plant species are projected to experience range reductions up to 80 percent before the end of this century under projected

---

[i] A megadrought is a prolonged drought lasting two decades or longer.

[j] One of the most obvious but almost incidental casualties will be economic damage to ski resorts and winter tourism from reductions in the snowpack and from increases in avalanches due to wetter snow.

warming." This staggering and imminent loss of native plants will in turn drive many animals to extinction—a huge blow to a many of the state's natural ecosystems—in less than 90 years. The ecological carnage will only worsen in the centuries thereafter. Such extensive damage will not just be costly to nature, but will also reduce economic activity.

In California's multibillion-dollar agricultural sector, economically important fruit and nut trees, such as almonds, apricots, avocados, walnuts, and olives, will be unable to produce in parts of central and coastal California—they require winter chilling. Grapes used in some of California's famous wines also will not flourish in the warmer days to come.

Sickened, insect-infested forest and grassland areas will have little appeal for outdoor recreation. Instead of experiencing "the gladness of nature," to borrow a phrase from conservationist John Muir, people will find ailing plants and dying ecosystems everywhere.

The Great Plains

Water availability will be the big issue as the weather gets hotter on the Great Plains of the Dakotas, Montana, Wyoming, Kansas, Nebraska, Oklahoma, Texas, eastern parts of Colorado, and New Mexico. In general, the northern plains will get wetter while the southern plains will become drier. Temperatures in the Great Plains are rising significantly and, by the end of the century, might rise by 5 to 10°F.

Seventy percent of the region is used for ranching and growing grain, fodder, and cotton, but virtually the whole region is withdrawing water from the High Plains aquifer faster than it is being replenished by rainfall.[k] The aquifer currently provides drinking water for 80 percent of the region's population and is used to irrigate millions of acres of land.[23] How long it can continue doing so is the question. As temperatures rise and rainfall decreases in the southern Great Plains, and as irrigation needs increase due to rising temperature, some areas will clearly need to shift from irrigated to less reliable rain-fed agriculture.

The Midwest

The big changes in the Midwest will be more extreme summer heat, more flooding, and a drop in the Great Lakes. All these trends are already becoming evident and will be aggravated as the climate heats up. Heat waves,

---

[k] Apart from rainwater that percolates in from above, the aquifer also contains nonrenewable glacial meltwater that permeated the ground at the end of the last ice age.

such as the unusual 1995 event that took 750 lives in Chicago and 90 in Milwaukee, will happen up to three times a year in a world with high emissions of heat-trapping gases. (See chapter 8 for more on the health consequences of heat waves.)

Higher temperatures not only mean more air pollution, but also more insect pests (ticks and mosquitoes) and greater evaporation from the Great Lakes. As lake levels fall, less water is available for transportation, fisheries, agriculture, industry, municipalities, tourism, and recreation, as well as for energy production and drinking. Once water levels drop, people will be farther from lakefronts and shoreline facilities. Docks, marinas, and harbors will be stranded. Wetlands and other fish habitat will be lost. Meanwhile, increased winter and spring precipitation with more heavy rains will lead to an increase in severe flooding, as in the Great Flood of 1993 along 500 miles of the Mississippi River.

Paradoxically, a reduction in summer rainfall will dry up streams and wetlands and reduce river flows and groundwater infiltration. That will hit fish and wildlife populations hard, and low flows on major rivers like the Mississippi will periodically make them impassable to barges and freighters.

What may seem like modest climate changes when viewed year to year will produce dramatic change in less than a century. By 2100, Illinois will have a summer climate resembling parts of south and central Texas. Michigan summers will resemble the summer climate now straddling Oklahoma and Texas. The hotter climate will enable plants native to the Southwest to get established in the Midwest. Native species adapted to the Midwest's twentieth-century climate will decline.

Coldwater fish habitats will shrink. Warming lakes will become depleted of oxygen as warm surface waters cap cooler waters below and the layering of water at different temperatures will become more pronounced. Oxygen-rich surface water will then not mix as well into lake depths. Warmer oxygen-poor water then will react more readily with mercury and other contaminants, introducing them into the food web. Some fish will become too toxic to eat.

Growing seasons will lengthen as the frost-free season is extended by a week or so, and some plants will benefit from increased carbon dioxide for photosynthesis. These gains for agriculture will be more than offset by less summer rain, higher temperatures, lower soil moisture, and more disease-causing organisms, pests and weeds, as well as by lower livestock productivity.

The Southeast

Much of what has been said about the Midwest—temperature increases, declines in agricultural production, increases in extreme heat and water scarcity, as well as oxygen depletion in lakes—also applies to the already steamy Southeast. But these problems will be compounded by an increase in drought and the drying up of lakes, ponds, and wetlands. The region will also suffer from sea-level rise, land subsidence, coastal flood risks, and exposure to increasingly powerful Atlantic hurricanes, along with insect outbreaks and wildfires in its forests.

Given that average annual temperatures in the humid Southeast have increased 2°F since the 1970s and are expected to rise 4.5°F by 2080 under a low-emission scenario and by as much as 10.5°F in the summer under a high-emission scenario, human health as well as livestock and wildlife health will be threatened during the peak summer heat. In places like North Florida, the temperature will be above 90°F for almost six months a year by the end of the century. Even beef cattle are harmed "at continuous temperatures in the 90 to 100°F range [especially as] the humidity increases,"[24] and other livestock will also suffer while crop yields will decrease. Many species of plants and wildlife will die or become endangered. Forests will decline, and more invasive plants will take over native ecosystems.

More waterborne diseases will spread (for example, in shellfish), and seas will rise off the southeastern United States by an average of two feet by the century's end. "More frequent storm surge flooding and permanent inundation of coastal ecosystems and communities is likely in some low-lying areas, particularly along the central Gulf Coast where the land surface is sinking. Rapid acceleration in the rate of increase in sea-level rise could threaten a large portion of the Southeast coastal zone."[25] Even intermittently flooded areas could experience persistent saltwater intrusion that would contaminate freshwater aquifers and increase the salinity of coastal estuaries, rivers, and wetlands. These partially submerged ecosystems will be eroded by hurricanes and storm surges that will have fewer and smaller barrier islands in their path to buffer beaches and other coastal areas.

The Northeast

Major climate changes are already evident in the Northeast. Just in the past 40 years, the climate has warmed an average of 2°F, and winter temperatures have risen by an average of 4°F. Within a few more decades, winter temperature increases could double, and temperatures could rise another 3.5°F in summer. These changes "could dramatically alter the region's economy, landscape, character, and quality of life."[26]

If high emissions continue, the frequency of heavy rains is expected to increase, with more rainfall occurring in intense downpours, and with more winter precipitation as rain rather than snow, and with earlier snow melting. That would reduce the snow season by half in the northern part of the region and to only "a week or two" in the southern part, GCCIUS found.

People who moved to the Northeast partly out of love for the crisp New England spring and fall weather or for relatively temperate summers will thus be in for a shock. By the latter part of this century, New Hampshire residents could find their climate similar to North Carolina's. The ski and snowmobile industries would decline, of course, but that would likely be a relatively minor problem compared with other expected impacts. Heat waves, now uncommon, would become more prevalent over the next few decades. Air quality would deteriorate. Dairy production would fall as heat-stressed cows give less milk in summer, and much of the Northeast would no longer be able to produce their traditional apples, blueberries, cranberries, or maple syrup.

The cod fishing industry would suffer economic losses as warmer ocean temperatures reduce cod growth and survival. Perhaps the most serious threat to the region will come from rising sea level, which will bring similar effects to the Northeast as in other regions. By the final years of the twenty-first century, floods that on the average only occurred once in a hundred years will be striking roughly every decade!

Inertia, Incredulity, and the Psychology of Climate Denial

It's worth pondering why, with information on the harsh impacts of climate change so readily available, many people, especially in the United States, seem so unwilling to heed the trenchant warnings. Certainly part of the story is the inherent complexity of understanding the climate system, and the difficulty of fully grasping it without scientific training. Maybe it's also because the impacts of climate change are so dire they produce a surreal sense of gloom. How much less disquieting it is to focus on more controllable mundane daily events.

People also often detach psychologically from unremitting bad news or deny it. It is easier to deal with pressing immediate concerns. And we welcome distractions. Facebook, Netflix, Reddit, or favorite TV shows beckon, as do movies, sports, music, and e-mail. How much more pleasant to escape from seemingly intractable global problems. How much easier to refrain from getting involved and to embrace the illusion that life can go on as normal if we ignore the gathering climate crisis.

But could there be more to it than that? Why else might people in the United States be so unwilling to fully face the realities of climate change? Why does much of the nation seem so lethargic in responding to the ever-more urgent warnings that scientists are delivering? I believe that the reasons are complex. The phenomenon known as cognitive dissonance may explain the tendency to deny climate change or its implications. In this way of thinking, one is so heavily invested cognitively or psychologically in an existing belief system that the beliefs are fortified by an often subconscious awareness of how damaging acceptance of a new belief would be to the older, deep-seated and familiar belief systems.

Cognitive dissonance operates even more powerfully when a new reality not only threatens old beliefs but also threatens to impose unwanted new economic costs, sometimes in the form of inconvenience, lifestyle changes, or threats to investments. Such changes are likely, given the pervasiveness of the carbon energy systems on which our society currently depends. Cognitive dissonance thus "kicks in" as we each consciously or subconsciously weigh the personal costs of more fully realigning our lives with the low-to-zero-carbon lifestyle that a complete commitment to climate protection requires.

Probably far more important than cognitive dissonance and scientific complexity is the impact of the prolonged campaign waged against climate science by representatives of the fossil fuel industry, a story I relate in my earlier book, *Climate Myths: The Campaign Against Climate Science*.[1] In essence, our current climate crisis reminds me of the words of the famous seventeenth-century French philosopher, scientist, and mathematician Blaise Pascal (1623–1662), who wrote: "We run carelessly over the precipice after having put something in front of us to prevent us seeing it."

People also tend to respond to ominous warnings of global heating with fatalism: there's no point in getting overly concerned; the issue is so enormous it's beyond anyone's control. Some people resolve the dissonance between the looming catastrophe and their sense of powerlessness by denying the disturbing evidence.[m] Instead of heeding warnings, they find it more reassuring to believe: "Nothing this extraordinary and nightmarish can ever

---

[1] John J. Berger, *Climate Myths: The Campaign Against Climate Science* (Berkeley, CA: Northbrae Books, 2011).

[m] For an account of an organized effort to capitalize on the natural tendency to deny the reality of climate change, see *Climate Myths*.

happen here—not to me, nor to my loved ones, nor the places I and my family hold dear."

Paradoxically, the more people seek refuge in illusions, science denial, reality avoidance, and magical thinking, the more self-fulfilling the ominous climate forecasts become. Drastic climate change *will* happen if we do not change the policies producing it. In fact, it would be a logical and consistent culmination to the broader pattern of environmental destruction that civilization has engaged in since industrialization.[n] As proof that the extraordinary can happen in one lifetime, allow me to share some of my own personal experience as an example of how rapidly and profoundly ecosystems are changing around us and yet how people have "normalized" the radical treatment of nature that our society practices.

Environmental Change in My Lifetime

I grew up in Riverdale, New York. As a child, I used to ramble in the summer picking wild raspberries in the woods beside the Hudson River. The breezes then had a unique scent in each season. The clean spring air coming off the river had a crisp nip to it that was actually thrilling and quickened the pulse. It also had a tangy, indefinable yet distinctive fragrance, a blend of new forsythia buds, young grasses, warming soil, and crocuses. One of my early memories, when I was about six years old, was digging in the empty lot next to our apartment building and unearthing the fragrant roots of wild sassafras. We would inhale their delectable aroma. They were at once sweet, pungent, slightly fruity, and yet earthy.

Although this was more than 60 years ago, I still remember the hilly field just across the street from our building. The milkweed and meadow grasses grew so tall they towered over us. We made secret tunnel-like paths through them to a little fort we had built called Grassy Hideout. In winter, when the tall grass and milkweed had died and the snows were deep, we went sleigh riding over the same hill.

Robins were everywhere in the springtime on the great lawn adjoining the building. Every few feet, another robust specimen, each with its own fiefdom and bright orange-red breast, pulled up large earthworms from the fresh, rain-sodden grasses that grew luxuriantly in pesticide-free ground. Insect life in Riverdale was abundant then. In the heat of the summer, when school

---

[n] See the introduction to my book *Restoring the Earth: How Americans Are Working to Renew Our Damaged Environment* (Alfred A. Knopf, 1985; Doubleday-Dell, 1987) and the environmental history sections of *Forests Forever: Their Ecology, Restoration and Protection* (San Francisco, CA and Chicago, IL: Forests Forever Foundation and Center for American Places at Columbia College, Chicago, 2008).

had long closed, we stayed out late after dinner. Sometimes our parents came out, too, for evening walks along honeysuckle-lined paths. At the right time of the season, the warm summer nights were alive with the sparkling lights of mating fireflies. Nature was alive and glowing with exuberance. We would sometimes capture a few fireflies in a jar to watch them closely as they winked on and off in the dark. The fields were also alive with grasshoppers, and, now and then, a stately praying mantis would appear. There were few pigeons in the neighborhood or raucous starlings.

The last time I returned to Riverdale, *à recherche du temps perdu*, houses stood where the sassafras and milkweed grew. Although it was spring, the air no longer smelled the same. Whatever spring scents remained were now commingled with the exhausts from millions of cars and trucks. Song birds and insect life seemed rare. Homeowners now spray herbicides and pesticides in their yards. The fireflies are probably long gone with the woodlands and meadows. Pigeons and other city dwellers have moved in. Most of the open space is gone, apart from parkland. The story is not unusual. Throughout the United States, much of our natural environment has been destroyed as open space has been converted to asphalt, cement, and houses, sacrificing native plants and animal life that gets in the way. Perhaps this is why reading the GCCIUS report sends shivers up my spine. I have seen environmental destruction. The report's projections thus seem very believable to me.

I now live in California where there are still large open spaces near my home. The desiccated summer woodlands of the Sierra foothills, however, can easily be torched by lightning as climate change brings megadroughts and massive fires back to the West and Southwest. After these very large, hot blazes, unprotected soil may well be reduced to powdery dust, easily washed away in the next heavy rain. As the summer environment grows more arid and less hospitable, thirsty, starving wildlife scatters or dies. California's emblematic giant sequoias (*Sequoia gigantea*)—trees that have withstood the storms of winter and the baking of summer sun for millennia—can nonetheless ultimately be vanquished by climate change. Experts fear that new seedlings will be unable to survive the dry season and that the old trees will succumb to the lack of soil moisture, their once-healthy, fog-shrouded, wispy green needles baked to dry brown. Many pine and spruce-cedar forests in the United States are already slowly drying, sick and teetering, attacked by insects, baking in summertime, and bursting easily into flame.

In Proust's *Remembrance of Things Past*, there might be a faint hope of return. In nature today, the nourishing womb of life still pulses and throbs.

But should the chance to save the climate elude us, we cannot ever go back and resuscitate landscapes laid waste by climate change.

No matter how understandable our as yet torpid and ineffectual response may be, climate realities are closing in on us. The climate system is unforgiving of human ignorance, political gridlock, economic expediency, shortsightedness, and hubris. Thus, the key question is: can we wake ourselves up quickly enough and rouse the rest of the world to prevent the scenarios in this chapter from happening? Time is exceedingly short, and bold, decisive action is required.°

As chapter 7 will show, however, an unwillingness to make major changes in our energy use may not just destroy ecosystems and cripple regional economies. Obstinacy about making needed policy changes can create conditions that push us over natural climate system "tipping points," triggering uncontrollable, irreversible climate catastrophe.

° The third book of this current climate series, *Climate Solutions: Turning Climate Crisis into Jobs, Prosperity, and a Sustainable Future,* (Berkeley, CA: Northbrae Books, 2014) documents the steps that need to be taken and indicates how to provide ample energy supplies for the world at an affordable price without destroying the climate.

CHAPTER 6

Tipping Point Perils

The climate is nearing tipping points.

—Dr. James Hansen

This chapter explains the risks of abrupt climate change and the notion of climate thresholds. It introduces the concepts of climate feedback, tipping elements, and runaway global heating. The major tipping elements discussed include the Greenland and West Antarctic Ice Sheets, Arctic sea ice, ocean currents, the Amazon rainforest, the Asian monsoon, and thawing methane deposits in the Arctic. Disturbances to these tipping elements could not only accelerate climate change but could also potentially cause a climate chain reaction that could bring about an irreversible, self-intensifying warming cycle.

The Climate System and Climate Feedback

T O ADEQUATELY EXPLAIN THE RISKS of climate tipping points, it is virtually inescapable to spend a few initial pages clarifying the basic terms and concepts applied in the rest of the chapter. If wading through this becomes too tedious, you can always simply skip to page 95 where I begin to apply these principles. But if you can soldier through the scientific theory, you will likely gain a deeper understanding of the mechanisms that cause climate change.

The climate system consists of the atmosphere, the hydrosphere (the planet's liquid water in all forms), the cryosphere (ice in all forms), the land, and the biosphere (the ensemble of all living systems) and all their complex interactions. At the start of chapter 4 (see page 51), I explained that a climate forcing agent is a substance, process, or event that produces a change in climate system. Climate, in turn, is itself a planetary-scale forcing agent. As defined by Barnofsky et al.,[1] global forcing agents are those which can produce a "state shift" in Earth's biological systems, meaning an abrupt departure from previous normal values and the establishment of a new mean. Other than climate change, the range of planetary forcing agents includes human population growth, habitat modification, as well as energy production and use.

With respect to climate forcing agents, a change in any component of the climate system will then induce changes in the others. Thus, changes in vegetation, moisture, ice, solar radiation, albedo (reflectivity) of the planet's surface, nutrient flows to ecosystems, and atmospheric composition provoke changes in the other components of the system and are all forcing agents. Some forcing agents are considered intrinsic to the internal operation of the climate system, whereas others are external to it, such as variations in solar radiation toward Earth, volcanic eruptions, and human interference in climate; these are known as external forcing agents.

Conceptually, climate system feedback occurs when the climate's response to a forcing agent serves to either reinforce or dampen the initial change. Feedback can thus be thought of as a self-perpetuating, continuous loop process—a shift in one climate process that affects another climate process that in turn responds by exerting a recursive effect on the first process. The climate system's response to a forcing—whether a feedback is induced or not—will invariably not just be expressed in a change in a single variable like temperature, but by interactive adjustments in precipitation, ice cover, ecosystems, atmospheric conditions, and a range of other biogeochemical variables.

Apart from the operation of climate feedback processes, some ordinary initial climate system adjustments in response to climate forcing will be linear, meaning that the effects or outputs are directly proportional to inputs so the relationship can be represented by a straight line on a graph. In that case, an increase or decrease in forcing is matched by a proportional increase or decrease in a climate system feature. Each step in the response happens at a rate that can be described as a series of simple additions or subtractions.

Other ordinary climate system responses, however, are nonlinear, meaning they are disproportionate. Thus, a one-unit change in a climate system variable will not produce a regular stepwise change in the other but instead a response that can only be described as an increasing or decreasing *multiple* of the original change; these changes are known as exponential and, when graphed, appear as curved lines of varying slope.

The power of feedback processes is that they can produce rapidly increasing or decreasing exponential responses from even linear processes. For example, if the first process in a feedback system is a linear increase in atmospheric greenhouse gas concentration that triggers rising temperatures that lead in turn to the release of more greenhouse gas, then it's an intrinsic property of the feedback process that it will force temperature to rise faster and faster for each additional unit of greenhouse gas released. In short, the climate sometimes responds to a forcing in a way that can greatly magnify a small climate change to produce a much larger effect. For example, if a climate forcing induces an amplifying temperature response in a feedback cycle, then the initial temperature change can ultimately propel the climate toward a warmer or cooler state at an accelerating rate.

The mathematical formula for calculating the total effect of a forcing on the climate is Total Effect = Direct Forcing Effect x $(1/(1-g)$, where g is the feedback factor.[2] Conventional estimates of the sum of all individual positive and negative planetary forcings, g_i, are in the range of 0.4 to 0.78.[3] As Lashof et al. have pointed out, the climate system is very sensitive to additional feedback. Thus, if the actual forcings are at the lower (0.4) end of this scale, then a one-degree (C or F) temperature gain resulting directly from an increase in atmospheric carbon dioxide would produce an eventual temperature gain of 1.6 degrees once feedbacks occurred.

By contrast, if the actual aggregate planetary feedback factor turns out to be at the upper end of the scale (0.78), then one degree of direct forcing is transformed by climate system feedback into a 4.5-degree rise in temperature. Moreover, as Lashof et al. also noted, if a still unknown feedback were to raise the total climate system feedback factor even very slightly above the high-end estimate—say by .05—then the impact of a 1-degree temperature increase would rise from 4.5 to almost 6 degrees! Similarly, if the forcing factor were 0.9, then the effect of the 1-degree direct forcing would be magnified tenfold to produce a 10-degree planetary temperature gain! A rise in g above 1 would render the whole climate system unstable.

Real World Examples of Forcing

The climate system is a mass of complex interactions and positive and negative feedbacks. Without knowing the strengths of countervailing processes, there is no way to tell what the effect of a change in the climate system will ultimately be. Computer models are essential to calculating the net effects as innumerable climate changes and feedbacks reverberate simultaneously through the entire climate system. Even in the following small and isolated example, things can quickly get very complicated and very confusing, making it essential to know the magnitudes of major effects and to distinguish between relatively small-scale local effects and global climate changes.

Let's say that vegetation is removed from a few square miles of land. That could happen in the course of a major logging operation. With dark green vegetation removed, the color of the ground would probably be lightened. That would increase the reflection of solar energy—a local cooling feedback—but it simultaneously reduces evapotranspiration (the movement of water from the ground through plant leaves to the atmosphere. With less moisture aloft, fewer clouds will form locally. That will reduce the interception by clouds of heat and sunlight radiated skyward from the Earth—a warming effect—and will reduce the reflection of solar energy from the tops of the clouds—a cooling effect. The reduction in the former effect will tend to cool the area and the reduction in the latter effect will tend to warm the area. (The net effect of clouds depends on their type and altitude.)

In addition to the processes above, with less transpiration there may be less regional rainfall, so in a generally warming world, the forest may not be able to grow back. The forest had removed carbon dioxide from the air before it was cut down, so now more greenhouse gas is aloft, another warming effect and one that will have a global effect because of the long atmospheric residence time of the carbon and the global circulation of the atmosphere. Now, too, because the atmospheric concentration of water vapor (a powerful heat-trapping gas) has been reduced locally by the loss of evapotranspiration, some cooling results. But it remains largely local because water's residence time in the atmosphere is short (a matter of days) compared with carbon dioxide's long residence time, and so the cooling effect diminishes rapidly as the air travels around the Earth and reaches thermal equilibrium with the rest of the atmosphere, but the carbon dioxide and the warming it induces remain.

Complicated as all this is, I have nonetheless omitted some processes occurring in the soil and leaf surfaces to avoid further complexity.[a] So, as noted, a computer model is needed to work out exactly what happens on balance, but it is likely that the net effect of all this would be a global warming and a local cooling. Many site-specific factors would need to be known quantitatively for

---

[a] In addition to all the effects of deforestation just noted, the ensuing reduction in evapotranspiration has yet another effect: since it takes energy to evaporate water, energy is removed from the surface of the plants during evapotranspiration and transferred to the atmosphere as latent heat. When the moisture later condenses as precipitation, the latent heat is released, warming the atmosphere.

the problem to actually be solved by determining the nature and magnitude of the temperature and other climate changes that would ensue.

The converse of the net local cooling feedback of devegetation also occurs when climate change expands the range of broad-leaved vegetation into the Arctic tundra. Environmental science professor Inez Fung at the University of California, Berkeley, has actually modeled this process. The trees' northward progression increases the absorption of solar energy, because tree leaves are relatively dark compared to the lighter-colored tundra, a powerful warming feedback. However, the trees also pump large amounts of moisture into the atmosphere, causing clouds that on balance create cooling, but not enough to overpower the trees' warming feedback. In fact, Fung's climate models show that replacing tundra with trees on a large scale will create so much warming (1 to 5.4°F) over the entire Arctic that it will melt sea ice.[4] When that happens, it causes still more warming, because the open ocean absorbs more heat than the light-reflecting ice. Thus the positive feedbacks in this example combine to intensify warming. Other feedback processes are going on simultaneously, but are not on balance of sufficient magnitude to nullify Fung's results.[b]

Climate Thresholds

The climate system is constantly buffeted by both positive and negative feedbacks as well as by chaotic internal behavior. Because it is a dynamic and nonlinear system with relative stability only under certain steady-state conditions, the climate system may undergo abrupt transitions.[5] The boundary to an abrupt climate transition is known as a climate threshold.

Forcing the climate to cross a threshold may bring large and irreversible climate changes because of the presence of ensuing climate feedbacks lurking beyond the threshold. When a critical threshold exists at the boundary of an irreversible and potentially catastrophic climate shift, the threshold is known as a tipping point. Thus, if climate change progresses past a tipping point, the climate will evolve or shift to a new state. As will be seen shortly, this abstract discussion has important implications for climate change.

A phase change from solid to liquid or from liquid to gas is a natural physical process that occurs when threshold temperature conditions are reached. In a living system, an organism's tolerance limits (such as the minimum amounts

---

[b] Trees, for example utilize carbon dioxide during photosynthesis, which removes carbon dioxide from the air and tends to cool the Earth. Trees, however, also release carbon dioxide during respiration, which adds to the carbon dioxide concentration in the atmosphere and thus tends to cause warming. Again, it is the net effect of the vegetation that is the focus of concern here.

of space, water, or nutrients it needs for survival or reproduction) are examples of its biological thresholds. Both physical and biological thresholds are important for understanding climate change. To cite a biological example, if a species' environment does not meet its required threshold conditions, the species goes extinct. If the species was vital to the ecosystem, its loss may then trigger the entire ecosystem's collapse, because of species' interdependencies.[c] In turn, the loss of an ecosystem on a large scale might further amplify climate change.

The Earth's climate may be relatively stable for a long time. Yet when an additional critical unit of energy is added or removed from the system, it may, by initiating one or more feedback processes, push a hospitable climate past a tipping point, causing an irreversible change far larger than the original disturbance. Rapid large temperature changes of 10.8 to 18°F occurred during the last ice age. During these cycles, known as Dansgaard-Oescher oscillations, the temperature in Greenland rose or fell as much as 18°F, sometimes within only a few decades.[6] The Dansgaard-Oescher oscillations left ice core and ocean sediment traces that scientists have now identified and studied. They provide physical evidence that disturbances in bygone eras tipped the climate into episodes of enormous and abrupt climate changes, sometimes within a human lifespan. Dansgaard-Oescher oscillations reveal how extraordinarily powerful positive climate feedback can be.

Very significantly, feedback processes also regulate the complex interrelationships of plants, organic matter in soil, soil microorganisms, nutrient availability, nutrient uptake, and temperature. Feedback, of course, as previously noted, also affects the concentrations of atmospheric gases, like carbon dioxide, methane, nitrogen oxides, and ozone. For example, temperature affects microbial respiration (the use by microbes of oxygen resulting in carbon dioxide production and the release of energy) and the microbial breakdown of soil organic matter. Coupled with the direct effects of temperature and moisture on the soil, the altered microbial activity affects nutrient availability and plant growth which, through photosynthesis and respiration, affects the concentration of carbon dioxide in the atmosphere, which in its turn produces a (warming) climate system feedback.

These processes are themselves reciprocally affected by the concentration of carbon dioxide (and other gases) in the atmosphere in yet other ways. For example, one result of these carbon-cycle linkages could be that,

---

[c] Individual ecosystems, like individual species, have climate-sensitive tipping points—for example, tolerance limits for heating, drying, water chemistry changes, or species loss—beyond which the ecosystem cannot recover.

as temperature increases above a certain threshold, soil nutrient availability drops, reducing plant growth and constraining oxygen production and carbon dioxide removal from the atmosphere by the plants, despite the increased availability of atmospheric carbon dioxide, which by itself tends to stimulate plant growth. The result therefore would be an overall increase in atmospheric carbon dioxide, pushing global temperature up still higher.[7] Without knowing the speed and strengths of these highly complex internal ecosystem dynamics, we are foolish to tamper with nature by heating the Earth.

The chemical behavior of methane in the atmosphere is another example of a positive climate feedback. (Methane is 84 times as potent as carbon dioxide at heating the Earth over a 20-year period.)[d] Because it is chemically reactive in the atmosphere, it simultaneously decreases the atmospheric concentration of the hydroxyl radical (OH^-) that normally removes not only methane but also many other heat-trapping gases and pollutants from the atmosphere.[8] Increasing the concentration of methane thus also increases the concentration of these heat-trapping gases. The release of methane thus initiates a heat-amplifying positive feedback cycle. (See "Hidden Carbon Bombs," page 102, for more about the implications of these feedbacks.)

Thinking about the Unthinkable

As explained earlier, at a tipping point, an incremental change in climate conditions provokes a disproportionately large and dramatic feedback. The temperature above which an ice cap or glacier would melt fits this definition. So does a temperature at which much of the long-distance transfer of heat by the oceans would halt, or at which permafrost would melt, releasing billions of tons of carbon dioxide into the atmosphere.

Tipping points are not reversible on time scales of interest to current generations. A climate threshold leading to a tipping point is thus a little like the spring-loaded steel spikes at a parking lot exit. You can roll forward over it, but you cannot back up. Similarly, passing that threshold is a one-way journey that leaves our climate at the mercy of self-reinforcing natural processes. They deliver ever-increasing amounts of heat-trapping gases to the atmosphere in response to global warming. The additional heat-trapping gas then generates

---

[d] Whereas the Intergovernmental Panel on Climate Change's *Fourth Assessment Report* (Working Group I, Chapter 2), estimated that methane was 25 times more potent than carbon dioxide over 100 years on a per-molecule basis and 72 times as potent over a 20-year period, the *Fifth Assessment* increased the estimate to 28 times over 100 years and 84 times as potent over 20 years. Older EPA estimates put the per-molecule effect of methane at 20 times the potency of carbon dioxide over 100 years. See Environmental Protection Agency, "Methane," Environmental Protection Agency, April 1, 2011, www.epa.gov/methane. Accessed March 20, 2012.

additional warming in a vicious, self-amplifying cycle. Eventually, this could significantly reduce the planet's ability to support human life and make the Earth a far less hospitable place for those who survived. Climate change can thus lead to civilization-altering conditions, as history reveals.

During the classic period of Mayan Civilization (about 300 to 1000 AD), millions of Mayans flourished in parts of what is now southern Mexico, the Yucatán, and northern Central America. Mayan classical civilization at first expanded and produced major cities like Tikal and Palenque during a long period of exceptionally plentiful rainfall. But then came a series of severe prolonged droughts. Food production fell. Social fragmentation and political collapse followed. Major Mayan cities were abandoned.[9] Based on very careful reconstructions of past temperatures in the region, climate change is the best explanation for the Mayan collapse.

Can Global Heating Be Kept Below 3.6°F?

Whether or not the Earth crosses a major climate threshold depends on how hot the planet gets. Therefore, it's important to have accurate forecasts of the future temperatures that will occur if we continue disturbing the climate. As explained in chapters 3 and 4, our impact on the climate is determined by the net climate forcing—the difference between the positive forcing of heat-trapping gases less the negative forcing of cooling aerosols from fossil fuel combustion plus natural sources.

Assuming that carbon dioxide emissions stay roughly at current levels but that, magically, the synthetic reflective aerosols are reduced to zero, the additional warming to which we are already committed from excess heat stored in the ocean would drive global temperature up not 1 to 2°F as discussed in chapter 5 but potentially by as much as 6.4°F. The most probable effect would be an additional 2.9°F of warming. Given the 1.4°F warming to date, the two increases combined imply a likely interim warming of 4.3°F relative to preindustrial temperature, and the planet wouldn't stop warming at that point. So, without current global levels of air pollution, the Earth *already* would be committed to an average temperature rise significantly above 3.6°F. And at some point in the future, the cooling aerosol pollutants are likely to be significantly diminished by more effective pollution controls and much more widespread reliance on carbon-free renewable energy. Then the additional heating now temporarily masked by those short-lived aerosols will occur.[10] Unfortunately, even with aerosols at current levels, it is not even clear that we can limit global heating to 3.6°F.

In a special 2009 issue of the *Proceedings of the National Academy of Sciences*, climatologist Hans Joachim Schellnhuber of the Potsdam Institute for Climate Impact Research reported on a recent study by the German Advisory Council on Global Change. He warned that, "Even a 2-in-3 chance of holding the 2°C [3.6°F] line *requires* that the industrialized countries achieve *almost-complete decarbonization* [emphasis added] by 2030 or purchase tremendous amounts of [greenhouse gas] permits from countries like India, Pakistan, and Ethiopia."[11] (Schellnhuber is alluding here to a provision under the Kyoto Protocol that allows industrialized nations to get credit for emissions reductions required under the Protocol by paying developing nations to cut their emissions instead.) More ominously, and at variance with Schellnhuber's guarded assessment, a recent climate projection by the Integrated Global Systems Model at MIT concluded that *the world now has a zero chance of avoiding a 3.6°F rise by the end of this century.*[12] That would be tragic. Unfortunately, the MIT study's midrange global temperature projection is for a totally unacceptable global average temperature rise of 9°F by the end of this century. The model also indicates a 10 percent chance that temperature will rise 12.6°F by century's end, and a 3 percent chance of a full 14.4°F rise by then.

In the same *Proceedings*, distinguished climate scientist Dr. Mario Molina and colleagues also wrote:

> Current emissions of anthropogenic greenhouse gases have already committed the planet to an increase in average surface temperature by the end of the century that may be above the critical threshold for tipping elements of the climate system into abrupt change with potentially irreversible and unmanageable consequences. This would mean that the climate system is close to entering if not already within the zone of 'dangerous anthropogenic interference.'[13]

It thus appears that if we fail to take these authoritative warnings with deadly seriousness and do not radically slash our heat-trapping gas emissions, we are destined not just to exceed, but to *greatly* exceed a 3.6°F average global heating through the prolonged action of the positive climate feedbacks I discuss in this chapter. The world will therefore be exposed to a very large risk of crossing a climate threshold from which the Earth will not recover for thousands of years.

It is only fair to add here that not all climate scientists agree with the pessimistic assessment produced using the Integrated Global Systems Model at MIT or with the ominous statements by Molina and Schellnhuber. Joeri

Rogelj and colleagues at four prestigious research institutions,[e] using detailed energy and climate models, concluded that it may still be possible to keep global temperature from exceeding 3.6°F or even 2.7°F under certain technologically feasible climate and energy scenarios[14] predicated on an early international consensus and decisive action to curb emissions. Their complex study examined the far-reaching technological, policy, and social changes that would be required, along with the costs and risks of various alternative energy demand and supply scenarios. The authors themselves acknowledge that their projected emissions "are at the high end of the literature range of 2°C-consistent scenarios." Their out-of-the-mainstream findings clearly showed, however, that delay in cutting emissions increases the risk of overshooting 3.6°F and, conversely, that reducing global energy demand significantly decreases the risk. Their work reinforces the conclusion of Schellnhuber and many others that large, rapid, and pervasive global reductions in carbon energy use will be necessary if we are to have a chance of staying below 3.6°F.

Whether or not their postulated low-energy scenario occurs, however, Rogelj's team found that hundreds of the world's 1,400 conventional coal power plants would need to be shut down by 2030 or sooner to protect the climate. The researchers also found that if developing countries refuse to join in global climate change mitigation efforts until after 2030, risks of a global overshoot of 3.6°F would rise. The world therefore urgently needs a political consensus that we are collectively facing a global climate emergency. That consensus would create the possibility for the multitrillion dollar global programs—in energy efficiency, renewable energy, agriculture, forestry, as well as carbon capture and storage—needed to head off massive climate change with a reasonable probability of success.

Dangerous Climate Tipping Elements

A climate tipping element is defined as a part of the climate system that is susceptible to responding to a climate forcing by causing the climate system to abruptly shift to another state. As explained, the new state is distinct in having a new mean condition outside the range of normal fluctuations seen in the existing state around its mean.[15] The precipitating forcing factor acting upon the tipping element could be a variety of agents. It might, for example, be a change in atmospheric composition, a change in the planetary surface,

---

e The Institute for Atmospheric and Climate Science, ETH Zurich; the International Institute for Applied Systems Analysis of Laxenburg, Austria; the National Center for Atmospheric Research in Boulder, CO; and Graz Univeristy of Technology in Graz, Austria.

a change in the oceans, a change in the Earth's biological systems, or other changes, each of which could culminate in a rise in global temperature that through the action of both fast and slow feedbacks could set irreversible heating processes or other climate changes into action.

Among the climate system's major tipping elements are the Asian monsoon, global-scale heat conveyance by ocean currents, ocean carbon cycling, the El Niño/Southern Oscillation (ENSO), the Amazon forest, the Greenland Ice Sheet, the West Antarctic Ice Sheet, Arctic sea ice, and the release of huge quantities of now-frozen seabed methane-hydrate deposits. Much of the ensuing discussion is based on the special 2009 "tipping point" issue of the *Proceedings of the National Academy of Sciences* cited on page 97.

The Oceanic Conveyor Belt Circulation

Vast ocean currents absorb heat in the equatorial oceans and transport that warmth toward the poles. This warms regions closer to the poles while cooling the equatorial region. To understand how this works, it is first necessary to understand something about ocean circulation patterns. Cold salty water is denser and heavier than freshwater. In the presence of winds that mix the ocean surface, the cold salty water sinks beneath warmer, fresher surface waters. At certain specific locations in the high ocean latitudes of the North Atlantic near the poles, cold, dense water descends to a depth of 6,500 to 10,000 feet at which it flows southward toward the equator. This flow is a part of the Atlantic meridional (north-south) overturning circulation (AMOC).

When the cold deep currents from the north reach warm equatorial waters, they gradually rise, absorb heat, and head northwardly again across the ocean, bringing tropical warmth with them. The AMOC is the dominant north-south heat transport mode in the Atlantic. In the Pacific, heat is transported primarily by a gyre-like mass movement of water northward on the western side of the Pacific basin past Japan, then across the ocean and southward along the eastern side of the Pacific. In the Atlantic, the AMOC is not the only transoceanic heat-transfer mechanism. Heat is also transferred by a massive wind-driven gyre-like circulation that brings warmth north via the Gulf Stream from the area around the Gulf of Mexico and then returns cooler water southward through the eastern Atlantic. (The Gulf Stream participates in both the gyre and the AMOC.) Winds also play an important role in producing an upwelling of dense subsurface water in the Southern Ocean that sends it northward along with other surface water, but warmed by the mixing and sloshing of winds and tides.

Because the meridional currents are driven by water density differences caused both by temperature and salinity, the phenomenon was until recently known as the "thermohaline" circulation, from the Greek words *thermē* (heat) and *hals* (salt). The powerful force driving this process is known as the "thermohaline pump." While its action propels billions of tons of water and massive amounts of heat around the whole ocean, including the Antarctic and Arctic seas, it is now understood to be only part of the oceans' heat-transport system.

Because of its enormous scale, the AMOC would at first seem entirely beyond human influence and control. But that is not quite the case. When the Earth's warming atmosphere accelerates the melting of Greenland and Antarctic glaciers, this adds tens of billions of tons of lighter freshwater to the oceans every year. A warmer climate also brings heavier rainfall and more river runoff into the North Atlantic, which also over time infuses vast volumes of lighter freshwater into the Nordic and Labrador seas where low temperatures normally increase the density of seawater.

Over geologic time, paleoclimatic records indicate that the AMOC has abruptly stopped on a number of occasions. As you may have already guessed, when the surface waters of the far North Atlantic become fresh and light enough, the freshwater lens formed prevents the sinking of vast amounts of heavier salty water that would otherwise propel those deep coldwater currents southward. In effect, it "cuts the engine" that drives this component of the ocean's circulation. Halting the AMOC would have profound effects on the world's climate and unknown ecological effects on the oceans and regions whose climates depend on it. Mathematical models indicate that the thermohaline pump has two stable positions: "on"—the current state—and "off"—a state it can swiftly switch into when infusions of surface freshwater reach a critical threshold.[16] Models used by physical oceanographers Stefan Rahmstorf and Matthias Hofmann indicate that when the volume of freshwater entering the North Atlantic exceeds a critical value, the AMOC begins to shut down. It doesn't come back on again until freshwater flows have fallen considerably below that critical value. The deep or abyssal ocean is slow to respond to atmospheric changes and therefore lags far behind the climate on a millennial scale.

Oceanographic research reported in the IPCC's *Fourth Assessment Report* indicated that the AMOC is likely to slow during the current century but is unlikely to totally break down completely, even if Greenland were to continue melting rapidly. If the circulation were to stall at some time in the more distant future, however, it is not hard to envision the pump getting stuck in the "off" position. That's because under current climate conditions, the AMOC pumps freshwater out of the North Atlantic basin, so the pump remains in operation by clearing away the freshwater that would cause it to stall. However, if it were overwhelmed by vast amounts of freshwater—for example, by the rapid melting of the Greenland ice sheet and large increases in precipitation and runoff in the North Atlantic—the pump would indeed stall. That would stop the export of freshwater from the basin. Then even more freshwater would accumulate, so the floating freshwater lid would become even larger and more difficult to penetrate. Thus the "off" state, like the "on" state, tends to be stable, making a restart of the AMOC even more difficult.

The future of the Greenland Ice Sheet therefore may have a critical bearing on the fate of the AMOC, though there is some divergence of scientific opinion on the strength of its effect. The optimists believe that rapid melting of Greenland's ice will merely slow the AMOC to 40 percent or so of its current value; other researchers fear that the melting will eventually cause the AMOC to stop.[17] Scientists at the American Geophysical Union meeting in 2012 presented images from NASA's Cloud-Aerosol Lidar and Infrared Pathfinder Satellite Observation (CALIPSO) showing clouds of smoke from Arctic fires—themselves linked to global warming—blowing over Greenland and coating the ice cap with soot. When darkened by soot, the ice cap becomes less reflective and so absorbs additional heat. Because of global warming exacerbated by soot particles, the entire surface of the Greenland Ice Sheet showed signs of melting in the summer of 2012. The darkening of the ice cap is not only a positive climate feedback. It also shows how some global warming impacts—Arctic fires and melting Arctic ice in this case—operate synergistically, their combined effects being worse than the sum of the parts.

Based on the expectation of heavier future rains and more river runoff into the North Atlantic along with a strengthening of winds, the Intergovernmental Panel on Climate Change (IPCC) has found that there is a one in ten probability that the AMOC may shut down during the current century. Not only would the climate be catastrophically affected, but the whole North Atlantic would get fresher and the entire ocean and its ecosystems would be severely disrupted. Given consequences like this, a 10 percent chance seems an unacceptably high probability. Rahmstorf and Hofmann point out that not only do most climate models make incorrect assumptions about AMOC stability, but most also fail to account for the increasing meltwater entering the North Atlantic from the Greenland Ice Sheet. Until models are refined, we will not be able to accurately predict when the AMOC may shut down, only that we are on a trajectory that—if unaltered—will eventually lead to a frightening result. If as a society we have underestimated this risk, however, the ocean's enormous inertia and slow response time imply that it will take a long time for society to rectify the miscalculation, if we can do so at all.

Hidden Carbon Bombs

Warming the ocean is a risky proposition, even if only by a few degrees. In many undersea areas, methane-containing ice lies frozen in undersea permafrost, but at temperatures close to their melting point. Slight warming of the water above these deposits is all it takes to melt the methane ice and release the stored methane. In a worst-case scenario, billions of tons of methane could be released with devastating impacts on Earth's climate. In addition to ocean warming, seabed mining, drilling, and dredging would also disturb these deposits and could release significant amounts of methane.[18]

Permanently frozen ground known as permafrost covers a quarter of the Earth's land area, including 60 percent of Russia.[19] The release of carbon dioxide and methane from permafrost and the ocean seabed is a classic tipping point risk. When permafrost thaws, the organic matter in it can be converted by bacterial action either to carbon dioxide or more potent methane.[f] For thousands of years, however—and in some cases about a million years—huge caches of methane have lain frozen and isolated from the atmosphere both in the northern latitude permafrost and in frozen seafloor sedimentary methane ice deposits known as hydrates or clathrates[20] in which gas is trapped in a crystal lattice cage of frozen water. The Arctic has now warmed to such an extent, however, that some permafrost and methane hydrate deposits are beginning to thaw. In fact, methane hydrates in some shallow seafloor areas are already bubbling up to the ocean surface—and into the atmosphere.[21] In deeper ocean areas, however, cold temperatures, high pressure, masses of overlying water, and low oxygen concentration hold the hydrates harmlessly in the icy ocean sediments. (The higher the pressure the hydrates are under, and the lower the oxygen concentration and temperature, the more stable they are.) As global surface temperatures rise, however, and gradually penetrate to the seafloor, frozen methane can thaw and form bubbles.

If the thawing occurs in the bottom of the deep ocean, the methane can be dissolved in the overlying seawater and oxidized to carbon dioxide by bacteria. While preferable to having the methane released to the atmosphere,

---

f Except in oxygen-poor environments, most of the thawed organic matter will be released to the atmosphere as carbon dioxide, the less potent of the two gases.

the dissolved carbon dioxide further acidifies the ocean. Modeling studies of methane bubble behavior have indicated that in the deep ocean, the depths of the overlying seawater do provide effective barriers against the transfer of methane bubbles to the atmosphere. The ocean's depth also delays the release of methane and carbon dioxide from frozen storage because of the length of time required for heat to reach the deep ocean sediments.

However, in very shallow seas of less than 328 feet in depth,[g] a large proportion of methane bubbles that emerge from once-frozen sediments are able to percolate up to the ocean surface without being dissolved and oxidized. They therefore escape to the atmosphere much as bubbles from a glass of soda water escape when the cap is removed. Substantial amounts of methane may also reach the atmosphere from thawing hydrates in shallow freshwater lakes, wetlands, and reservoirs."[22]

About a third of the Arctic region is covered by millions of freshwater lakes, and their numbers are increasing as higher Arctic temperatures melt more and more permafrost. Ecologist Katey Walter Anthony from the University of Alaska Fairbanks has studied lakes from Alaska to Greenland and Siberia. Her measurements of gas releases from northern lakes and soils show that fossil deposits of methane that were sealed away for thousands of years are now escaping and entering the atmosphere. Bubbles of long-dormant methane gas are already clearly visible as they break on the surface of these far northern lakes. In some areas, the surface looks as if it were boiling. Anthony now commonly finds hundreds of spots on a typical lake in which roughly 10 to 30 quarts of methane are being released per day.[23]

As Arctic seas, lakes, and tundra continue to warm, and more and more permafrost melts, scientists are not certain precisely how quickly methane and carbon dioxide from permafrost and clathrates will reach dangerous levels. Some methane experts affiliated with the US Geological Survey (USGS) view global oceanic methane releases as insignificant on a planetary scale. Referring to them as "chronic releases," the researchers project that the releases will only cause significant warming in 1,000 to 100,000 years.[24] I hope their assessment is correct, because if they are in error and the permafrost and shallow seas release more methane than anticipated, it would be too late to stop the process and an unimaginable catastrophe would unfold. That is because ocean sediments contain huge amounts of carbon—1.8 to 2.2 trillion

---

[g] It has also been hypothesized that the formation of gas bubbles in seafloor sediments from thawing methane could destabilize certain ocean sediments, causing landslides that could release large amounts of methane. Ancient seafloor evidence exists for these cataclysmic releases. (See discussion on page 107 of the Paleocene-Eocene Thermal Maximum [PETM].)

FIGURE 6-1. Whitish circles reveal where methane bubbles emerging from thawing permafrost in the sediments of an Arctic lake in Greenland in November 2010 became trapped in the lake surface ice when the lake refroze after the spring surface thaw. Arctic lakes used to freeze solid from top to bottom but most no longer do due to global warming. This allows decaying organic matter in the thawing permafrost of the lake bottom to release methane. The dark circle of open water indicates an area located over a methane release powerful enough to keep the lake from completely freezing over. Photo courtesy of K. W. Anthony and NASA's Earth Observing System Data and Information Website: https://earthdata.nasa.gov/featured-stories/featured-research/leaking-lakes.

tons (1.6 to 2.0 trillion metric tons)—in methane hydrates.[25] If these as-yet imprecise estimates of the methane deposits' magnitude are roughly correct, the deposits would hold two to two and a half times the amount of carbon in the Earth's atmosphere, and several times as much as all the carbon in all the oil, gas, and coal ever burned to date.[26] Thus, were significant fractions of these hydrates to thaw and reach the atmosphere, they would without question superheat the climate.

The known geologic record indicates that there have been no abrupt releases of methane hydrates from the deep ocean for at least the past 650,000 years. This would be reassuring were it not for the following disturbing facts. First, average global temperature is rising fast and is beginning to affect both permafrost and methane hydrates.[27] By 2100, average global temperatures will be warmer than at any time in the past 5 million years, if current emissions trends continue. As global temperatures rise, more and more extensive areas of permafrost and clathrates will begin thawing. And because average increases in global temperature are magnified in the Arctic as previously noted, the Northern Hemisphere could be as much as 21°F warmer by 2100.[28]

Second, in addition to the ocean methane deposits, roughly 2.1 trillion additional tons (1.9 trillion metric tons) of carbon are stored on land in permafrost in the Northern Hemisphere.[29] Most of it lies in the upper, more vulnerable layers of the permafrost.[30] These enormous Arctic deposits extend from the far north to about 60° north latitude. Some of this permafrost is much more susceptible to thawing than the carbon in the deep ocean and, as noted, has already started melting. Moreover, a recent report by the United Nations Environmental Programme found that even the Intergovernmental Panel on Climate Change's estimates for future global warming are underestimates because their models did not adequately consider the accelerating releases of carbon from permafrost.[31]

Third, as we have discussed earlier, when permafrost soils thaw in an aerobic environment, bacteria gradually decompose the carbon and release it as carbon dioxide to the atmosphere. Then the warmer climate that results thaws even more permafrost and releases even more greenhouse gas. The process tends to accelerate and at some point becomes unstoppable. Fourth, whereas temperatures in the bottom waters of the deep oceans change relatively slowly, methane deposits beneath shallow seas along continental margins warm more quickly and so can escape to the atmosphere much more readily. Thus, in the Arctic's shallow East Siberian Sea, researchers have been finding extensive evidence that millions of tons of seabed methane are already bubbling out of clathrate deposits.[32] Professor Igor Semiletov, head of the International Arctic Research Center's Siberian Shelf Study, told BBC News in 2009 that methane in the atmosphere above the Arctic Ocean "is 100 times higher than normal background levels and, in some cases, 1,000 times normal."[33]

The East Siberian Sea is the world's largest sea lying on a continental shelf and is less than 50 meters deep. While methane in frozen permafrost on land in places like Siberia is protected against thawing by icy temperatures, the ocean keeps the temperature on the bottom of the East Siberian Sea 22 to 31°F warmer than the surrounding land surface containing permafrost.[34] The seafloor here contains submerged Siberian peat deposits that were once part of the frozen Arctic coastal plain before they were flooded 7 to 15 thousand years ago.[35] They contain large quantities of methane.

Researcher Natalia Shakova and her colleagues are now finding that the water here is perennially supersaturated with methane from top to bottom from all the gas already bubbling out of the seafloor. Methane concentrations above the sea surface are ten times normal open-ocean levels.[36] The East Siberian Sea alone is venting an estimated seven million metric tons of

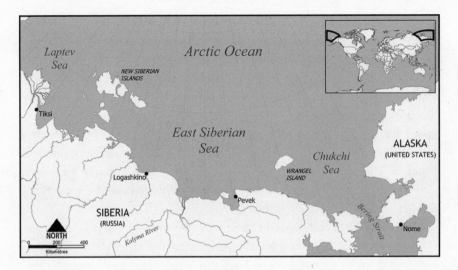

FIGURE 6-2. Map of East Siberian Arctic Sea. Source: Google Images/Wikipedia Commons.

carbon dioxide per year to the air—a process that until recently was largely ignored in climate models. The new research by Shakova and others reveals that, in effect, the "lid" holding methane in the permafrost of shallow seas has opened and that undersea frozen methane—which until recently was thought stable—is now leaking.[37] The quantities of methane potentially at risk are substantial as the East Siberian Arctic Sea has an amount of carbon comparable to that of the Siberian tundra.[38] At the moment, these releases are still not "alarmingly altering the contemporary global [methane] budget," according to the researchers, but need to be closely monitored to determine whether these releases might be signaling the start of a period of massive releases.

Clearly, much still needs to be learned about the release of methane from the seafloor. Contemporary models of the process can only provide educated guesses about how methane hydrates will behave when the oceans get warmer. Studies by David Archer and colleagues, however, suggest that a warming of 5.4°F could result in release of half the gargantuan amount of methane stored in seabeds.[39] In view of all the uncertainties, methane hydrates and land-based permafrost must be regarded as another climate tipping point that could potentially deliver enormous amounts of carbon to the atmosphere. Methane releases from these sources might conceivably contribute to an irreversible cascade of positive feedback events, especially if the deep oceans warm enough to trigger methane releases over large areas of the ocean.

Even if these releases did not accelerate, the Arctic has been warming much faster than predicted and twice as fast as anywhere else on the planet, putting its permafrost at risk. The National Oceanic and Atmospheric Administration has reported that springtime air temperatures in the Arctic from 2000 to 2007 were an average of 7.2°F higher than from 1970 to 1999.[40] Thus, not surprisingly, release of methane from Arctic wetlands (still a relatively small source) has increased by a third from 2003 to 2007.[41] Longer-term measurements are needed to better assess the risk from vast areas of tundra permafrost, Arctic lakes, and wetlands with billions of tons of additional carbon.

More research is also needed to track the destabilization of methane hydrate by shifting ocean currents. One recently published study in *Nature* found that rising temperatures in the Gulf Stream at "intermediate depths" off the East Coast of the United States had already destabilized 2.5 billion tons of frozen methane hydrate under 4,000 square miles of the seafloor.[42] "A changing Gulf Stream," the authors wrote, "has the potential to thaw and convert hundreds of gigatonnes of frozen methane hydrate trapped below the sea floor into methane gas, increasing the risk of slope failure and methane release."[43] (One gigatonne equals one billion metric tons.)

Undersea slope failures are believed to have played a role in initiating an extreme warming known as the Paleocene-Eocene Thermal Maximum (PETM) 55 million years ago in which the ocean became acidified and the entire climate warmed dramatically. The amount of clathrate destabilized so far represents a mere "0.2 percent of that required to cause the PETM," the authors of the *Nature* study say, and the Gulf Stream has been warming for millennia. However, if temperatures have now risen enough to cause clathrate destabilization over other large areas of the ocean, that could have serious consequences for the climate and ocean. It calls for further inquiry, especially since, as the authors point out, the study area is unlikely to be the only part of the ocean experiencing changing ocean currents and temperatures.

Another significant explanation for the extreme (9°F) warming in the PETM has been provided by DeConto et al. who linked the warming with cyclical changes in the Earth's orbit that increased both the Earth's obliquity (tilt) and the eccentricity (noncircularity) of its orbit.[44] This caused a warming of the polar regions, both in the Arctic and Antarctic, and that triggered the decomposition of soil organic carbon that had been frozen in permafrost. This process then released billions of tons of methane to the atmosphere.

In conclusion, concentrations of atmospheric carbon dioxide are already above any concentrations for the past 800,000 years and, in just a few decades, could be more than double the highest levels known during those years, if current emissions trends continue.[45] If we persist in continuously adding more and more carbon dioxide and methane to the atmosphere, we will at some point reach Arctic temperatures that will cause thousands of square miles of permafrost and shallow clathrates to thaw. This would in effect set off the "carbon bombs" lying quietly in the permafrost and the marine clathrate deposits. Earth would then begin a progressive planetary overheating from which civilization as we know it would not recover.

Unfortunately, the implications of uncontrolled permafrost melting and the release of carbon from northern latitudes do not seem to be on most world leaders' minds. They, in turn, are not helping to raise public awareness of this danger. Until leaders comprehend the implications of this phenomenon and begin factoring it into their decision making, scientists need to be much more assertive in bringing it to their attention.

Permafrost and clathrate carbon bombs are not the only major tipping point hazards on the horizon. Another potentially devastating climate-change trigger depends not on the persistence of ice but on the presence of adequate sunlight.

Monsoon Disruption

Climate history reveals that over periods lasting from years to centuries during past ages, there have been large variations in East Asian monsoon rainfall. Today, hundreds of millions of farmers depend on this rainfall for their livelihoods. Changes in the transport of heat and moisture to monsoon regions seem to control the monsoon's strength, which is also influenced by vegetation and other land surface features. A simplified model of rainy season monsoon circulation suggests that a threshold value of solar radiation over a given area is required during the rainy season to maintain the monsoon. Otherwise, a lack of solar radiation—or its redistribution by aerosols from the Earth's surface to the midtroposphere (Earth's lower atmosphere)—can cause a breakdown in monsoon circulation.[46] Based on this model, local cooling due to aerosol pollution could trigger a monsoon collapse[47] if China and India fail to impose effective pollution controls that would limit aerosol production in a timely manner. However, it is expected that they eventually will impose air quality standards for health reasons. Nonetheless, the risk of a monsoon collapse requires further attention and scientific investigation because of the gravity of its consequences.

Were it to occur, loss of seasonal monsoon precipitation plus the concurrent predicted loss of high-elevation glacial meltwater from the Himalayas due to climate change would drastically curtail some Asian water supplies and could leave hundreds of millions of Asians without sufficient food or water. The results could be great hardship, economic distress, social chaos, and political instability that would spill across borders and destabilize other parts of the world as refugees contended with their neighbors for the necessities of life. Terrorist groups might exploit the situation to recruit new members among the desperate and impoverished.

Another vulnerable aspect of the global climate system—and another possible trigger for dangerous, self-reinforcing climate change—can be found in the equatorial Pacific Ocean. Cyclical events here, when perturbed, could seal the destiny of the Amazon rainforest.

A Fickle El Niño

The El Niño/Southern Oscillation (ENSO) is a sequential fluctuation (usually on a four-year cycle) in the location of a vast pool of warm ocean water in the tropical Pacific. The eastward expansionary phase of the warm water is known as an El Niño. Its contraction (when a large tongue of unusually cool water spreads from east to west in the equatorial Pacific) is called a La Niña. The El Niño phase of the cycle is marked by unusually warm water in the equatorial Pacific. The La Niña phase is associated with a strong contraction of the warm pool and an unusually cold equatorial Pacific. The ENSO cycle not only affects ocean temperatures but winds and upwelling.[h]

The position of the warm and cool water masses alters ocean currents and global atmospheric circulation, and thus profoundly affects the weather of large regions of the Earth, warming or cooling them and also making them wetter or drier. Thus ENSO fluctuations thus "drive substantial variability in rainfall, severe weather, agricultural production, ecosystems, and disease," according to a recent study by Powers et al.[48] Major shifts in the timing or intensity of the ENSO "will therefore have serious climatic and ecological consequences."[49] One of these could eventually be the destruction of the Amazon rainforest.

Powers, the senior research scientist of the Australian Bureau of Meteorology, found using four recent climate models that ENSO cycles are likely to intensify in the latter half of the current century to produce increased drying

---

[h] Although models can reproduce important aspects of ENSO behavior and impacts, its complex relationship to climate change is still not well understood.

in the western Pacific and increased rainfall in the eastern and central Pacific. Powers et al.'s results are consistent with those of Shayne McGregor of the University of New South Wales ARC Centre for Excellence for Climate System Science.[50] McGregor found that the ENSO cycle seems to be becoming more variable, more active, and more intense when compared with historical records extending hundreds of years into the past.[51]

ENSO arises through a coupling of the ocean and atmosphere. This occurs as a consequence of ocean basin geometry; sea surface temperature differences along the equator; resulting wind patterns; and ocean currents driven by the Earth's rotation. The winds and currents in turn affect the upwelling of cold deep ocean water and hence differences in sea surface temperature. Global climate conditions both set the stage for ENSO and are in turn influenced by ENSO through large-scale air-ocean interactions, a form of feedback to the climate system.

In a normal year, the equatorial Pacific is warmer on the western side, and the easterly trade winds actually pile water about two feet higher on this side of the ocean. During an ENSO cycle, the pool of warm water slowly expands or contracts on its east-west axis across the Pacific. This phase shift produces abundant rainfall in regions above the warm water and drought in areas near colder water, while also affecting ocean temperatures, winds, and upwelling.

ENSO is an inherent property of the climate system. Unlike a climate forcing agent, such as a greenhouse gas that heats the entire Earth and climate system by trapping more solar energy from space, ENSO primarily redistributes heat. Its geographic impact is determined not only by the location of the warm water pool, but by the depth of the boundary, known as the thermocline, between the ocean's warmer surface water and its colder depths. ENSO can affect both monsoons and storm tracks.

Normally, east-west trade winds across the entire Pacific near the equator push ocean surface water to the west and therefore produce the continental upwelling along the western coast of South America that brings up nutrients and supports rich commercial fisheries. In crossing the ocean, these winds absorb and transport warm water from the ocean surface as part of a larger cycle called the Walker Circulation (figure 6-2).

As warm humid air rises in the warm, low-pressure areas over the western Pacific, it expands. The resulting drop in pressure pulls in more air from the east, much like a chimney draws air into a fireplace. As the air column cools, it releases excess moisture as rain. Once high aloft, the dried air is pushed

eastward above the Pacific. It then descends over or near the American continents before again traveling west at low elevations back over the Pacific.

The immediate effects of the ENSO are starkly evident.[52] The El Niño
phase, which occurs irregularly every three to seven years, is correlated with
heavy rainfall in Southern California and western South America. Because
the western Pacific is cooler during an El Niño, Australia and Indonesia are
subject to drought. By contrast, the displacement of the warm equatorial
water farther to the west during a La Niña event interferes with this pattern
so that the eastern equatorial Pacific is cooler and less moisture descends over
the Amazon and western parts of South America. More moisture absorbed
from the Pacific by the east-west trade winds is transported to west, and then
deposited in the area of Indonesia and Australia in the far western Pacific as
the warm, moist air rises.[i]

Because ENSO affects the location of the ascending and descending
moisture-laden winds as well as the precipitation they bring, ENSO has a
strong effect on national economies through its effects on farming, graz-
ing, and forestry. It also affects commercial fishing through marine nutrient upwelling, which impacts the distribution and abundance of marine
organisms.

As El Niño Goes, So Goes the Rainforest

Investigations of ENSO by climate modelers have not proven that ENSO
definitely is a tipping element capable of pushing the global climate into
an irreversible, self-amplifying heating mode in this century.[53] However,
ENSO nonetheless could conceivably trigger a climate "excursion," meaning
a self-amplifying global warming, and it could certainly dry out the Amazon
rainforest. Since ENSO can transfer heat and moisture patterns to the larger
climate system, if global heating perturbs ENSO, then ENSO, in turn, could
disrupt regional ecosystems sufficiently to push the climate across a threshold
detrimental both to biodiversity and human welfare.[54] To understand how
huge ecosystems like the Amazon rainforest could be destroyed by disturbances to ENSO cycles brought on by climate change, see figure 6-2, which
show how much ENSO affects ocean-atmosphere interactions in the eastern
Pacific. Normally El Niños are followed by La Niña conditions. However, if

---

i More detailed analysis of El Niño reveals that it has an eastern Pacific and a central Pacific variant.
Eastern Pacific El Niño events are associated with a year-round reduction in rainfall in the northern,
eastern, and central Amazon. Central Pacific El Niño events decrease rainfall in the Amazon in the
austral summer but may increase it or have no effect in other seasons.

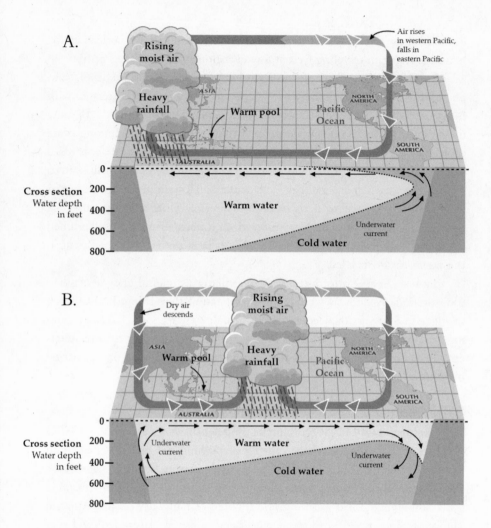

FIGURE 6-2. The El Niño Southern Oscillation (ENSO) cycle is depicted in diagrams (A) and (B). (A) shows how the Walker Circulation normally drives trade winds from east to west across the equatorial Pacific, sweeping warm surface waters west so they accumulate east of Australia and Indonesia. There, as the warm, humid trade winds rise and cool, they release their moisture as rain. Now drier and cooler, the air mass returns at higher elevations back across the Pacific. Under El Niño conditions (B), the trade winds weaken, allowing the warm water pool to drift back in an easterly direction across the Pacific. As warm water accumulates in the mid-Pacific, warmed air rises above the area and splits the trans-Pacific Walker cell into two cells. Rain falls beneath the area of rising air. A deep layer of warmed water accumulates off the coast of South America, where it blocks the upwelling of nutrient-rich waters that support marine life. An exceptionally strong Walker Circulation produces La Niña conditions (not shown) in which moisture-laden winds release their moisture farther west than normal.

La Niñas were more frequent or prolonged, that would reduce rainfall in the Amazon. A protracted reduction in the rainfall reaching the Amazon rainforest, for example, could set in motion a progressive drying and conversion to grassland that could destroy the forest, releasing vast amounts of carbon now stored as living plant matter in the forest.

The Amazon has 25 percent of the world's biodiversity and is a major part of the Earth's climate system, because of its effects on heat, moisture, and the cycling of carbon dioxide.[55] Some 15 percent of all the photosynthesis on Earth takes place in the Amazon.[56] The carbon dioxide that would be released as the forest dried and burned would contribute to global warming, and the ensuing ecological changes would exterminate the forest wildlife. Climate scientists haven't predicted how ENSO will respond to increased global temperatures, but some modelers expect that the Amazon could be significantly affected as early as 2050.[57]

As the climate heats up in the Amazon basin (as it is doing by almost half a degree Farenheit every decade), the different regions of the rainforest will respond differently to decreases in water supply and changes in the length of the dry season brought on by a La Niña.[58] In the eastern Amazon, climate models indicate that the forest will be able to withstand some drying while still maintaining its character without turning into tree-studded grasslands. However, if fires accompany the decrease in moisture, the viability of the forest is doubtful.[59]

More than one-quarter of the Amazon is "facing incipient fire pressure, now being within 6 miles (10 km) of an ignition source."[60] By the end of this century, according to some climate change scenarios, the Amazon will be as much as 5.4 to14.4°F hotter. This expected warming is 30 to 50 times faster than the natural warming of the Amazon following ice ages.

Whereas natural fire is rare in the Amazon basin, it is more likely in the vast areas of the Amazon where the forest canopy has been destroyed by logging and clearing for ranching or for agriculture and settlements. The western Amazon generally suffers very little water stress in dry years, but its northern margin and southwestern portion resemble the eastern Amazon.[61] If fires accompany the decrease in moisture there, the viability of the eastern Amazon forest is doubtful.[62]

Climate models projecting forest survival are still imperfect in their estimates for evapotranspiration,[j] their exclusion of deforested areas, and their

---

j Water transported from ground to atmosphere through the plant stem and leaves.

uncertainty in how much global rainfall and temperature will vary as the world heats up. Consequently, they are still unable to accurately determine how much regional forest temperatures will rise and how vegetation will respond. (The models also vary in their accuracy in simulating current climate and rainfall patterns.) Most models required recalibration to observed rainfall patterns of the twentieth century for more realistic forecasting of the rainfall patterns to come in the twenty-first century.[63]

Studies that attempt to assess forest stability from rainfall are focused mainly on projecting how vegetation boundaries will shift over the next century rather than the region's ultimate fate. Rainfall changes will be occurring at the same time as changes in the length and duration of the wet and dry seasons as well as changes in temperature, atmospheric aerosol content, local carbon dioxide levels, transpiration, soil moisture, and human-induced fires.

All these changes will eventually have large ecological effects on the Amazon for intrinsic reasons and because of their influence on the forest's vulnerability to fires. While the Amazon rainforest is clearly in grave peril and is being subject to rapid degradation from many causes, experts believe that its destruction is not inevitable and that proper protection against fire, logging, roads, and human incursion could buffer the region somewhat against the increasingly likely water shortages.[k]

If climate change leads to more frequent or more intense El Niños as new scientific research suggests, [64] the Amazon rainforest would not be the only casualty. Another could be degradation of the Galapagos Islands off the coast of Ecuador. Surrounded by the world's second largest marine reserve, the islands are a world heritage site, an international biosphere reserve, and an Ecuadorian national park.[65]

Stronger and more common El Niños could lead to the impoverishment of the marine reserve surrounding the Galapagos. During El Niños in this part of the Pacific some 620 miles off the Ecuadorian coast, the trade winds slacken and so does the upwelling of cold, nutrient-rich water from the ocean depths.[66] The lack of nutrients leads to a breakdown in the marine food chain and to the starvation of marine species and land-based creatures that depend on them, like the marine iguanas and other rare species found nowhere else on Earth. El Niños that occurred between 1982 and 1998 brought such warm water to the Galapagos that the coral reefs around them turned white

---

[k] Ways of protecting the Amazon and other tropical forests from clearcutting and other flagrant abuse are discussed in an earlier book of mine, *Forests Forever: Their Ecology, Restoration, and Protection.*

and died.[67] Under normal conditions, the time between El Niños is sufficient for species to recover from starvation and rebuild their populations. However, if El Niños return more frequently, some species will go extinct.[68]

Dust in the Wind

While we are accustomed to considering temperature, moisture, sunlight, ocean currents, wind, and atmospheric gases as important climate variables, dust is sometimes overlooked.

Yet dust particles from the Saharan Region of Africa today have surprisingly far-reaching and complex climatic effects on the planet.[69] Dust increases the scattering of sunlight back to space, promotes cloud formation, and also—by absorption and reradiation of long-wave radiation—blocks the escape of heat from the Earth to space, as does carbon dioxide. Dust also contains and transports nutrients, such as iron and phosphorus, to nutrient-poor ecosystems where they increase productivity. On land, this fertilization increases the rates at which carbon is removed from the air during photosynthesis and is later released back to the air as living things breathe. Dust also provides important nutrients such as iron to ocean ecosystems and coral reef islands and even to the Amazon rainforest. The ultimate influence of dust on climate is the overall net effect of the countervailing heating and cooling processes.

To find the origin of the Sahara's dust, we need to go back seven thousand years to the time when Lake Megachad in what is now Chad was the largest lake on the planet. Part of the ancient lake's bed is known as the Bodélé Depression. Today, it is the planet's largest dust source, producing some 772,000 tons (700,000 metric tons) per year—more than half the Sahara's output.[70] Variable ground-level winds erode the lake's dusty sediments and loft them across entire oceans and continents.[71] These dust particles become airborne more quickly and easily where the ground is dry. When the ground is wet after rain, however, dust production diminishes. As the climate warms, the position of prevailing winds shifts and precipitation also changes. Both phenomena affect the magnitude and rate of dust transport. Scientists are now studying the effects of the Chadian dust to determine whether the Bodélé Depression will act as a dangerous positive feedback to amplify global warming, or whether it will exert a cooling influence.

De-icing the Planet

Over the ages, the amount of ice on the planet in polar ice caps, high-altitude glaciers, and sea ice has varied greatly. During the current warm period between ice ages, ice masses were relatively stable—until the late twentieth-century and early twenty-first-century warming. Loss of ice is occurring rapidly today and is a positive climate feedback with a strong warming effect on the planet. The question now is: at what point does the melting of ice on the planet become unstoppable? The answer is quite different for land-based ice sheets than for floating sea ice.

Sea Ice. No evidence from studies of climate over the past several million years suggest the existence of an irreversible tipping point in the melting of polar sea ice. However, sea-ice melting does produce a positive feedback as reflective white ice is replaced by less reflective and more heat-absorbing water.[72] Earth then warms and more ice melts until, eventually, no more polar sea ice remains. But the real story is not quite so simple.

We now know that negative feedback effects are also built into the Arctic sea-ice melting and re-formation process. These operate to speed up the recovery of lost sea-ice cover, thereby reducing the positive feedback effect described above. One form of negative feedback is the faster growth of new, thin ice which can form more rapidly than thicker, older ice. Even when sea ice melts extensively in summer, the absence of insulating snow cover on newly formed ice allows it to freeze more quickly than older, snow-covered sea ice. Thus, sea ice has the capacity to recover area quickly. While it may not be as thick as the ice it replaces, it can reflect just as much light.

Changes in the amount of heat transmitted to polar regions could partly compensate for warming induced by accelerated sea-ice melting.[73] A recent study found no tipping point inherent in sea-ice interactions that could cause irreversible loss of sea ice as long as the climate does not become so warm that ice cannot re-form.[74]

Ice sheets. The Greenland and West Antarctic Ice Sheets behave quite differently from sea ice, however. These ice masses may indeed have tipping points beyond which intermittent melting results in unstoppable melting, even if temperatures fall below the level that initiated the tipping point behavior.[75] The retreat of these ice sheets appears to be strongly influenced by the lubricating effects of meltwater accumulating at their base and by the loss of floating ice shelves. When the point of attachment of an ice shelf to the shallow seafloor just offshore melts, and the ice shelf breaks away from the land, land-based glaciers dammed in place by the ice shelf are then free to flow more rapidly seaward. This loss of ice-shelf mooring is an important accelerator of glacial outflow from the West Antarctic Ice Sheet.

Commenting on risks to this ice sheet, Dr. James Hansen has warned:

The West Antarctic ice sheet is particularly vulnerable to removal of its ice shelves.... Loss of the entire West Antarctic ice sheet would raise sea level 6 to 7 meters (20–25 feet) and eventually open a path to the ocean for part of the much larger East Antarctic ice sheet. Once the ice sheets' collapse begins, global coastal devastations and their economic reverberations may make it impractical for humanity to take actions to rapidly reverse climate forcings. Thus if we trigger the collapse of the West Antarctic ice sheet, sea-level rise may continue to even much higher levels via contributions from the Greenland and East Antarctica ice sheets.... Global chaos will be difficult to avoid if we allow the ice sheets to become unstable. Sea level changes to heights at least several meters greater than today's level occurred in interglacial periods that were at most 1 to 2 degrees Celsius warmer than today.[76]

Although we don't know the exact tipping points for either the Greenland or the West Antarctic Ice Sheet, global average temperature increases are amplified at high latitudes. Scientists have forecast that—because of the doubling of average global temperature increases in the Arctic—the Greenland Ice Sheet will begin melting when local temperature has reached 4.9°F above preindustrial levels, long before global average temperatures rise 3.6°F. Contemporary observations indicate that the melting of the Greenland Ice Sheet has already begun and that both the Greenland and West Antarctic Ice Sheets are losing mass. If we continue our present greenhouse gas emissions, the melting will continue and accelerate.

This chapter was meant to provide an overview of risks to the Earth and life as we know it from major tipping elements in the climate system. These included the release of frozen methane in the seabed and permafrost; the disturbance of monsoons critical to agriculture; drought and fire threats to the Amazon rainforest; and other lurking tipping points in the climate system. The next chapter will discuss some estimates of the economic costs of climate change that could tip regions and nations from solvency to economic hardship or beyond.

CHAPTER 7

Economic Perils

*"Global warming pollution, indeed all pollution, is now
described by economists as an "externality." This absurd label
means, in essence: we don't need to keep track of this stuff
so let's pretend it doesn't exist. And sure enough, when it's not
recognized in the marketplace, it does make it much easier
for government, business, and all the rest of us to pretend
that it doesn't exist. But what we're pretending doesn't exist
is the stuff that is destroying the habitability of the planet."*

—AL GORE

This chapter outlines the unacceptable costs of abrupt climate change
and explains why the benefits of a global climate rescue plan would
exceed both the out-of-pocket costs and the costs of inaction. It shows
that in the United States and other nations, greenhouse gas emissions
could be slashed while actually raising economic output and boosting
national productivity, rather than stifling it.

Guesstimating the Costs of Climate Change

UNLIKE THE DIRECT, short-term impacts of costly but discrete climate
disasters, such as hotter heat waves or more intense droughts, the coming broad economic impacts of rapid climate change—unprecedented in
human experience—are more complex and difficult to meaningfully quantify
in economic terms. The authors of the IPCC's Fourth Assessment in 2007
acknowledge under the heading "Advances in Knowledge" that little progress

was made between it and the 2001 Third Assessment in advancing knowledge on the costs of climate change (for impacts, mitigation, and adaption) and on the critical question of our "proximity to thresholds and tipping points." They add, "The coming disruptions and degradations of major natural resource systems, such as the world's oceans, forests, and biodiversity, clearly fall into this broad category of hard-to-quantify impacts. In addition, governments will face major challenges and costs in trying to maintain order amid chaotic, climate-induced events while coping with chronic food insecurity, public health challenges, international security, and public safety issues."[1]

Conventional economic analysis is sorely tried to convey an accurate picture of the true costs to present and future generations of the extraordinary devastation that rapid climate change will bring.[2] One distinguished effort, a report for the British government led by former World Bank chief economist and Chancellor of the Exchequer Sir Nicholas Stern, nonetheless concluded in 2006 that, based on formal economic models, the global cost of unmitigated climate change would amount to "losing at least 5% of global GDP [Gross Domestic Product] per year, now and forever."

Stern's widely cited report went on to say, however, that if a wider range of impacts not included in those models were considered, the cost of global climate "could rise to 20% of GDP or more." Looking back on the report in 2013, however, Lord Stern now says, "I underestimated the risks. . . . Emissions are rising pretty strongly. Some of the effects are coming through more quickly than we thought then. . . . I think I would have been a bit more blunt. I would have been much more strong about the risks of a four- or five-degree [Celsius] rise."[3]

Among that "wider range of impacts" Stern may have had in mind are the climate externalities described in one of the report's most often-quoted conclusions that, "Climate change is the greatest market failure the world has ever seen, and it interacts with other market imperfections." Stern meant that the market economy generally fails to assign costs to climate-disturbing activities commensurate with the hidden (nonmarket) costs those activities impose on the marketplace and society.[a] Such activities are commonly known

---

[a] The market fails not only to properly attribute the costs of climate change to those who caused it, but also to properly reward those who create public benefits by helping avoid those costs through creating cleaner and more efficient technologies. These efforts benefit society as a whole, beyond any economic benefit to their developers. Public benefits are the opposite of externalities.

in economics as "externalities" and can be positive as well as negative.[b] A whole range of activities—like driving a car or flying a plane, heating a home or business, burning fuel to make electricity or produce cement, flaring natural gas, cutting down a forest, or raising livestock—all create negative external effects on the climate. (Market externalities are not limited to climate change impacts but also include the general production of air, water, and noise pollution and public health effects.) What all these activities have in common is that those responsible for them generally are not charged for them sufficiently or at all. Then, in the absence of correct price signals to market participants to account for externalities, the market fails to efficiently allocate societal resources. In *Hidden Costs of Energy: Unpriced Consequences of Energy Production and Use*, the US National Research Council found that just the readily quantifiable external damages of energy production in the United States cost $120 billion annually in 2005.[4,c]

Negative externalities produce many problems. For example, they often lead to the inequitable distribution of costs to individuals who experience no benefits from the activity generating the costs. A special case of this highly relevant to climate change is economic unfairness over time, in which one generation's combustion of fossil fuel in the present imposes costs on future generations. Risk, imperfect information, and transaction costs are other examples of factors that lead to market inefficiencies. When externalities and other market inefficiencies are ignored in attempting to analyze the full economic impacts of climate change, conventional economic analysis produces a misleading rather than an accurate, meaningful picture of climate change's true costs.

Research Fellow Jonathan Koomey of Stanford University has succinctly summarized other important reasons why the conventional cost-benefit analysis approach for understanding the economic impacts of climate problems is deeply flawed, and why comparing the cost-effectiveness of alternative pathways to a needed emissions-reduction goal makes more sense.[5]

---

[b] A positive externality—a special type of public benefit—is the effect of an action or decision that benefits those not part of the decision or action. Examples include the construction of a flood-control dike by a property owner to keep a river in its banks and thus to prevent his property from flooding while thereby also protecting neighbors. Another example is a vaccination program that not only immunizes the vaccinated but prevents illnesses among the unvaccinated by suppressing disease.

[c] This estimate, for example, excludes the cancer and other serious health effects (such as reproductive disorders and birth defects) produced by hazardous (toxic) air pollutants. It did assess the costs of conventional air pollutants and estimated the damages produced by greenhouse gases by assuming that each ton of $CO_2(eq)$ produced $30 in damages.

The latter approach abandons the pretense that the costs of future climate damages can be accurately forecast, and it draws attention to the fact that, as emission-reduction technologies are deployed, their costs decline through "learning by doing," an important benefit often overlooked in conventional cost-benefit analysis.[6] Koomey also notes that using this "working toward a goal" approach also focuses attention on the enormous additional costs of delaying the adoption of meaningful emissions reductions, a cost that the International Energy Agency has estimated is currently at half a trillion dollars or more a year for every year of delay.[7] (Delay raises costs because emissions must then be cut more radically later.)

Conventional economic studies nonetheless are useful for indicating the general nature of the economic effects of climate change, even when they may be unreliable in gauging their magnitude. For example, studies show that as temperature rises in poor developing countries, industrial output, investment, and per capita income will all fall, relative to their expected performance under normal climatic conditions.[8] But the studies are less convincing when they claim that a precise percentage decline in output or income will occur with every additional degree of temperature increase. Some of the most advanced modeling tools used to estimate the economic effects of climate change have gaping flaws. For example, although in ordinary business and private life, insurance is purchased against risks, some models exclude insurance costs and trade across regions (except for trade in fossil fuels).[9]

Even if models attempted to include insurance costs, no insurance policy will suffice if, for example, the ocean that feeds billions of people is disrupted and its fish stocks plummet or disappear. Yet climate change is obviously putting the oceans at great peril: if carbon emissions continue their relentless increase, by the end of the century, the oceans will likely be more acidic than at any time in the past 20 million years.[10,11]

Likewise, no effective insurance exists to adequately compensate the United States if a catastrophic megadrought returns for decades or centuries to the Southwest as in the past. Similarly, who will insure Everglades National Park—much of which lies only a few feet above the Atlantic—against sea-level rise? Would the loss of this World Heritage Site, the largest tropical wilderness in the United States, be insurable or adequately valued in economic analyses if lost?

The Costs and Widespread Ripple Effects of Drought and Agricultural Losses

In contrast to the uncertainties about how an entire economy will fair under climate change, agronomists can indeed make more restricted forecasts of how individual crops will respond as temperature rises or precipitation changes.[12] When these studies are aggregated, a good picture can be developed of how a particular nation's agricultural sector is likely to perform under various climate scenarios. However, often the agricultural sector accounts for only a relatively small portion of a country's national output. Much larger economic impacts may occur throughout the economy, including its industrial and manufacturing sectors, through a loss of labor productivity due to climate-induced heat stress or health effects.[13] In addition, the indirect effects of a drop in agricultural productivity can reduce the output in other resource-dependent sectors. For example, suppose temperature rises and rainfall decreases significantly in a developing nation dependent on agriculture and tourism. Grain, forage, and livestock production are then likely to decline. The prices of meat, milk, eggs, cereals, bread, and hides are thus likely to rise. If so, hunger, malnutrition, and illness will increase, especially among the poor. Political instability and crime will likely also rise. Tourism will then likely decline, and hotels and restaurants will then probably see their business dry up.

As global temperature rises, agricultural yields in drought-affected areas, such as the African Sahel, East Africa, and southern Africa, may drop by half in some countries, according to the IPCC.[14] Crop yields are already declining in Asia, "The frequency of occurrence of . . . heat stress [on crops] in Central, East, South and South-East Asia has increased with rising temperatures and rainfall variability,"[15] according to the IPCC. Each 3.6°F rise in average temperature reduces yields of wheat, rice, and corn by 10 percent.[16]

Drought combined with strong winds can also lead to dust storms that sicken people and blow away valuable farmland topsoil, driving marginal farmers off their lands. Food then has to be imported. Drought also leads to the expansion of deserts. Deserts caused by the overintensive exploitation of land are already destroying cropland in Central Asia, China, Saharan Africa, and the Middle East.[17] Climate change will make the problem worse. The economic costs are unknown. This scenario is already playing out in Iran where precipitation is down 43 percent over the past decade compared with the previous 30 years.[18] According to a recent account by Joel Brinkley, a Pulitzer Prize–winning former *New York Times* reporter:

Iran ·is, quite literally, blowing away. Lakes and ponds are drying up. Under-ground aquifers that supply most of the nation's water are emptying fast. . . . Massive dust storms sweep across the country almost daily, afflicting 23 of the nation's 31 provinces, making it hard to breathe and killing thousands of people a year. . . .[19]

China ·and India, the world's number-one and number-two wheat producers—and the world's first and second most populous nations—both depend heavily on meltwater from glaciers in the Himalayas and the Tibetan plateau, many of which are disappearing.[20] An authoritative study of glaciers there and in the Pamir Mountains (Tajikistan) found that the glaciers studied there had lost 9 percent of their ice (by area) just in the 30 years from the early 1970s to the early 2000s.[21] As major glacial-fed rivers like the Brahmaputra, Indus, and Ganges diminish, China and India with some 2.5 billion people will struggle to buy enough food and land abroad to keep their ever-growing populations fed. That will drive up world grain prices and make food more expensive (by unknown amounts) for consumers in the United States and elsewhere.[22]

Both China and India, along with other Asian and Middle Eastern coun-tries, are already buying and leasing agricultural land abroad in efforts to keep pace with their growing populations' food requirements.[23] Western investors, including US universities and pension funds working through British hedge funds and European speculators, are also buying up land in Africa, but for bio-fuel production.[24] While the scale of these purchases may not be large relative to the billions of acres in global farmland, such purchases tend to drive up land as well as food prices in the countries selling or leasing their resources, and sometimes local farmers are displaced in the process.[25] As Lester Brown notes in Plan B 4.0, "Countries selling or leasing their land are often low-in-come countries and, more often than not, those where chronic hunger and malnutrition are commonplace."[26]

At an increase in global mean temperature of 3.6°F above 1990 levels, the portion of the world's surface affected by extreme drought will increase from 1 percent of all land area to 30 percent, according to a "medium confidence level" estimate by the Fourth IPCC Assessment.[27] As drought increases and agricultural productivity declines, farm and agricultural commodity export incomes fall in the expected regions. Those farmers able to stay in business will have less money to spend on seed and farm implements. Less capital will also be available for domestic investment. Livestock and dairy production are likely to fall. Leather goods, including shoes, are likely to get more expensive.

Food processors, breweries, and manufacturers will likely see higher raw materials costs and raise their prices. In addition, if drought makes forests across a region more susceptible to insect infestations and fire, timber will be more expensive for the housing and furniture industries and, ultimately, fewer people will be able to buy or rent homes. Higher housing costs mean that people will have less disposable income.

In the summer of 2012, 80 percent of the agricultural areas of the United States were stricken by the most extensive drought since the 1950s. Corn and soybean production were hit particularly hard, reducing supplies for both people and animal feed. Ranchers were forced to cull herds. California, which produces close to half of all fruits, nuts, and vegetables grown in the United States,[28] had its driest year on record in 2013, and a drought emergency declaration was in effect in the state in 2014. Although market impacts often lag drought by several months, drought eventually increases food prices.

The costs of drought extend to the energy sector. A large Sierra Nevada winter snowpack accumulates in a state like California during years of normal precipitation. During the spring and summer, snowmelt fills the state's reservoirs and is later discharged from dams to spin turbines to generate hydropower. In a normal year, that power is available to meet the state's peak summer demand. But in a drought year like 2013, little snow fell in the mountains of Claifornia. Without its usual snowpack, the state in 2014 will have to replace the hydropower with more expensive electricity from other sources, primarily natural gas generators. Consumers and businesses will therefore have to pay higher electricity bills, and combustion of the replacement fossil fuel will create air pollution and release more carbon dioxide, further heating the climate.

Aquatic Impacts. Droughts that cripple agriculture can also damage shipping and service industries. In the United States in 2013, Lake Michigan and Lake Huron declined to record lows, and other Great Lakes were abnormally low, all because of climate-related drought and high temperatures.[29] Not enough winter snow has been falling in the watershed for some years now to replenish water evaporating from the lakes during very hot, dry summers. Thus, deepwater freighters carrying coal and iron ore have had to lighten their loads to operate in shallower water. Some harbors have had to close entirely. Tourists who once came to fish stayed away. Lakeside docks, restaurants, motels, fishing charters, and grocery stores all felt the pinch.[30] Low water levels in rivers and lakes can also affect the operation of power plant and factory cooling systems and sewage treatment plants. Cheap hydro-power becomes less available. Higher electricity costs then ripple through the economy's manufacturing and industrial sectors.

Coastal Assets in Jeopardy

Sea-level rise and the flooding of coastal areas and coastal infrastructure are other major economic impacts of climate change. In California alone, a study by the University of California and the research firm Next 10 found that $2.5 trillion of the state's $4 trillion in real estate will be damaged by climate change.[31]

In addition, as sea level rises, estuaries and the lower reaches of rivers flowing into them will become increasingly saltier, lowering the quality of irrigation and drinking water supplies derived from those sources. Invest-ments of unknown magnitude will need to be made to ensure that public and agricultural freshwater needs can be met. Also, the related extensive loss of wetlands that serve both as storm buffers and as valuable habitat for commer-cial fish will drive up the cost of seafood and increase the cost of protecting coasts against flooding.

The Costs of Extinctions

As chapter 10 explains, climate change is likely to cause very high global rates of extinction. A study that compared European plants under various climate change scenarios between now and 2080 found that half or more would be classified as vulnerable or threatened with extinction by then.[32] Another longer-term study from that up to half of all European flora "is likely to become vulnerable, endangered or committed to extinction by the end of this century."[33] The Amazon rainforest and its biodiversity may also be devas-tated by climate change, as described in chapter 6. While the loss of a species of life is tragic from an ethical perspective and may undermine the viability of an ecosystem, these consequences are often ignored by economists. Even if we wanted to put a price tag on a species' existence, how would we go about it? How much is the loss of a beautiful flower or a capuchin monkey worth? What is a jaguar's value? Is it a few hundred dollars for a pelt multiplied by the dwindling number of jaguars, or is it, perversely, the amount of money people are willing to pay to see a jaguar? If setting a price on an endangered plant or animal is problematic and counterintuitive, how are we to put a price on the incomprehensible yet expected loss of millions of species due to cli-mate change? Even if the vast majority of species are invertebrates, this raises even broader questions about how to ascribe an economic value to natural resource systems destroyed by climate change. What is a coral reef habitat or an ancient tropical forest wilderness worth? How can we reckon in dollars and cents the almost magical natural harmony that exists within the ancestral wildness of a complex, healthy, intact forest ecosystem?

Sea Ice and Sea-Level Economics 101

In chapter 6 and elsewhere, I discussed the ecological consequences of sea ice loss and how, through positive feedback effects, it leads to further global warming.[d] How does one look at the loss of sea ice from an economic perspective? On the positive side of the ledger, sea ice loss will facilitate the transit of ships from the Atlantic to the Pacific through an ice-free Northwest passage, saving shippers money.[34] On the negative side, by accelerating warming, the effects of sea ice loss exacerbate all the other effects of global warming. The loss of sea ice has profound effects on Arctic ecosystems as noted in chapter 1. As sea ice dwindles and disappears, it leads to starvation for ice-dependent marine mammals and to the loss of traditional subsistence lifestyles for native peoples reliant on hunting, fishing, and gathering. Current trends indicate that all Arctic summer sea ice will be gone well before the turn of the century.[35]

As sea levels rise from the combined effects of land-ice loss and the thermal expansion of seawater, the rice-growing river deltas and coastal floodplains of Asia will gradually be submerged. "Even a 3-foot rise," writes Lester Brown, "would devastate the rice harvest in the Mekong Delta, which produces more than half the rice in Viet Nam, the world's number two rice exporter . . . and would inundate half the rice land in Bangladesh, home to 160 million people."[36] How many people will be malnourished, famished, or impoverished by the loss of agricultural land on this scale? What will that do to the price and availability of rice on which so many people in the world survive?

Collateral Economic Damage: Coldwater Fish and Northern Ecosystems

Earlier in this chapter I discussed the impact of drought and lower water levels in lakes and rivers. Higher water temperatures and reductions in river and stream flow will also affect fish and other aquatic life, as well as birds and mammals that depend on the fish. The IPCC's *Fourth Assessment Report*

---

[d] From 1979 to 2008, sea ice disappeared more than four times faster than predicted by the IPCC. Just in the past 30 years, Arctic sea ice has declined in extent by half.

contains the following dire forecast: "In the continental US, cold-water [fish] species will likely disappear from all but the deeper lakes, cool-water species will be lost mainly from shallow lakes, and warm-water species will thrive except in the far south, where temperatures in shallow lakes will exceed survival thresholds.... Species already listed as threatened will face increased risk of extinction."[37]

Trout and salmon will be most affected, especially in the high northern latitudes. Abundant salmon runs are not only important for feeding people and countless creatures, but they play a vital ecological role in forests transporting nutrients from the oceans to the land. These nutrients—some in short supply on land—are brought by predators of salmon, such as bears and wolves, into the tundra or forest, where their droppings provide essential nutrients that enrich the soil and nourish vegetation. Thus, when coldwater fish such as trout and salmon vanish, a high-quality, delicious source of protein disappears and nutrient cycling that is vital to the health of the entire forest food chain is interrupted. Without the decaying bodies of spawned-out salmon in the shallow headwaters of river systems, these ecosystems will have fewer insects; this will reduce fish and bird populations that feed on insects. To quote John Muir, a giant of the American conservation movement, "When we try to pick out anything by itself, we find it hitched to everything else in the Universe."

Health-Related Costs of Extreme Weather

The effects of climate change on human health are discussed from a medical perspective in chapter 8. Health effects also have economic impacts, however. Especially in vulnerable developing lands, global warming will increase deaths and disease from malnutrition, starvation, heat waves, floods, storms, drought, fires, air-quality degradation, civil disorder, dislocation, and civil strife (including conflicts over increasingly scarce water and other resources).[38] More money will then need to be spent to provide health care, emergency services, and to preserve public order. In addition, as noted earlier, tropical cyclones like Typhoon Haiyan that hit the Philippines in 2013 are likely to increase in severity. That will bring more deaths, injuries, and costly damage to businesses, homes, personal property, infrastructure, and crops.

Large public health and, specifically, mental health costs are likely. Some climate change victims will be unable to endure the mental stresses from diseases, hunger, unemployment, poverty, and refugee life. Some may suffer breakdowns associated with a loss of fresh water, sewage, and power services, or the inability to continue with traditional lifestyles. Earnings will be reduced or eliminated if victims are unable to work. People who are seriously malnourished in childhood due to the effects of climate change may suffer lifelong reductions in mental function due to nutrition-related developmental impairment and thus may be unemployable, dependent on social services, or eligible only for low-wage jobs. The human casualties, especially in developing countries, will include all those with less secure access to food, water, and employment, or whose family members are suffering these ills.

Massive coastal flooding related to sea-level rise and storm surges is likely to damage schools, hospitals, legal and employment services, and public safety. In the wake of the flooding, local and regional economies may wither. As noted in volume 2 of the IPCC's *Fourth Assessment Report*,[39] severe disasters not only have an immediate effect but may also slow or halt economic development insofar as the developing country's economic surplus is entirely absorbed for an indefinite time in responding to climate-related catastrophes. Social and political order may be adversely affected. The loss or decline in public order and morale may then produce still more stress, mental disorders, social chaos, and economic impacts, as in the chaotic aftermath of Hurricane Katrina. Yet future climate-related disasters won't be confined just to Louisiana and Mississippi. They may occur next in Florida, Panama, Bangladesh, Australia—or again in New York City, New Jersey, and Washington, DC.

Costly Yet Imponderable Cascading and Interacting Effects

I've alluded previously in this chapter to secondary and tertiary "ripple effects" of climate change, as when an economic crisis precipitates a mental health crisis that causes an economic impact. Wildfires can have similar ripple effects. When wildfires burn near populated areas, they can cause severe air pollution which adds public health damages to the fires' direct costs in lives, property, and fire fighting.[e] In addition, the cost of wood pulp and lumber may be affected, which would then affect paper and wood products. Finally, silt from burned-over and erosion-prone lands will clog waterways and require costly dredging.

The melting of glaciers also sets in motion a cascade of economic consequences that may not only eventually lead to regional water scarcity, but will impact wildlife populations, biodiversity, tourism, and even public safety,

---

[e] When state and federal costs are combined, wildfire fighting in the United States is a multibillion-dollar expense. Not only has the fire season lengthened, but fires are larger, hotter, more damaging, and more extensive. Over eight million acres burned in 2012. See Darryl Fears, "US Runs Out of Money to Battle Wildfires," *The Washington Post*, October 2012.

because of increased flooding, rockslides, and avalanches. Additional economic losses will occur should people be hurt by any of these processes. As these examples remind us, it is unlikely that anyone can realistically assess all the costs of climate change when so much about its cascading effects is unknown.

To cite a planetary-scale example of prospective cascading impacts, what will the economic effects be of altering the Atlantic Meridional Overturning circulation (AMOC)—an enormous current that transfers heat from the equator to the poles? Disturbing the current will disrupt ocean ecosystems and alter the productivity of both the ocean and the land masses where temperature is modulated by the current. This may jeopardize marine fisheries. Evidence exists that the currents are already slowing.[40] Who can say what the ultimate price tag may be? How, too, shall we put a price on the impact of changing the ocean's chemistry or its ability to take up heat or carbon dioxide? Should we guesstimate the cost of these impacts by putting a price tag on the ecosystem services that the oceans provide us, for example, the service of absorbing carbon? Then, as the ocean's capacity to absorb carbon diminishes over time (as it becomes more and more saturated), do we tally damages in dollars per ton of unabsorbed carbon? Do we then use the economic methodology known as discounting to translate the future economic costs over the next 500 or 1,000 years back to an estimated present value? (If we select a high enough discount rate—a percentage related to current interest rates—we may succeed in shrinking the stream of future costs to a trivial contemporary sum.) Does it really make any sense, however, to apply this narrow economic approach to an ecosystem as vast and basic to survival as the oceans? Suppose we performed such a foolish calculation and concluded that we could indeed economically afford to sustain trillions of dollars of damage by impairing or destroying a vital ocean service, like carbon absorption or indefinitely sustaining marine fisheries. It would still be an enormously unwise thing to do because of the nonmarket environmental effects and the equity effects: harm to certain groups of ocean-dependent low-income regions and people. In the case of using up the ocean's carbon- or heat-absorption capacity, that would then lead to amplification of future heating, committing the world to more intense global heating and risks of runaway climate change. This would make little sense no matter how large our current GDP might be.

Investing in Adaptation: Needs and Pitfalls

Climate change will hit poor and developing nations especially hard. Often they are already dealing with extreme poverty, hunger, disease, crowding, water and energy shortages, and inadequate government. They often lack adequate housing, education, health care, employment, social services, and strong institutions. Preoccupied with day-to-day survival and maintenance of public order, they have few resources to make costly, long-term investments in flood prevention and village or city relocation. Massive outside aid will be required to assist such regions with climate adaptation.

Adaptation, of course, won't actually avoid or reduce climate change. In fact, it is quite conceivable that the more intensely we pursue adapting to climate change, the more resigned or complacent some people may become about it and, more importantly, the fewer resources we may have left to deal with its fundamental causes by reducing greenhouse gas emissions.

Adaptation necessarily focuses on protecting people and their livelihoods. In the process, some plant and animal populations will incidentally be helped. Some wildlife may be saved from flooding by a seawall built to protect a populated area, for example. But nature is generally left to fend for itself. Investing in adaptation has unfortunately become necessary because climate change is occurring. But in contrast to building dikes, flood gates, or other engineered structures that buy time, slowing the rate of climate change by reducing heat-trapping gas emissions addresses the core problem and benefits both human and nonhuman life.

Adaptation and Complacency. Believing that government has what appears to be a sound adaptation plan may lull some people into assuming that they are protected and that public officials have everything under control. This could be a dangerous illusion. As Typhoon Haiyan bore down on Tacloban in the Philippines in 2013, for example, the government reportedly failed to order an evacuation. Many lives that could have been saved were lost as people drowned in their homes or went to unsafe buildings designated by local officials as public shelters. However, even if sound adaptation plans exist, there is no assurance that they will ever be fully or properly implemented. As the IPCC's *Fourth Assessment* points out:

> There are significant barriers to implementing adaptation. These include both the inability of natural systems to adapt to the rate and magnitude of climate change, as well as formidable environmental, economic, informational, social, attitudinal and behavioral constraints.[41]

The authors conclude:

Unmitigated climate change [pursuing business-as-usual emissions scenarios] would, in the long term, be likely to exceed the capacity of natural, managed, and human systems to adapt.[42]

Several states created climate change adaptation plans in the United States between 2006 and 2009.[43] However, according to a 2012 evaluation of state climate change and water preparedness by the Natural Resources Defense Council, only four US states had a comprehensive plan.[44] California's 2009 climate adaptation strategy report was approved by then-governor Arnold Schwarzenegger in a public ceremony on Treasure Island in San Francisco Bay. If unprotected, the island itself ironically will be underwater by the end of the century if average surface temperatures increase 9°F and local sea level responds by rising 4.5 feet; in that event, San Francisco International Airport would also be underwater.

The Oakland-based Pacific Institute, a nonprofit research group, estimates that counties just in the San Francisco Bay Area alone face tens of billions of dollars in costs to replace property inundated by rising seas or storm surges.[45] Even in the relatively small area studied, however, it is not known how, whether, or by whom the huge adaptation costs would be paid. California has struggled for years with multibillion-dollar budget deficits and only recently managed to produce a balanced budget after making deep cuts to education and to services for the aged, blind, and disabled.[46]

As other geographic areas and states complete their impact studies and adaptation plans, more comprehensive estimates of the total US costs expected from sea-level rise will become possible. Many states, however, face major economic challenges even without the future costs of climate change. Unfortunately, these climate bills will be presented by nature whether we like it or not.

Can We Afford to Cut Carbon Emissions?

To stave off the worst future impacts of climate change, large investments are going to need to be made in mitigating climate change by changing the energy systems and energy-use patterns that create greenhouse gas emissions and by reducing emissions from agriculture and forestry. Even though reducing carbon emissions by gradually transforming energy systems will require substantial upfront investments for decades, some studies have found that these investments will require only a small annual fraction of our GDP and will ultimately generate net benefits for the economy, both in terms of new economic activitiy, new employment, new tax revenue, and avoided climate damages.

The 2006 Stern report agrees with a number of other major studies that have concluded that the world could reduce the worst impacts of climate change by making deep emissions cuts for a small fraction of global GDP. The Stern report concluded this would cost 1 percent of global GDP per year. Looking specifically at the United Kingdom, the study projected that 80 percent cuts in heat-trapping gas emissions could be accomplished there with very minor impacts: an almost unnoticeable 1 percent increase in food and clothing prices, 1 percent in car purchases, and 2 percent in household goods and electronics over a 40-year period.[47] Referring to the price increases expected in gasoline and electricity, the Stern report stated, "this need not lead to big changes in lifestyle." Electricity, for example, was forecast to rise by 15 percent by 2050; most people would not consider a 15 percent price increase spread over 40 years as a major problem, especially because energy only accounts for about 2 percent of the cost of British consumer goods.

The UK Committee on Climate Change established by Parliament as an impartial source of expert advice has confirmed recently, after studying the costs of Britain's Climate Change Act of 2008,[48] that meeting its goal of cutting British greenhouse gas emissions by 80 percent by 2050 would indeed cost only 1 to 2 percent of GDP. That estimate does not even include credit for the economic benefits that would ensue, such as improved air quality, more secure energy supplies, new employment, or global export of low-carbon goods and services.

Similarly, a study commissioned by the British publication *New Scientist*[49] and conducted by Cambridge Econometrics found that these deep cuts in heat-trapping gas emissions could be accomplished with "barely noticeable increases in the price of food, drink and most other goods by 2050."[50] The insensitivity of consumer goods to energy prices also may well apply to the US economy. A study in the journal *Energy Economics* found that halving heat-trapping gas emissions in the United States between now and 2050 would add less than 5 percent to the cost of consumer goods.[51] A related study by the Netherlands Environmental Assessment Agency found that by investing a mere 0.2 percent of GDP in 2020, developed countries could raise their carbon reduction targets (as outlined at the 2009 Copenhagen climate conference) so their heat-trapping gas emissions would fall 25 to 40 percent below 1990 levels by 2020.[52]

Another detailed study by the Dutch Ecofys group for the World Wildlife Fund in 2011 asserted that it would be possible to create a fully renewable global energy system that would be 95 percent "sustainably sourced" by
2050.[53] Ecofys also projected that this would require an investment of 1 to
2 percent of GDP in coming decades but that by 2040, the investments would
produce a net positive rate of return and that by 2050, the return would equal
2 percent of global GDP. Even if these kinds of estimates were off by 100 or
200 percent, the sums are far from prohibitive. Using examples from the US
economy, this chapter will offer many possible constructive ways that a few
percent of GDP could be reallocated to climate protection.

Putting a Price on Carbon

The most practical and powerful way to protect the climate is by raising the economy-wide cost of releasing greenhouse gases. That can be done
either by passage of a carbon tax, as eight nations and some Canadian provinces have done, or by creating a carbon market in which participants trade
carbon emission allowances.[f] In emission trading schemes, government sets a
cap on the total quantity of carbon that can be emitted and then allows the
emissions market to set the price of emissions. The European Union, New
Zealand, Switzerland, and various regions throughout the world have adopted
this approach. While a carbon tax is administratively simpler and provides a
clear, transparent price signal that is not easily evaded, passage of a carbon tax
in the United States has thus far proven politically difficult.[54,g] Resistance to
establishment of carbon markets, by contrast, has been less vehement; nine
US states already participate in carbon markets. That includes eight northeast
and Mid-Atlantic states which are participating in a regional carbon market
known as the Regional Greenhouse Gas Initiative.[h]

---

[f] The nations with a carbon tax are Australia, Kazakhstan, the Scandinavian nations, Switzerland, and
the United Kingdom. Nations with a carbon market are members of the European Union, New
Zealand, Switzerland, and parts of the United States, parts of Canada, and parts of India and China.
Carbon pricing is also under consideration in Brazil, Chile, Mexico, Turkey, and Ukraine, and it is
scheduled to go into effect in South Korea in 2015, per The Climate Group (see reference 52).

[g] Unfortunately, an excessive number of permits were issued in the ETS and so in 2013, the price of
carbon fell to a new low of $3.50 per ton (€2.63). Low prices are also attributable to the recession and
to the availability of cheap carbon offsets in developing countries where it is usually much cheaper
to save a ton of carbon emissions than in the more industrially advanced nations.

[h] The initiative began in 2003 and, after Maryland joined in 2007, included nine states until New
Jersey withdrew in 2011. See Environmental Defense Fund and International Emissions Trading
Association, "Regional Initiative, The World's Carbon Markets: A Case Study Guide to Emissions
Trading," May 2013. http://www.ieta.org/assets/Reports/EmissionsTradingAroundTheWorld/edf_
ieta_rggi_case_study_may_2013.pdf

More than 20 carbon-pricing systems now exist around the world, with the largest being the somewhat dysfunctional European Union's Emissions Trading Scheme (ETS), which includes more than 30 nations and covers about half the EU's carbon emissions. Australia recently passed a carbon tax but will be moving to a carbon market system.[55] China and India are trying both carbon taxes and carbon markets. In the United States, California has been a national and global leader in the field of energy efficiency standards since the 1970s.[i] More recently, it has also taken a leading role in establishing a comprehensive market-based cap-and-trade program to reduce greenhouse gases and has a separate program to increase the state's reliance on renewable energy.[j] The program requires California's electric utilities to increase the share of the power they generate from renewable resources to a third of their total generation by 2020.

California's cap-and-trade program was established under the Global Warming Solutions Act (AB 32), which former California Governor Arnold Schwarzenegger signed in 2006. Under AB 32, the state issues a fixed but gradually shrinking number of carbon emission allowances each year. They entitle holders to discharge an equivalent quantity of carbon in that year. Holders are then allowed to sell these pollution permits, thereby setting a market price on carbon in the state. The state gradually reduces the number of available permits, thereby raising the value of each emission allowance and increasing each company's incentive to reduce its emissions. AB 32 applies to 85 percent of all greenhouse gas emissions in the state, including emissions from industry, transportation, agriculture, forestry, waste management and recycling, not just the electric-generating sector. All large carbon emitters in California have to report significant heat-trapping gas emissions to the state. Each year California lowers its annual statewide emissions cap to gradually coax the state's emissions to 1990 levels by 2020. The state initially gave large carbon emitters a fixed number of carbon emission allowances that they could either use or trade with each other. If a company can reduce its emissions for less than the market value of an emission allowance, it will have an incentive to make the emission reduction rather than use its permit, since it can sell the permit and pocket the difference.

---

i For more details about California's leadership in energy efficiency and the development of Federal energy efficiency standards, see "Efficiency: The Sleeping Giant" in John J. Berger, *Charging Ahead: The Business of Renewable Energy and What It Means for America* (New York, NY: Henry Holt, 1997 and Berkeley, CA: University of California Press, 1998).

j California also has notable state incentive programs for installation of residential and commercial solar panels as well as for low- and zero-emission vehicles and cleaner-burning transportation fuels.

Conversely, if it is very costly for a company to reduce its own emissions, it can buy the permits it needs from another company. The seller presumably can reduce its emissions more cheaply than the buyer, so statewide emissions are reduced more economically than if each company had to actually reduce its emissions by an equal percentage. The cap-and-trade system is thus consistent with the state's intent to minimize the cost to businesses of complying with the state's carbon reduction mandate. Although it could have charged carbon polluters for permits from the outset and set challenging targets, California initially gave away 90 percent of the permits for free to utilities and industries. The rest of the permits were sold at an auction that in effect set a price on carbon in the state.

Based on the latest auction results, the price for 2016 will be a mere $11.10 per ton of carbon dioxide,[56] which is close to the regulatory price floor. (The market also has a $40/ton price ceiling.)[57] Since permits are still so inexpensive, the program as yet does not provide a strong incentive for reducing carbon emissions. (Any measure that costs the polluter more than $11.10 per ton will be looked on with disfavor since it will be cheaper to simply purchase a permit than to reduce the pollution.) Eventually, permit prices will rise to levels that will constitute a greater emissions deterrent. Yet even if the state achieves its goal of returning to 1990 emissions levels by 2020, California companies in the aggregate will still be allowed to emit 427 metric tons of greenhouse gases a year as they did in 1990, which will continue warming the planet. Since carbon emissions do not respect state or national boundaries, California is endeavoring to coordinate its efforts under AB 32 with several other western states and Canadian provinces. An agreement linking California's carbon market with Quebec's is due to go into effect in 2014.[58]

Although business interests objected to AB 32, a University of California, Berkeley, study of the statute's expected economic impacts found that achieving the state's emission reduction goal would *add* more than $70 billion to the Gross State Product (GSP) and create close to 100,000 new jobs by 2020.[59,60] If a program of gradually and moderately controlling heat-trapping gas emissions is not only profitable for the California economy, but good for the environment, it probably would also be profitable and good for the United States as a whole. After all, California has the largest GSP of any state in the nation, and if it were a nation, it would be the world's ninth largest economy (as of 2010).[61,62]

The economic benefits that occur when new emission-reducing energy systems are gradually phased in to an economy include increased employment, reduced air and water pollution, health benefits, fuel savings, and the avoidance of fossil fuel subsidies and tax exemptions.[63] On the other side of the ledger, costs would arise if subsidies had to be provided to compensate those investors who were obliged to prematurely retire energy systems made uncompetitive by the new systems. Given that studies have shown that the adoption of a cap-and-trade carbon market system creates net economic benefits, the argument that it places undue burdens on the broad economy is without merit. Similarly, results from five years of experience with carbon taxes in the province of British Columbia, Canada, indicate that they, too, are "kinder and gentler" than their opponents have charged. The revenue-neutral carbon tax adopted in British Columbia in 2008 rose from $10 a ton in 2008 to the current $30 a ton (at which it has been capped). Emissions during the past five years fell 15 percent in British Columbia while they rose elsewhere in Canada. The BC economy, however, outperformed the rest of Canada; and corporate and individual taxes were reduced to the lowest in Canada for corporations and for individuals earning less than $119,000 (CAD).[64]

These studies of modest emissions reductions do not demonstrate, however, that the steep emissions reductions actually required to prevent global warming of more than 3.6°F can be achieved without large costs for writing off fossil fuel infrastructure and natural resources. Wrenching the economy suddenly from fossil fuels to renewable energy would create economic dislocations. If these costs were not properly mitigated by expansionary government policies—heavy government investments in energy efficiency and renewable energy infrastructure—energy prices would probably rise, disposable income would probably fall, and growth might suffer during the transition. Yet even sacrificing economic growth for a period of time would have a silver lining, as it would modernize the economy, reduce long-term energy costs, protect the climate, reduce pollution, and—of critical imiportance—avoid the disastrous long-term economic costs that uncontrolled temperature increases of 7 to 10°F or more would inflict.

According to professors Kevin Anderson and Alice Bows at the University of Manchester, "Economic growth cannot be reconciled with the breadth and rate of impacts as the temperature rises towards 4°C and beyond—a serious possibility if global apathy over stringent mitigation persists."[65]

Finding the Money to Protect the Climate

Despite these indications that both carbon taxes and carbon markets can be quite benign, suppose that in the worst case, that were untrue. Could the US government nonetheless afford the 1 to 2 percent of GDP that studies say is required as investment capital to reduce carbon emissions and put the nation on a path to deep emissions reductions? Even if you believe that these studies underestimate the final cost of transforming the US economy by a full 100 percent, it would still be eminently affordable.

The US GDP is currently about $16 trillion. One to two percent of GDP is $160 to 320 billion, not a great deal of money for a nation as wealthy as the United States of America in a cause as preeminent as combating climate change. Changes in our tax code, including the elimination of some tax loopholes for the wealthy, and adjustments to national spending priorities could readily make those funds available. Let's begin by considering what some might consider the most unlikely way to raise this money: reducing bad habits. Every year Americans spend about $90 billion on cigarettes that annually kill 443,000 people.[66,67] Tobacco use of all kinds costs the economy about $200 billion a year in public and private health care costs and productivity losses.[68] One hundred billion dollars are spent each year on illegal drugs, which cost the nation another $93 billion in drug-related law enforcement, drug treatment, and prevention.[69,70] Americans also spend $125 billion a year on gambling.[71] Obesity adds another $190 billion a year nationally to health-related costs.[72] Alcohol abuse causes 75,000 excess deaths a year and costs the economy $224 billion.[73] The unhealthy habits and addictions just cited cost well over $1 trillion—some three to six times what it might take to launch a meaningful effort to combat climate change. No, of course, addictions, illnesses, and alcoholism cannot just be wished away and should not be neglected. But given the gravity of the climate crisis, finding ways to reduce the prevalence of these conditions and then repurposing some of this partially avoidable spending might not be a bad idea.

Even without laying a finger on cherished vices like smoking, the United States could surely find enough cash to deal with the climate crisis by reforming the federal tax code to close tax loopholes that disproportionately benefit the affluent, and to discourage US corporations from hoarding cash, especially abroad, where "offshoring" allows them to defer payment of US taxes. Closing just four such loopholes would generate more than $200 billion a year.[k] Ordinary taxpayers today are paying more in taxes to make up for the uncollected revenues.

---

[k] While I believe there are better ways to raise these funds than across-the-board elimination of the following four loopholes, the elimination of special low tax rates on dividends and long-term capital gains would yield $90 billion per year; ending the tax exemption of income from life insurance and annuities would produce $40 billion per year; eliminating the income exclusion for interest earned on state and local bonds would also produce about $40 billion; and eliminating the exclusion of capital gains at death would provide another $52 billion for a total of $220 billion, which is in the range of 1 to 2 percent of US GDP.

With respect to hoarded cash, sixty of the wealthiest corporations deposited $166 billion offshore beyond the IRS's reach in 2011, according to a *Wall Street Journal* analysis.[74]

Based on Federal Reserve data, in 2012, US nonfinancial companies had on hand more than $1.7 trillion in cash and easily traded securities.[75] When the worldwide holdings of these corporations is included, IRS data show that the 2009 liquid assets of nonfinancial corporations were more than $5 trillion. The sum is likely greater in 2013, since corporate profits have grown almost a trillion dollars from 2009 to 2011.[76] As David Cay Johnston wrote for Reuters, "The 2009 cash reported to the IRS equaled America's entire economic output that year from New Year's Day through May Day."[77] Imagine how much the United States could slash its carbon emissions if these funds, instead of sitting idle, were used to put millions of unemployed Americans to work creating the clean, efficient, and carbon-free renewable energy system needed to protect the climate and provide for a sustainable future. Given that the nation is thus awash in underutilized capital, policymakers don't actually have to go after spending on tobacco, illegal drugs, junk food, or gambling in the United States to fund our energy transition. They could eliminate tax loopholes for the wealthy by means of testing tax loopholes too important to dispense with, and they could provide tax credits to induce corporations to invest their idle cash in jobs creating clean energy systems and other greenhouse gas–reducing investments in agriculture and forestry.

Changing How We Tax and Spend

With suitable tax incentives to properly direct the cash companies and investors have, the dollars now overseas or on the economic sidelines—parked in tax-free treasuries, for example—could be channeled into public and private clean-energy infrastructure investment, providing a flood of capital that could kickstart the US economy. Some of those funds could also be invested in a new federal environmental bond program to protect and restore badly damaged natural resources systems necessary to a healthy economy. These resources include soils, forests, mined land and rangelands, surface waters, aquifers, and the oceans. Making soils, forests, wetlands, and rangelands healthier has the added climate benefit of increasing the rate at which they naturally remove carbon dioxide from the atmosphere. Moreover, a healthy natural resource base provides a stream of long-term public benefits, so it is not unfair to fund them with borrowing that will be repaid in part by future resource users.

In short, investing heavily in energy efficiency, new energy technologies, as well as in renewable natural resources and human capital, would sharply cut carbon emissions, boost employment, and advance technological innovation. The economy would prosper, the vast majority of Americans would see their living standards rise, and the environment could be nurtured.

Income inequality and social justice issues are tightly intertwined with society's current failure to fund an adequate response to climate change and are therefore highly relevant to any discussion of the affordability of mitigating climate change. Without the enormous income and wealth disparities in America today, ordinary people would not experience so much privation. They would thus have far less reason to resist the allocation of a few percent of a fairly divided GDP to climate protection. The hoarding of cash mentioned earlier is an indication of the increasingly unequal distribution of income and wealth in America. The United States in 2009 had some $54 trillion in wealth, but 93 percent of it was held by just the top 20 percent of all Americans.[78] Conversely, the bottom 80 percent had only 7 percent.[79] The nation's wealthiest 1 percent owns 40 percent of all the nation's wealth,[80] including 50 percent of all the stocks, bonds, and mutual funds. The bottom half of all Americans, however, owns a mere half of 1 percent of those investments.[81] The even poorer bottom 40 percent of all Americans have virtually no wealth or investments.[82] Income in America is also extremely unequally divided, with the wealthiest 1 percent annually earning nearly a quarter of the nation's entire income—a situation that has only grown more unequal over the past 30 years along with the increasing inequality of wealth.[83]

With concentrated wealth goes concentrated political power, and the ability to create laws and policies that enhance one's income. Some of the super wealthy who are benefiting from the unequal division of income and wealth are also deeply vested in supporting policies that further entrench the fossil fuel economy that aggravates the climate crisis. Raising tax rates on superwealthy individuals and wealthy corporations would not only be equitable but would make a massive infusion of capital available for investment in transforming the energy sector and developing substitutes for fossil fuels.

Reexamining Our Spending Priorities. Even without dipping into the national wealth squandered because of addictions and preventable illnesses, and even without reversing the egregious pattern of income and wealth inequality in America, the nation could still afford the energy transition needed to protect the climate. We would just have to reorder some other priorities, for example, eliminating fossil fuel subsidies and trimming the

$701 billion budgeted in FY 2012 for military spending.[84] The United States spent more on military defense in 2011 than the next 13 countries combined, not even including an additional $178 billion spent for "defense-related" outlays.[85] US military spending thus accounts for 41 percent of the entire world's annual military budget of $1.7 trillion.[1]

Having such an enormous military budget and the industries and jobs it supports might—by perhaps fostering overconfidence—increase the temptation to use military force precipitately in costly, ill-advised campaigns. The United States, for example, ignored the advice of international weapons inspectors who affirmed that Iraq had no weapons of mass destruction—the ostensible reason for the US invasion—and invaded anyway.[86] The Iraq War then killed hundreds of thousands of people, cost the United States well over $3 trillion, and wreaked havoc on Iraq.[87] Needless to say, these $3 trillion could have been far more wisely invested in clean energy and energy efficiency, which would have helped protect the climate and would have raised American workers' living standards by creating large numbers of new jobs. Ironically, at the same time as the United States maintains an outsized military establishment which it justifies on national security grounds, the nation is failing to protect itself and the world against the immediate looming national security threats presented by climate change. (Studies cited in chapter 8 have documented that climate change increases the odds of human conflict.)

A Giant Spigot of Untapped Cash

Another source of capital that could be tapped to address climate change is the enormous river of cash flowing through the stock, bond, and commodity exchanges of the world each day. In the course of 2011, some $30.7 trillion worth of stocks changed hands, just in the United States,[88] where on an average day, more than two billion shares of stock are traded. Many of these trades are by wealthy speculators, hedge funds, and other large institutional investors who often try to game the market, increasingly with sophisticated computer systems capable of executing high-frequency trades in fractions of a second. These "flash" transactions provide few public or social benefits and can actually destabilize the market. Imposing a half-percent transaction fee on the value of each stock bought or sold would help discourage flash trading and other speculation and would produce $150 billion a year in revenue; if bonds,

---

[1] *SIPRI Yearbook 2012: Armaments, Disarmament and International Security* (Oxford University Press: Oxford, 2012) cited in "World Military Expenditure Stops Growing in 2011, Press Release, Stockholm International Peace Research Institute, April 2012. http://www.sipri.org/research/arma ments/milex/sipri-factsheet-on-military-expenditure-2011.pdf.

futures contracts, derivatives, and credit default swaps were also assessed a small fee, another 1 percent of GDP or more could easily be collected to fund an energy transformation.

Removing Perverse Fossil Fuel Incentives

The amount of money spent subsidizing fossil fuels is actually a significant expense, domestically and globally. Moreover, these subsidies are a needless and imprudent use of capital to subsidize mature and profitable technologies that do not require subsidies and that damage the climate. The funds instead ought to be diverted toward a clean-energy transformation. Just from 2002 through 2008, the Environmental Law Institute calculated that, despite economically difficult times for most Americans and massive budget cuts, the federal government still provided $72 billion in fossil fuel subsidies.[89] Globally, the world spends about $700 billion a year on fossil fuel subsidies.[m] This practice cumulatively wastes of trillions of dollars in global investment capital, a very substantial portion of the capital needed to create a 100 percent renewable energy economy. Not only does this money subsidize industries that are mortally damaging the planet, but it helps create and maintain fossil fuel infrastructure with a 30- to 50-year operating lifetime, thereby locking in future carbon emissions. Finally, Americans spend $1.08 trillion each year buying oil and coal products, most of which quite literally go up in smoke.[90] Much of that can gradually be eliminated by investing purposefully in national energy efficiency, electrified transport, public transport, and an electrical grid powered mainly by fuel-free wind, solar, hydro, geothermal, and energy storage. Much of the residual liquid transportation fuel requirements could then be met in the future by increasing fleetwide fuel efficiency and by biofuels that include ethanol from nonfood cellulosic material, biogas from methane generators, and biodiesel made from vegetable oil and animal fats. (Biodiesel made with oils from oil-rich microalgae and fungi is also under development.) As the fuel economy of cars and trucks improves through regulations and technological progress, overall demand for liquid fuel can be reduced in the United States while the supplies of biofuels can be greatly expanded.

The Affordability of Climate Protection Programs

US economic studies are broadly consistent with the UK studies (cited earlier) on the cost of lowering carbon emissions. When the US Congressional

---

[m] As reported by the International Energy Agency and the Organization for Economic Cooperation and Development.

Budget Office (CBO) studied the prospective economic effects of the proposed American Clean Energy and Security Act of 2009 introduced by Congressman Henry Waxman and Senator Edward Markey,[n] the CBO similarly estimated that the bill would cost the United States only 0.25 to 0.75 percent of GDP in 2020 and only 1 to 3.5 percent of GDP in 2050, by which time real (inflation-adjusted) GDP would be roughly 250 percent of 2009 GDP. Yet the bill was defeated in Congress. CBO also pointed out that the effect of the law on household well-being would be less than the minuscule effect on GDP. Most importantly, CBO pointed out that the economic analysis of the bill did not include any benefits of averting climate change. To ignore the benefits, however, is absurd, given the enormous costs that reducing climate change would enable us to avoid.

Another way to view the costs of climate protection is through a framework presented in the IPCC's *Fourth Assessment Report*. There the IPCC forecast that a carbon tax of just $80 per metric ton of carbon dioxide would stabilize global temperature by 2050.[91] That tax would raise $440 billion a year, which is 2 to 3 percent of US GDP, and would only raise the price of gasoline by about 70 cents. Economist Robert H. Frank has pointed out that—even if a $300 per metric ton tax were imposed (something few people are proposing)—such a fee would still leave gasoline a lot cheaper than in many European nations today.[92] Moreover, much of the impact on drivers could be offset by increasing average fuel efficiency standards, as many European nations have done, and by rebating a portion of the tax to individuals, especially low-income people. In addition, as hybrid and electric vehicles continue to improve and decline in price, many drivers would find it cost-effective to shift to those types of vehicles, both of which can be largely fueled from renewable sources.

Those who still doubt that the nation can afford to combat climate disruption may recall what happened when the balance sheets of large financial institutions were threatened in 2008 by the subprime mortgage crisis and the associated credit defaults it triggered. Absolutely without hesitation, the doors of the US Treasury and the Federal Reserve and the FDIC flew open. The federal government then transfused hundreds of billions of dollars into the "too big to fail" institutions as soon as their financial troubles surfaced. No federal rescue measure was too costly. No loan guarantee was too high.

---

[n] The bill passed the House of Representatives, then under Democratic control, but failed to pass the Senate.

No sacrifice of taxpayers' money was too great to protect the banking system. Shouldn't a climate emergency be treated with *at least* the same urgency?

Former Goldman Sachs managing director Nomi Prins makes clear in her telling analysis of the federal bailout[93] that the Troubled Asset Relief Program (TARP) disbursements to banks were but the most visible currents in a quiet ocean of taxpayer money that flooded from the Treasury and the Fed to large financial institutions. According to Prins, through a convoluted web of programs and agencies, the federal government made more than $14 *trillion* dollars in bailout funds available on short notice.[94] If implied guarantees to financial institutions are included (most of which were not ultimately spent), Prins calculates that the total amount of taxpayer money pledged in the course of the bailout was more than $20 trillion. Ordinary Americans who lost jobs and homes and health insurance by the millions in the crisis, however, got about 15 cents on the dollar in economic benefits from the federal government's financial rescue package, Prins noted.

In the same era, however, the US government pinched pennies when funding the renewable energy research, development, and deployment that could help prevent climate disaster.[95] (It also stinted on helping the millions of struggling American families hurt by the financial tsunami, as Special Inspector General for TARP Neil Barofsky pointed out in *Bailout: An Inside Account of How Washington Abandoned Main Street While Rescuing Wall Street*.)[96] Bottom line: while the federal government hastily stepped in to provide generous, "no strings attached," taxpayer-funded aid to the multibillion-dollar financial firms that caused the financial debacle—even allowing their wealthy executives to keep lavish bonuses—the government balked at funding far less costly efforts to protect the climate.

The Need for Bold Action

As shown, through adjustments to the federal tax code or reductions in the military budget or small fees on securities transactions, the United States with its $16 trillion 2012 GDP could readily afford to invest in a gradual long-term transition to an energy-efficient, climate-safe, renewable energy economy. Given the gravity of the climate threats we face, however, a slow and gradual approach is no longer enough. We need to put the nation on an emergency footing to combat both climate change and ecological collapse, much as we suddenly geared the nation up industrially to fight World War II. Climate protection is far more important to the world over the long term than rescuing the financial system was in 2008. The financial system can be repaired in a few years of prudent management or, in exceptional circumstances, over a decade or two. But we have no proven technology to repair a damaged climate system. Unlike the transitory and reversible financial crisis, the climate crisis is a global survival struggle in which we must very quickly gain the upper hand or face the risk of bringing on a virtually permanent and uncontrollable climate disaster.

As soon as possible, the president of the United States should set up a new task force, perhaps under the direction of the National Renewable Energy Laboratory in collaboration with other national laboratories, to create an effective, credible, fully-funded national climate protection plan. It should describe in detail how the nation can get the greatest possible carbon reduction and economic benefit for each dollar invested. And it must aim toward an eventual zero-net-carbon, clean-energy economy, however long and difficult it may be to attain.°

The goal of protecting the climate cannot be achieved just by burning less fossil fuel. Agriculture, forestry, deforestation, landfills, mining, and fracking collectively also release billions of tons of carbon dioxide a year. The national climate protection plan therefore must reduce emissions from these activities as well. For example, the plan must include ways of protecting vulnerable forests from logging and clearance by intentional burning in order to prevent release of the forest's stored carbon to the atmosphere. Efforts to proactively use biological systems for removing excess carbon dioxide also will have to be on the agenda. This means establishing wetlands, planting more forests, and reestablishing deep-rooting native prairie grasses in the plains states and elsewhere to store large amounts of carbon in the soil, for example. In addition to harnessing natural carbon removal processes, however, other means will also need to be developed to begin drawing down atmospheric carbon levels because, as shown in figure 1-1, even if we could suddenly zero out emissions, carbon levels would remain elevated at dangerous levels for millennia.

The Stern report already foresaw in 2006 that: "stabilization [of atmospheric CO_2 concentrations and hence global climate] is likely ultimately (well beyond 2050) to require complete decarbonization of all other activities and some net sequestration of carbon from the atmosphere (e.g., by growing and burning biofuels, and capturing and storing the resultant carbon

---

° A zero-net-carbon economy will require a high economy-wide standard of energy efficiency; maximum use of renewable technologies; use of non-food-based biofuels to meet unavoidable liquid fuel needs; excellent land-use practices (elimination of deforestation; carbon capture and storage for any residual fossil fuel use; and active largescale carbon sequestration by augmentation of natural carbon removal processes, discussed above); and other means.

emissions, or by afforestation and [reforestation])."[97] Unfortunately, even the reduction of carbon emissions has stuck in our collective craw and thus, as a society, we appear to be a very long way from investing in the management of natural capital like trees, soil, and wetlands that can accelerate carbon removal. The developed nations today are even reluctant to provide adequate funding to protect the forests that still stand in tropical developing nations.

Stinting on Climate Protection

In the 1992 United Nations Conference on Environment and Development (also known as the UN Earth Summit or Rio Summit), participants signed the Convention on Biological Diversity and a Declaration on Global Forests, but still have not invested much to implement the agreements. Whereas the UN estimated that a mere $30 billion annually was needed to protect tropical forests, only half a billion was spent in 1991.[98] A generation later, in 2010, spending had reached only $1 to 2 billion a year.[99] Worse, little of the money was reaching the forest or forest dwellers, and little was going to challenge corrupt, ineffectual institutions that condoned or were complicit in deforestation. Much of it was and is spent on government agencies that failed to stop deforestation in the past[100] and have little control over what happens on the ground.[101] Forest protection needs to be elevated to the presidential and prime ministerial echelon of diplomacy. It must there be integrated with broader bilateral and multilateral treaties that provide trade and other very generous financial incentives to host nations, tied to their performance in fulfilling forest protection and restoration objectives.

The effective management or large-scale restoration of cropland, grasslands, and forests to a healthy condition can store meaningful amounts of carbon in soils, thereby taking carbon out of circulation for long periods of time. Professor Douglas Kell of the University of Manchester, for example, measured the benefits of breeding crops to have longer, bushier root systems. He found that increasing the depth to which roots grow could double soil carbon storage compared to the amount stored by common grain crops today.[102,103] Additives like biochar (charcoal produced by heating plant material in a zero- or low-oxygen atmosphere) can also help damaged or depleted forest soils recover.[104]

Despite numerous meetings and official statements, the industrialized world provides little support for developing countries to minimize or adapt to climate change or to reverse their largely carbon-intensive development paths. In a 2010 meeting at the United Nations' Cancun, Mexico,

Climate Summit, the international community again confirmed the 2009 UN Copenhagen Climate Conference's goal of providing only $30 billion in fast-start funding until 2012 in order to support programs addressing climate change in developing countries.[105] A Green Climate Change Fund funded by the conference's industrialized nations is intended to ultimately provide $100 billion per year by 2020 for the developing countries. Meanwhile, by contrast, most foreign investment in developing countries currently tends to be for the extraction of fossil fuels, timber, manufacturing, or other economic development activities that increase rather than decrease carbon emissions. Much of this extractive investment is facilitated by publicly supported export credit agencies. For example, while industrialized countries supported more than $103 billion worth of energy-intensive projects or exports during the mid- to late-1990s, their export credit agencies supported only $2 billion in renewable energy projects.[106] Thus, from the perspective of climate change, developed nations supported 50 times as much economic activity devoted to aggravating the climate problem than to fixing it.

The situation had improved considerably by 2011 as international lenders provided more support for renewable energy, but some still lent primarily to conventional fossil fuel projects. The US Export-Import Bank, for example, provided about 12 times as much financing ($8.6 billion) for traditional energy projects, such as oil and gas development, as for renewable energy exports ($771 million) in 2011.[107] The World Bank Group's renewable energy portfolio grew to almost $3 billion in 2011, with another $1.5 billion for energy efficiency, but it still spent $3.5 billion on conventional energy sources in 2011.[108]

The Global Environmental Facility (GEF) (created in 1991 by the United Nations Development Fund, the United Nations Environment Programme, and the World Bank) is expected to coordinate international support for addressing climate change and other environmental challenges. But the GEF is constantly underfunded and from 1991 to 2004, GEF funding for climate change projects amounted to only $11 billion (including the outside cofinancing which GEF leveraged at roughly a 5:1 ratio; the GEF contribution therefore would have only been about $2 billion). Meanwhile, the GEF simultaneously continued to support "CO_2-intensive fossil fuels projects."[109] As of 2011, the GEF had provided $3.8 billion in funding for projects aimed at addressing climate change, which together drew $21 billion in cofinancing. Its 2011 global funding, however, amounted to only $366 million, a very small amount for the entire developing world.[110] The European Bank for

Reconstruction and Development's Sustainable Energy Initiative, the Asian Development Bank's energy efficiency initiative, and the World Bank's Clean Energy Fund all together only provided on the order of $4 billion per year circa 2006, reflecting the low priority that developed nations actually accord to the global climate threat.[111] The Stern report found the world spending only $34 billion on incentives for using low-emission technologies and only $10 billion internationally for all such energy R&D.[112] By contrast, the world is expected to spend around $20 trillion by 2020 on mostly carbon-intensive energy infrastructure.

Working Group III of the IPCC's *Fourth Assessment Report* noted that the chances of stabilizing carbon in the atmosphere "will be heavily constrained by the nature and carbon intensity of this [fossil fuel] investment." But the report held out hope, stating that the early estimates indicate that the massive investment could be largely redirected toward zero- or low-carbon energy "with net additional investments ranging from negligible to less than 5%."

Expectations of Gradual Change. Most economic studies of climate change extend for no more than a century and assume that climate will alter in a gradual, orderly manner providing society and ecosystems with a period of gradual adjustment. Yet recent paleoclimatic research has found irrefutable evidence of abrupt past climate changes that have occurred within as little as ten years, a virtually instantaneous change on a geological timescale (see chapter 6).

The IPCC *Fourth Assessment Report* uses the technical term *singularity* to denote an extraordinary "game changing" geophysical event—a "blockbuster" climate disruption materializing as a result of what were thought to be routine positive feedback effects:

> Very few studies have been conducted on the impacts of large-scale singularities, which are extreme irreversible changes in the Earth system, such as an abrupt cessation of the North Atlantic Meridional Overturning Circulation, or rapid global sea-level rise due to Antarctic and/or Greenland ice sheet melting. . . . *Due to incomplete understanding of the underlying mechanisms of these events, or their likelihood, only exploratory studies have been carried out* [emphasis added].[113]

These unknowns imply that even the best economic studies today could still be vastly underestimating the eventual costs of climate change. In *Hidden Costs of Energy: Unpriced Consequences of Energy Production and Use*, the National Research Council authors caution that, "Although a number of the possible outcomes have been studied—such as release of methane from permafrost that could rapidly accelerate warming and collapse of the west Antarctic or Greenland ice sheets, which could raise sea level by several meters—*the damages associated with these events and their probabilities are very poorly understood* [emphasis added]."[114]

The good news this chapter brings is that, rather than exposing the world to these potentially catastrophic and still poorly understood risks, the world can afford to lower carbon emissions and protect the climate, a course safer and cheaper than inaction. Moreover, as the next chapter shows, following this more prudent course also will produce important public health and environmental benefits that will save millions of lives. [115]

CHAPTER 8

Health Perils

*Climate change currently contributes to the global burden
of disease and premature deaths. . . . Adverse health impacts will
be greatest in low-income countries. Those at greater risk include,
in all countries, the urban poor, the elderly and children,
traditional societies, subsistence farmers, and coastal populations.*

—WORKING GROUP II, FOURTH ASSESSMENT REPORT OF
THE INTERGOVERNMENTAL PANEL ON CLIMATE CHANGE[1]

Rapid climate change not only impacts the natural environment and
the economy; it also kills and sickens people. Although the media
focuses on the immediate effects of extreme weather, such as floods
and storms, these disasters have a long tail, meaning they cause many
pernicious health and environmental effects that only appear long after
the emergency crews have left and the TV news cameras have been
turned off.

Climate Lethality

WORLD HEALTH ORGANIZATION statistics indicate that over the past
35 years, more than five million people have died from increases
in disease and malnutrition brought on by climate change. They perished
from the effects of extreme weather, increased disease transmission, diminished food production, or some combination of those causes.[2] The number of nonfatal illnesses and injuries was at least several times greater. Thus,
in 2000 alone, climate change would have caused 150,000 deaths and the

loss of 5.5 million "disability adjusted life-years,"[3,a] according to the 2007 *Fourth Assessment Report* of the Intergovernmental Panel on Climate Change (IPCC).[4] If elevated rates of death, disease, and disability continue on a similar trajectory into the foreseeable future—or rise steeply as temperatures climb and extreme weather increases—many millions more will die or experience illnesses and injuries as a result of climate change.

True, in some northern climates, cold-related fatalities have declined, but the decline is small relative to the global increase in heat-associated deaths and disease. In addition, where increased temperatures or altered rainfall patterns chance to make life unsuitable for parasites or disease carriers, some local reduction in the spread of disease will be found. "However," the IPCC concluded, "the balance of impacts will be overwhelmingly negative."[5]

Conflict and Violence

Underscoring that conclusion, researchers from the University of California and Princeton lead by Solomon M. Hsiang recently found a powerful connection between climate change and the frequency of war, conflict, and interpersonal violence, all of which, shall we say, are prejudicial to human health.[6] As climate deviates from historical norms as evinced by rising temperatures and abnormal rainfall patterns or El Niño events, Hsiang et al. found that interpersonal violence and crime, and group violence, became more common along with breakdowns in social order and political institutions. To measure how much climate at a particular location was deviating from normal in order to determine its effect on conflict, Hsiang et al. expressed the difference in terms of a common statistical measure known as a standard deviation. I will briefly explain this concept and why it's so useful in the present context. I will begin with a few key statistical concepts that provide needed context for understanding the notion of a standard deviation.

The term *data* here represents a collection of climate variables, such as temperatures, or rainfall frequencies, or intensities over time. Most collections of data in which the data are randomly generated will have a classic pattern of variability around the data's average value. Known as a normal distribution, the pattern has a characteristic bell-like shape when plotted on a graph. (On the graph, the individual climate data are arranged on a scale of increasing

---

[a] A disability adjusted life-year is a measure of years of life lost prematurely plus years of healthy life lost due to disease or disability. For more details, see World Health Organization, Health Statistics and Health Information Systems, Metrics Disability Adjusted Life-Years, http://www.who.int/health-info/global_burden_disease/metrics_daly/en.

magnitude on the bottom axis and the likelihood of each value's occurrence, also on a scale of ascending magnitude, is on the vertical axis.) (See figure 9-2.) The average value of all the data in the collection or set of data, its mean, will be found in the middle of the graph from left to right. The standard deviation is defined as the square root of the variance or, in other words, as the square root of the average of the squared differences of each value in the population (or sample) from the mean. Calculated in this way, the standard deviation of a normal population is a particular value that is equidistant above and below the mean and that includes about 68 percent of all the values in the data. The magnitude of the standard deviation indicates the variability of the data on which it is based. When data are highly variable and widely dispersed from their average value, the calculated standard deviation will be large. Similarly, when the variability is low and most values are tightly grouped around the mean, the standard deviation will be small.

Just as we noted that in a normal population about 68 percent of all the values in the data will be within the limits of one standard deviation above and below the mean, so two standard deviations from the mean in a normal population will include about 95 percent of all the data, and more than 99 percent of all data in a normal distribution will be within three standard deviations of the mean. This implies that if on the average climate change has produced a new average temperature or rainfall pattern that is a standard deviation or more away from the previous one, the new value is unlikely merely to be a chance occurrence; and the farther it is from the historic mean, the less likely the difference can be explained by chance.

Using standard deviations as a yardstick for indicating how much climate has changed in a locality over time is a clever move—a bit like converting fractions to a common denominator before working with them. It enabled the researchers to analyze and compare the results of many studies from different times and places in what is known as a meta-analysis. The use of standard deviation units thus freed the researchers from relying on temperature changes or even on percentage temperature changes as the basis for their analysis. (Neither choice would have been satisfactory, since absolute and percentage climate changes in one locality cannot easily be compared with those in another where temperature and rainfall patterns likely differ in magnitude and variability.) By expressing the new altered climate in units of standard deviations from the average of past climate values, the researchers could easily tally how much the climate had changed in a locality, so that any ensuing changes in the level of conflict, violence, or social breakdown could be quantitatively related to antecedent climate changes. Here is what they discovered after controlling for other possible explanatory variables besides climate change.

For each standard deviation of warming from a previous climate mean, they found that the chance of individual violence rose 4 percent and the chance of group violence increased by 16 percent. The connections between warmer temperatures and abnormal rainfall patterns on the one hand and violent conflict on the other are extremely important, because millions of individuals are already involved in violent crime, interpersonal conflicts, and war around the world. Thus, even a small percentage increase will create large numbers of additional crime and conflict victims.[b]

By 2050, many parts of the world are expected to warm by two to four standard deviations, if current emissions trends persist. Based on the new research results, this amount of warming implies that social conflict is likely to increase by more than 50 percent and the odds of interpersonal conflict are likely to rise by 16 percent.[7] In particularly hot, vulnerable, or violence-prone areas—such as Somalia, the Sudan, western Pakistan, Yemen, and elsewhere—an eventual breakdown in social order can easily be envisioned.[8]

In the United States, where 1.2 million violent crimes were committed in 2011,[9] 96,000 to 192,000 additional violent crimes per year by 2050 would be expected. Globally, where the direct and indirect deaths caused by war from 2004 to 2007 are already on the order of a quarter million a year,[10] the warming cited could add 80,000 to 160,000 additional war-related deaths annually by 2050. Under similar assumptions, the global murder rate could rise by 160,000 to 320,000 from about half a million a year.[c]

Based on their advanced statistical reanalysis of data from 10,000 BCE to the present in some 60 published studies, Hsiang et al. concluded that the connection between climate stress, conflict, and violence held "almost everywhere: across types of conflict, across human history, across regions of the world, across income groups, across the various durations of climate changes, and across all spatial scales."[11]

Thus, people in the United States and other developed nations will not be immune from the expected increase in war and violence given that the

---

[b] Violent crime in this context refers to murder, rape, robbery, and assault.

[c] From 1945 to 2000, an average of 745,000 deaths a year occurred in war and conflict. See Milton Leitenberg, "Deaths in Wars and Conflicts in the twentieth Century," Cornell University Peace Studies Program, Occasional Paper #29, 3rd ed. http://www.cissm.umd.edu/papers/files/deathswarsconflictsjune52006.pdf.

researchers found that the effects of climate change applied to high-income countries as well as to the developing world. Given the magnitude of the effects seen, the researchers forecast that by 2050, climate change could have a "large and critical impact" on the frequency of human conflict.

Garden-Variety Hazards to Health

In previous chapters we have already indicated that, as the world gets hotter, heat waves, droughts, fires, cyclones, and floods become more frequent and severe and take an immediate, obvious toll. However, climate disasters have myriad ways of undermining human health and welfare, sometimes causing death and illness months or even years after the extreme event.[12] For example, domestic violence and mental illnesses, including depression and post-traumatic stress disorder, may arise in the aftermath as disaster-related hardships, disruption of normal life, and stress persist, long after the world has forgotten the disaster.

Intense Heat

Heat waves increase the incidence of heatstroke, which is a potentially life-threatening ailment. A European heat wave in August of 2003 thus caused 35,000 deaths, with 14,800 deaths in France alone.[13] A US heat wave in July 2011 that raised temperatures 20°F in large parts of the United States affected 200 million Americans, killing dozens and breaking heat records in many places. Nighttime temperatures were also greatly elevated—a pattern characteristic of global warming[14] and associated with the urban heat island effect.

The human body generates metabolic heat, which it must dissipate to the atmosphere through the skin (and breath) in order to maintain an internal temperature of 98.6°F. But if ambient conditions of temperature and humidity make it impossible for the skin to stay below 95°F to cool the body, the body's internal temperature begins to rise. If skin temperature gets above 98.6°F, the body's internal temperature will rise in a matter of hours into the 107 to 109°F range, at which point irreversible tissue damage begins to occur, leading to death or chronic illness. Heat stress and heat exhaustion are less serious conditions than heat stroke (the final stage of a continuum of disorders), but they also lead to short-term physical and mental impairment that makes victims more accident-prone and thus can increase injury rates. The very young, ill, or elderly are most susceptible to extreme heat and drought, since they are less able to maintain a safe body temperature.

Urbanization, which is increasing rapidly in many parts of the world, can accentuate heat waves through the "urban heat island effect," in which solar heat is absorbed by stone and other building materials and reradiated. In confined spaces close to buildings, the reradiation of heat intensifies the heat island effect. The concentrated combustion of fossil fuels in urban areas also concentrates the discharge of waste heat there.

When air temperature is above the body's external temperature, the body does not naturally lose heat to the environment and needs to sweat or drink cold liquids in order to keep from overheating. But when ambient temperature and humidity are both high, it gets harder and harder to sweat, and heat waves become more intolerable. Therefore our ability to survive during a heat wave or other extraordinarily hot conditions depends both on temperature and humidity. "Wet-bulb temperature," the temperature of a thermometer wrapped in a wet porous material, provides a single measurement that is based on both of these critical variables. Wet-bulb temperatures of about 93°F already can result in dangerous elevation of body temperature, and humans cannot survive for more than a matter of hours when wet-bulb temperatures are above 95°F, even though they would have no trouble surviving much higher temperatures in the desert, where humidity is low.[15]

Currently, wet-bulb temperatures around the world peak at about 80°F during heat waves, while the maximum peak wet-bulb temperature on Earth is about 86°F. However, each one-degree increase in global temperature increases the wet-bulb temperature by about three quarters of a degree. So if average global temperature ever increases by 12 to 13°F, then typical maximum wet-bulb temperatures would be around 90°F and peak maximum wet-bulb temperatures would be 95°F, which not even fit and healthy people could survive. This would render parts of the world uninhabitable, unless people at times took shelter underground or in air-conditioned spaces. These findings demonstrate that if global warming continues unabated for long enough, human survival itself would be increasingly difficult or impossible in the overheated regions. Although not very likely, some experts believe it is possible that continuing with unabated greenhouse gas emissions could lead to a 12 to 13°F temperature increase by 2100.[16] Even if that didn't happen for centuries, it would still be impossible to justify present actions that would eventually precipitate these conditions.

Drought's Complex Effects

Higher temperatures increase the severity and extent of droughts, as described in previous chapters. Droughts then lead to food scarcity that in turn brings hunger, malnutrition, starvation, and increased susceptibility to infectious diseases. Low-income people in poor, warm, agricultural nations, especially in sub-Saharan Africa and parts of Asia, are likely to be hardest hit by drought and unpredictable rainfall. Subsistence farmers, pastoralists (who rely on animal herds), and traditional societies are among the most vulnerable groups in these areas. Droughts also reduce rural incomes and, by driving marginal people further into poverty, increase their susceptibility to a wide variety of maladies. In addition to killing crops, droughts also bring a scarcity of water for domestic use (bathing and hand-washing), sewage transport, and power generation. Water contaminants become more concentrated during drought. Small rivers and drainage canals may become stagnant, providing ideal habitat for disease-bearing mosquito larvae.

In certain locales, droughts bring dust storms that cause respiratory problems. Although the transmission mechanism is not completely understood, dust storms may carry disease-causing organisms. Meningitis, for example, has spread geographically in West Africa following drought.[17]

More Frequent Fires

Fires tend to become more common as temperatures rise, and as heat and drought reduce soil moisture. Although long-term global lightning records do not exist, hotter temperatures have created more frequent and violent storms in specific regions,[18] and large storms are often accompanied by lightning. Higher temperatures also appear to be associated with a shift in lightning from the tropics to middle latitudes, where fires are more likely to result. Climate models generally indicate that higher temperatures increase the number of lightning strikes, but some uncertainties about it remain.

We do know, however, that when lightning strikes tinder-dry forests or drought-stricken grasslands, it can ignite massive wildfires that burn hot enough to kill trees and harm the soil. These intense fires thus cause serious ecological damage (far beyond the routine impacts of normal, low-to-medium intensity fires that tend to remove fire-adapted brush and grasses without killing root systems). The large, ultrahot fires also produce massive volumes of smoke and other airborne contaminants that can cause cardiovascular and respiratory problems in susceptible people.

Airborne Irritation

Ozone is a corrosive allotrope of oxygen composed of three oxygen atoms instead of the usual two of a normal oxygen molecule. In the lower atmosphere, ozone is formed by a chemical reaction of nitrogen oxides, methane, and volatile organic compounds (VOCs) in the presence of sunlight. (Lightning increases ozone production in the atmosphere by creating nitrogen oxides.) The primary sources of nitrogen oxides and VOCs are the production and combustion of fossil fuels. While ozone in the stratosphere protects the Earth by absorbing damaging ultraviolent radiation, ground-level ozone, to which auto and truck exhausts contribute mightily, is a serious health hazard.

When inhaled, ozone irritates the eyes, lungs, and respiratory tract, and damages lung tissue. Long-term exposure reduces lung function.[19] Exposure to elevated ozone concentrations is associated with an increase in the incidence of asthma, pneumonia, obstructive lung disease, and other respiratory disorders, as well as heart attack and premature death.[20] Ozone concentrations are currently increasing in most regions of the world due to an increase in fossil fuel production and use in conjunction with rising temperatures that accelerate ozone-producing chemical reactions in the atmosphere.

The Union of Concerned Scientists has determined that by 2020, an increase in ground-level ozone production due to global warming could trigger 2.8 million serious respiratory illnesses and cause 5,000 infants and seniors to be hospitalized for serious breathing problems, generating costs of more than $5 billion dollars a year.[21]

As both temperature and carbon dioxide rise, so do pollen-related conditions and allergic reactions. Some pollen allergens are now becoming more concentrated in the atmosphere, since the growth of ragweed and other weedy plants is stimulated by elevated atmospheric carbon dioxide and they have longer growing seasons. Chronic allergic reactions can then open the door to additional respiratory infections.

Flooding and Health

Paradoxically, a hotter climate produces both more droughts and more floods, as rainfall tends to be concentrated in fewer but more intense events. Flooding—the most common type of natural disaster—has devastating health effects on vast numbers of people by causing accidents, injuries, illnesses, drownings, and contamination of water, food, and soil.[22]

Serious flooding can also trigger landslides in steep, deforested areas above towns and villages, along with injuries and drowning. It also can lead to contamination of water, air, soil, and sediment with carcinogenic or toxic chemicals, such as insecticides, fuels, hazardous wastes, and heavy metals. Individuals exposed to these substances will have an increased risk of later developing cancer, neurological disorders, and genetic defects. Latent impacts will not be immediately apparent, and the cause may later be difficult to correctly attribute. Cancer, for example, can take years or decades to appear, and its cause may be difficult to pinpoint.

Unsafe water, poor sanitation, and lack of effective sewage treatment already bring about almost two million deaths worldwide from diarrhea-causing diseases.[23] The poorest nations and their lowest-income residents in crowded substandard housing are most vulnerable. Higher temperatures and more infrequent but heavier rainfall can increase the spread of waterborne infections and diseases, including cholera and other diarrheal diseases as well as hepatitis, *E. coli* infections, and cryptosporidium infections. Untreated sewage can flow from overloaded sewer systems, contaminating waterways, bays, and freshwater supplies. Higher temperatures also increase food spoilage and thus food poisoning.

Heat and deluges also alter the range and seasonal occurrence of diseases transmitted by mosquitoes, ticks, and rodents. Following deluges, mosquitoes reproduce in the leftover standing water. Rodent populations, which can carry plague or hantavirus, may also expand after a flood.

Floods in China seriously disrupted the lives of 130 million people in 2003.[24] Exceptional floods also occurred in the United States in 2010 in southern New England, Tennessee, Minnesota, and Wisconsin.[25] In 2011, the Mississippi River flooded, and other major flooding occurred in the Mid-Atlantic states.[26]

Oceanic Health Risks

High surface-water temperatures, particularly in coastal zones, can lead to "algal blooms," a heavy overgrowth of algae, which can produce harmful toxins. When ingested by shellfish and other filter-feeding organisms, the toxins concentrate in their flesh and contaminate them. People consuming tainted shellfish or reef fish can become seriously ill. Plankton blooms in certain locales, such as the Bay of Bengal, are thought to serve as a summer "reservoir" for cholera bacteria, *Vibrio cholerae.*

Elevated ocean temperatures can also increase the abundance of a related bacterium: *Vibrio vulnificus* that produces a disease with a 50 percent mortality rate and is responsible for 95 percent of all deaths from eating seafood.

This emerging human pathogen secretes a toxin in the small intestine that causes vomiting, acute diarrhea, and abdominal pain. *V. vulnificus* can also infect wounds and cause blood poisoning.[27] Most common in warmer waters during summer, *V. vulnificus* can be found in oysters, crabs, and clams, as well as in seawater, sediments, and plankton. Another relative, *Vibrio parahemolyticus*, also transmitted in raw, undercooked, or poorly refrigerated seafood, can cause various abdominal symptoms, as well as fever and chills.[28]

Dreaded Diseases

A few years ago, people in the Italian village of Castiglione di Cervia, on the Adriatic Sea near Ravenna, began getting gravely ill. They had high fevers, nausea, weakness, and horrible pain in their muscles and joints. The disease soon reached epidemic proportions. After weeks of fear and dread, experts finally isolated and identified the virus. To everyone's surprise, the disease turned out to be chikungunya, a tropical disease found in large parts of Africa, Southeast Asia, and Indonesia. Chikungunya means "that which bends up" in Kimakonde (a language spoken in Mozambique), since the disease can cause contortions.[29] The pain from chikungunya can last for weeks or even months, and there is no vaccine to prevent it. It turned out that the disease had arrived in Italy in the blood of an Italian who had visited India. Chikungunya had then been spread locally by the tiger mosquito, an alien insect first found in Ravenna in 2004, and that has since spread within southern Europe and as far north as Switzerland and France.[30]

The appearance of tropical diseases in Italy and elsewhere in the developed world may not be unusual for long. As Europe's weather warms, the continent becomes increasingly hospitable—especially during the summer months—to mosquitoes formerly unable to thrive because of chilly nights and cold winters. People in developed northern countries typically have no immunity to tropical viruses, and therefore are highly susceptible to them when infected. Global warming is increasing the range of all mosquito-borne diseases, including malaria and dengue (breakbone fever), a viral disease with symptoms similar to chikungunya's but afflicting some 100 million people every year. This is not surprising, since about 2.5 billion people live in areas where they may be exposed to dengue.[31] The *Stegomyia aegypti* mosquito is the most important transmitter or vector of dengue viruses.[32]

When it comes to tropical diseases, however, malaria trumps dengue and chikungunya. The incidence of malaria is about three times higher than chikungunya. It afflicted 190 to 310 million people in 2008 and caused 700,000 to a million deaths that year.[33] Some 3.3 billion people live in regions where they can contract malaria. It is caused by a microscopic parasite carried and transmitted by certain species of mosquitoes. The parasite infects various organs, such as the liver and blood cells, producing a wide range of symptoms, from unpleasant to lethal. Mosquitoes with the malaria parasite, like other mosquitoes, can survive only where winter temperatures exceed 61°F. Chillier temperatures keep them in check in temperate zones and at higher elevations. But higher temperatures also speed up the parasite's maturation and facilitate the infection's spread.

Because of global warming, the area of potential malarial infection is expanding, allowing the disease to spread north and to higher elevations. From 50 to 80 million *additional* cases of malaria are therefore expected worldwide by the latter half of this century. Due to the complexity of predicting malaria's response to climate change, this estimate may change. Some areas will see an increase in malaria, while others will see a decline. What seems clear, however, is that certain areas now on the margin of where malaria is normally found may see the disease appear locally for the first time. Heavy rainfall, which may occur more frequently in a warming world, also can increase the prevalence of mosquitoes due to flooding. Ironically, drought can also increase mosquito-related illness, when people in developing countries respond to it by storing water around homes in open containers.

For the many reasons mentioned, global warming thus increases the incidence of waterborne as well as mosquito-transmitted diseases. These diseases include river blindness, cholera, yellow fever, Rift Valley fever, and various intestinal diseases. Rising temperatures, which speed up spoilage, also will cause increases in salmonella poisoning from eating contaminated food.

Other diseases affected by climate include tick-borne encephalitis, Saint Louis encephalitis (spread by migratory birds), meningitis, leptospirosis and hanta virus (transmitted by rodents), Nipah virus (transmitted by bats), Lyme disease (carried by ticks), and schistosomiasis (hosted by snails). Lyme disease, for example, is likely to expand in Canada, while tick-borne encephalitis is heading farther northeast in Europe. Global heating will also hurt human health through its little known harmful effects on medicinal plants.

Threats to Medicinal Plants

As many as 70,000 plant species are used for medicinal purposes in Western medicine and traditional therapies.[34] Two prominent medicines for malaria, for example, come from plants: artemisinin from the wormwood

(*Artemisia annua*) native to China (where it is known as Qinghao) and quinine from trees of the *Cinchona* genus native to tropical South America. Half of all prescription medicines come from plants, and less than 2 percent of these species are cultivated, so almost all medicinal plants are still collected in the wild, many in tropical rainforests and alpine areas where they are very susceptible to climate change.[35]

Medicinal plants are often not only in danger from a changing climate, but also from development, habitat loss, overexploitation, illegal trade, and incursions of invasive nonnative species. Despite all these threats, only 4 percent of all plants have even been assessed for their medicinal value.[36] We thus risk losing plant species of potentially great but still unknown medical value and scientific interest along with plants of known use.

The rosy periwinkle of Madagascar, for example, provides a treatment for leukemia, and taxol and paclitaxel from yew trees provide medication for cancer. Devil's claw (*Harpagophytum procumbens*), a plant native to southern Africa, is used to produce a time-honored treatment for fever, joint pain, and digestive disorders. The Himalayan snow lotus (*Saussurea laniceps*) grows above 13,000 feet in the eastern Himalayas and is used by Tibetan healers to treat heart disease, high blood pressure, and reproductive ailments. The Arctic *Rhodiola rosea* has traditionally been used to treat infections and strengthen the immune system.[37] Both Himalayan and Arctic ecosystems are experiencing rapid warming.

Certain animal species also have great medicinal value. The painkiller Prialt—which is 1,000 times more powerful than morphine and can effectively treat severe, opiate-resistant chronic pain—is based on a compound produced by a marine cone snail.[38] The cone snail's coral reef habitat, however, is being destroyed. Much could be learned from the world's 700 species of cone snail, each of which, according to Dr. Eric Chivian, produces 100 to 200 distinct toxic chemical compounds with which it attacks its prey "by firing poison-coated harpoons at them."[39] Other threatened animals, from polar bears to crucifix toads, may offer biochemical therapies for osteoporosis and type 2 diabetes, as well as adhesives to treat torn tissues.[40] Puffer fish venom may help chemotherapy patients with chronic pain; the toxin of certain sea anemone may lead to a medication for autoimmune diseases; the fire-bellied toad's toxin promotes wound healing; and the waxy monkey frog produces a toxin that suppresses angiogenesis, the growth of blood vessels, so it may lead to a drug that can control cancerous tumors or save the eyesight of patients with diabetes-related retinal disease.[41]

Poverty, an Additional Risk Factor

As stated earlier, (see "Flooding and Health"), the poorest peoples of the world are likely to be the most vulnerable to climate change. They live in the poorest-quality, most crowded housing, often in hot urban slums where they are most susceptible to communicable diseases, as well as to flooding and mudslides. They are also least likely to be able to afford air conditioning or to enjoy modern sanitation or good health care. Ironically, the disadvantaged, who bear the worst consequences of global heating, are least responsible for its onset, as they historically have used the least amount of fossil fuel energy per person. So climate change clearly raises many social justice issues.

In Bangladesh, 57 percent of the population (about 52 million people) could be flooded in the event that global temperature rises 7.2°F and monsoon runoff swells major rivers.[42] Enormous floods have already happened in Pakistan (2010) as discussed earlier. Other huge floods have also occurred in China (1998, 2011) and Thailand (2011).[43]

While I have previously described the ecological effects of climate change on the Arctic, I did not stress how adversely climate change will affect the health of Arctic peoples who have subsisted for thousands of years by hunting marine life, birds, and caribou and by gathering wild berries and other plants. As the wild game and fish dwindle in the warming climate—and as the people are forced to abandon their traditional lifestyles and ancestral homes—they will be forced to adopt unfamiliar diets. Their traditions, culture, and much of their cultural identity may disappear along with the wildlife. Tragically, when native peoples lose their traditional ways of life and subsistence foods, health problems such as alcoholism, drug use, heart disease, and diabetes often become more prevalent.

Although this chapter focused on the health effects of climate change, the issue will return in the next chapter, "Extreme Weather Perils," because extreme weather threatens human health, along with welfare and the economies of affected regions.

CHAPTER 9

Extreme Weather Perils

*Climate change is a risk-multiplier. It has the potential to take
all the other critical issues we face as a global community and
transform their severity into a cataclysm.*[1]

——HRH CHARLES, PRINCE OF WALES

As the world gets hotter and more energy is available to feed storms and
cyclones—and as normal rainfall patterns are disrupted—violent weather-related disasters have become more common, severe, and costly. To
convey what it's like to experience these hitherto unimaginable disasters, the chapter looks in depth at life amidst two recent major climate
events.

Mega Disasters

WHEN THE YANGTZE RIVER FLOODED IN 1998, it inundated 60 million
acres in China, displaced more than 200 million people, and killed
thousands. Americans barely noticed. Seven years later, Hurricane Katrina hit
the United States, killing more than 1,800 people and causing more than
$81 billion in property damage. It got the nation's attention, but not every-
one saw a climate connection, and the storm damage soon receded in society's collective memory, supplanted by a succession of news stories du jour.
While hurricanes are weather phenomena, they *are* related to climate and
climate change.

Hurricanes develop in the tropical ocean between 5° and 20° north and
south of the equator where trade winds prevail and the water is warm. The
ocean needs to be almost 80°F at a depth of 160 feet, and the greater the

difference between air and water temperature, the easier it is for a hurricane with fierce winds and torrential rains to develop. Hurricanes begin as
a cluster of rain storms that coalesce into a single larger tropical depression
that strengthens into a tropical storm. As warm moisture-laden air rises from
the ocean surface, it creates a low-pressure area that draws more warm air
in toward the center of the storm system. Cool high-pressure air above the
storm system causes moisture in the warm air to condense, releasing heat to
the atmosphere and energizing the storm.

Gradually, the Earth's rotation begins to spin the stormy air mass coun-
terclockwise, organizing the storm system into the familiar spiraling mass of
clouds with a clear eye at the center that you see on weather maps. As the
storm travels over warm ocean water, it "feeds" on heat that it draws from the
ocean, strengthening as more and more warm, moist ocean air condenses. As
the energy in the whole system increases, winds are drawn toward the eye
of the hurricane with increasing force. Meanwhile, cool, dry air descends
into the eye from above, suppressing condensation, leaving the eye clear and
cloudless. The eyewall immediately around the eye contains the strongest
winds. Radiating out from the eye in concentric circles are bands of rain
storms that drop wave after wave of torrential rain as the storm travels.

Once the storm winds reach 74 mph, it is classified as a hurricane. The
weakest of them are Class 1 and have winds from 74 to 95 mph; the strongest hurricanes, Class 5, have winds of 155 mph and above. Scientists from
the National Oceanic and Atmospheric Administration's Geophysical Fluid
Dynamics Laboratory believe that as the climate grows warmer during the
rest of the twenty-first century, the world will see an increase in the intensity
of rainfall from hurricanes and that the number of very severe Class 4 and
Class 5 hurricanes will increase.[2]

Even after Hurricane Katrina, not everyone was prepared to make a connection between climate change and megastorms, however. Then in 2012,
Superstorm Sandy arose in the Caribbean and grew into the largest Atlantic
hurricane ever, with winds swirling over an area 1,000 miles in diameter and
the lowest central barometric pressure of any Atlantic hurricane ever to hit
the United States north of Cape Hatteras, North Carolina. Suddenly, it was
hard to disregard the possibility that Superstorm Sandy was exacerbated by
climate change.

The giant storm deluged the entire Eastern seaboard and 24 states, some
as far from the coast as Michigan and Wisconsin. Ten states and the District

of Columbia declared states of emergency. The hurricane caused an estimated
$68 billion in damage. Hundreds of thousands of homes and businesses were
destroyed or damaged. As if in answer to the question, "Was Superstorm Sandy
a rare freak of nature or likely to become more common in an overheated
world?," Super Typhoon Haiyan hit the Philippines just a little more than a
year later, the third Class 5 hurricane to do so since 2010. (*Typhoon* is the term
used for hurricanes in the western Pacific. Hurricanes in the Indian Ocean
and Bay of Bengal are called *cyclones*.)

Typhoons in the Philippines have been getting stronger since the 1990s,
too. Haiyan's wind gusts reached 235 mph with sustained winds of 195 mph
over water at landfall.[3] It may have been the most powerful hurricane ever to
hit land (but necessary ground measurements were not taken to confirm the
record).[4] Tropical cyclones have killed nearly 170,000 people globally from
2000 to 2010, affecting more than a quarter of a billion people and doing
$380 billion in damage, according to calculations by the World Meteorologi-
cal Organisation.[5] All this is not coincidental. As explained, tropical cyclones
derive their energy from warm ocean waters, and as ocean temperatures have
increased, so has the energy available to drive these storms. Ocean data indicate that the subsurface ocean was about 5.4°F warmer than usual beneath
Super Typhoon Haiyan's storm track. When a typhoon roils this deeper water,
bringing abnormally warm water to the surface, additional energy becomes
available to power the hurricane.

Because higher temperatures mean oceans are warmer, particularly at and
near the surface, they provide more energy to sustain more powerful storms.[6]
Tropical Atlantic sea surface temperatures for the past 50 years are statistically
very closely correlated with Atlantic hurricane activity as measured by the
Power Dissipation Index, which measures storm intensity, frequency, and duration. In the summer of 2005, for example, when Hurricane Katrina occurred,
sea surface temperatures were at record highs—more than 1.6°F above their
70-year norm.[7] (They have been above average since the late 1970s.) Not
coincidentally, the 2005 tropical storm season had 15 hurricanes—the most
ever recorded. The same season produced six of the eight most damaging
storms on record for the United States. The IPCC's *Fourth Assessment Report*
concluded that there has been a definite global increase in Category 4 and
5 hurricanes since 1970.[8] Severe storms, incidentally, can not only bring torrential rains, high winds, and storm surges, but baseball-sized hail that can fall
at 100 mph.[9] Global climate models consistently project that a future warmer

climate will have more intense mean and peak rainfall from tropical cyclones. Some models also show higher peak winds.[10] They suggest that there may be fewer weak tropical cyclones, however, so the total number of tropical cyclones worldwide may drop.[11]

In the midlatitudes, extreme waves are expected to be bigger, increasing risks to people at sea.[12] Hurricanes can cause enormous ocean swells up to 20 feet high and 50 to 100 miles wide.[13] After emerging from a tropical ocean, a big storm can then spawn devastating inland floods and even tornadoes. Storm surges are responsible for most of the lives lost and damage caused by hurricanes. As discussed earlier, Hurricane Sandy's 13.8-foot storm surge flooded parts of Manhattan. Flooding associated with surges has killed hundreds of thousands of people in Bangladesh.[14]

Severe hurricanes can also produce ecological disasters. Hurricane Katrina blew down mature trees over more than five million acres of Alabama, Florida, Louisiana, Mississippi, and Texas. A Tulane University researcher calculated that hundreds of millions of trees were killed and even more were damaged.[15,16] Instead of helping to protect the climate by absorbing atmospheric carbon dioxide as they did while living, dead trees contribute to global heating by releasing carbon dioxide. Trees killed by Katrina will release

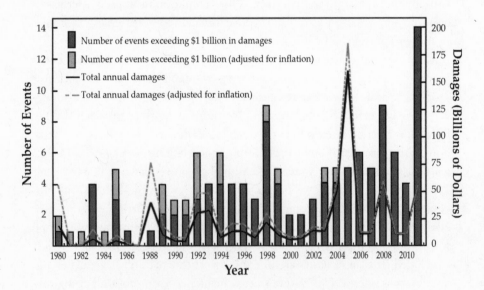

FIGURE 9-1. The costs and frequency of extreme weather are rising steeply as this graph reveals. Source: National Climate Data Center and National Oceanic and Atmospheric Administration.[17]

an estimated 105 million tons.[a] Other ecological damage and wildlife losses probably went unrecorded.

The Rising Costs of Extreme Weather

In the United States, data reveal that the economic costs of weather- and climate-related disasters are increasing dramatically and have become far larger than the costs of all other natural disasters in the past half century. Weather-related disasters account for nearly 90 percent of all presidentially declared emergencies.[18] Just since 1980, the United States has experienced more than 107 such disasters costing more than $1 billion each, for a total of more than $750 billion.[19]

As population density, property values, and coastal development all increase along with climate disturbances, damages from single storms by 2020 could reach unprecedented levels.[20]

The year 2011 was a record for extreme weather and climate disasters in the United States. US National Weather Service Director Jack Hayes summed things up this way:

> In my weather career spanning four decades, I've never seen a year quite like 2011. Sure, we've had years with extreme flooding, extreme hurricanes, extreme winter snowstorms and even extreme tornado outbreaks, but I can't remember a year like [2011] in which we experienced record-breaking extremes of nearly every conceivable type of weather.[21]

The United States in 2011 had more than a dozen disasters each costing $1 billion or more.[22] In 2012, the insured losses from these natural disasters were even greater, reaching $58 billion—more than twice the average from 2000 to 2011.[23,b] The year 2012 thus was the second costliest in weather-related disasters.

It is not surprising that as global average temperatures have risen over recent decades, the number of extreme weather events has increased alarmingly. That's because, according to the principles of statistics, when the average of some variable, such as temperature, shifts slightly, the incidence of extreme

---

[a] Some experts think Chambers's estimates may be high. See Daniel W. Stolte, "Dead Forests Release Less Carbon Into Atmosphere Than Expected," University of Arizona Communications, March 22, 2013. This research is not a critique of Professor Chambers's findings but basic research that probably applies broadly to prior published estimates of carbon released by dead forests. Louisiana officials estimate that 20 percent of the downed trees were harvested, so they will be turned into lumber instead of decomposing.

[b] Naturally, not all losses are insured.

values of the same variable has a tendency to rise markedly.[24] Imagine a normal bell curve plot of temperature or rainfall values on one axis plotted against their likelihood. Clearly, those closer toward the mean are the likeliest to occur. Now imagine that the mean of the distribution increases slightly. It can be proven mathematically that this will cause a disproportionate increase on the occurrence of extreme values. (See figure 9-2.)

Scientists now expect more frequent and expensive climate-related disasters in the future: disaster experts forecast that these disasters will cost more than $7 trillion in the next 75 years. Yet a recent survey developed by the National Association of Insurance Commissioners found that almost nine out of ten insurance companies have no comprehensive plan for dealing with climate change.[25] That may be why insurance regulators in five states are requiring large insurers to respond to a survey on climate change risk.[26] The questions require insurers to begin thinking about managing climate risk and disclosing their carbon footprint reduction plans. While many insurance companies have yet to realize the implications of climate change for them, the reinsurance industry, which insures other insurance firms to help them manage their risks, is definitely concerned. J. Eric Smith, the CEO of Swiss Re Americas, one of the world's largest reinsurers, recently declared, "What keeps us up at night is climate change. We see the long-term effect of climate change on society, and it really frightens us."[27]

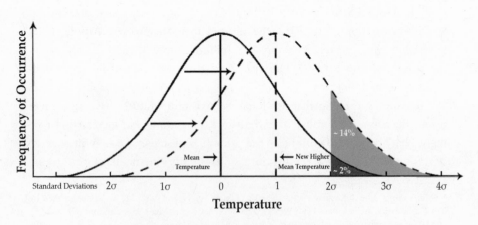

FIGURE 9-2. This diagram of a normal distribution, mean, and standard deviations from the mean illustrates that when the mean of the values shifts, the incidence of extreme values is disproportionately increased.

While no individual extreme event can be taken out of context and attributed solely to human-induced global heating, rising global temperatures do increase the statistical likelihood of extreme events. So the rising number of disasters and insured losses is consistent with the fact that the past decade was the hottest on record since about 1850 when modern instrumental weather record-keeping began.[28,29] (Examples of the increase in previously rare extreme events that occurred during this decade were given in chapter 2 and elsewhere.) This is not to deny that some of the increase in damages from extreme weather is due to population growth and movement into coastal areas. The number of people living in coastal areas of the United States, for example, grew from 120 million to 153 million between 1980 and 2003.[30]

Obviously, an increase in weather extremes is of tremendous global concern. Whereas natural systems and human societies have adapted to the normal range of climate variation, they are much less adapted to highly unusual events. So an increase in extreme events is very damaging both to ecosystems and to the built environment. In an authoritative 2008 report on extreme weather and climate in North America, distinguished scientists from the National Oceanic and Atmospheric Administration and the National Center for Atmospheric Research found many important changes in weather extremes over the past several decades.[31]

They observed more frequent and powerful hurricanes in the Atlantic Ocean and the northward migration of both North Atlantic and North Pacific storm tracks outside the tropics, along with a strengthening of the strongest storms. They also found more unusually hot days and nights; fewer unusually cold days and nights; and fewer frost days as well as more intense and frequent downpours. The report went on to forecast that North America will have a continuation of these trends along with more frequent heat waves and intense droughts, more powerful hurricanes, and more large storms outside the tropics, where they normally occurred.

The higher global temperatures of the past 50 years are, in turn, a major factor responsible for more heavy rains, since higher temperatures increase evaporation, thereby raising atmospheric humidity. More violent winter storms with higher winds and waves are also expected. The IPCC's *Fourth Assessment Report* also documents an increase in the strength of intense cyclone systems in temperate zones outside the tropics.[32]

EXCEPTIONAL FLOODING

During 2007, one of the hottest years on record, China, Indonesia, Mexico, Uruguay, and elsewhere had exceptional or record flooding. Some 25 million people were impacted by floods in Southeast Asia that killed 2,500 people. Floods also struck 15 countries across Africa. May to July 2007 was the wettest on record for England and Wales since 1766 and both areas experienced widespread flooding.[33]

The year 2009 produced torrential or heavy rainfall over parts of Brazil, Colombia, Peru, and Argentina, causing floods and mudslides.[34] In western Africa, heavy rainfall—more than ten inches in Burkino Faso—caused flooding that affected 100,000 people.[35] Torrential rains in southern Africa caused rivers to flood in Zambia and Namibia, disrupting nearly a million people's lives.[36] And in coastal Queensland and New South Wales, Australia, heavy rainstorms dumped nearly a foot of rain daily in January and February.[37]

In the United States, the director of the National Weather Service issued an unusual urgent public alert in 2010 warning that a third of the

FIGURE 9-3. Muddy Cedar River floodwaters cover large parts of Cedar Rapids, Iowa, in mid-June 2008 during a massive Midwestern flood that hit five states hard enough to trigger a Federal Emergency Management Agency voluntary homeowner buyout program. Qualified homeowners were allowed to sell their home to their city or county on condition that they demolished it and agreed not to rebuild in the floodplain. Photo © Jim Roberson, Associated Press.

FIGURE 9-4. A duck swims by the window of a building in downtown Grand Rapids, Michigan, in April 2008 when the Grand Rapid river rose to a record of almost 22 feet above flood stage. Weeks of exceptionally heavy rains caused the Mississippi, the Illinois, the Missouri, and many other rivers to flood in seven Midwestern states. Photo © Cory Morse, Associated Press.

country faced an elevated flood risk. States from New Mexico to Maine were in harm's way, in part due to El Niño, which had displaced normal storm tracks. Flood-related states of emergency were declared in Massachusetts and Rhode Island. North Dakota faced the second consecutive year of near-record flooding, an event unprecedented in 110 years of Red River Valley flood records.[38]

Big floods have become almost familiar to Americans over the past two decades. The great Midwestern flood of 1993, now known as "[o]ne of the most devastating floods of modern times,"[39] wreaked havoc across nine states. So did the 2008 Midwest flood, which took lives, burst levees, and caused billions of dollars in damages.[40] (See figures 9-3 and 9-4.) Climate models now consistently project that the kinds of intense precipitation events that caused these great floods will become more frequent, particularly over northern Europe, Australia, New Zealand, and South Asia (during monsoon season), as well as in other areas. The models also consistently show an increase in the number of wet days over northwest China and at high latitudes in winter, while projecting decreased rain in the Mediterranean area and South Asia.[41]

Droughts and Their Shocking Impacts

Drought is a function not only of rainfall and temperature but also of local soil, water availability, wind speed, solar radiation, and humidity. The Palmer Drought Severity Index measures soil moisture deficits based on temperature, precipitation, and available water content. Maps using the index can show moisture conditions across the globe relative to average moisture patterns for the past century. Maps like this can be used to draw attention to significant drought conditions, such as the major droughts in the developing world depicted in this chapter. Figure 9-5 illustrates the severe and widespread African drought that began in the 1980s, afflicted at least nine countries of the sub-Saharan Sahel region into the early 1990s, and returned in 2010. The severe Kenyan drought of 2011 was so severe it killed 150,000 livestock and affected 23 million people.[42] (See figure 9-6.)

Drought has also gripped parts of Alaska, Canada, the Mediterranean, eastern South America, eastern Australia, and large parts of Eurasia since the mid-twentieth century.[43] China in February 2009 had the worst drought in 50 years.[44] And as mentioned in chapter 8, more than two thirds of Iran is turning to desert and, according to the country's agricultural minister, the Iranian plateau is becoming inhospitable to human habitation.[45]

FIGURE 9-5. A woman carrying two malnourished babies pauses on her way to a distribution of maize in her village in Somalia, ca. 2000. Courtesy of UNICEF.

FIGURE 9-6. A herd of cattle succumbs to drought in northern Kenya in 2011.

While individual weather extremes can always be dismissed as aberrations, these droughts in the aggregate constitute a trend consistent with global climate change driven both by rising temperatures and reductions in precipitation over land. Other previously discussed factors that affect atmospheric conditions include the North Atlantic Oscillation (which affects westerly winter wind strength across the North Atlantic), El Niño, and shifts in normal storm tracks.[46]

In brief, droughts since the 1970s are more intense, last much longer, and are occurring over larger geographical areas.[47] Numerous respected climate models project that as global warming intensifies, summer drying will likely increase in the interior of many continents in midlatitudes, including areas such as central and southern Europe and the Mediterranean. More dry spells can also be expected over southern parts of Australia and New Zealand.[48]

Among other things, drought leads hundreds of millions of people to increasingly depend on limited and often waning supplies of water from underground reservoirs that often have taken ages to accumulate. These fossil groundwater resources should be carefully managed, especially as important emergency water supplies because, once depleted, they are for all practical purposes nonrenewable. Ever-more-frenetic efforts to overpump groundwater faster than it can be replenished will eventually make pumping too costly or physically impossible.[49] That can be catastrophic for those who depend solely on groundwater, as well as for vegetation and wildlife dependent on springs. (Groundwater causes springs to flow and moistens the root zones of vegetation, keeping it alive when rainfall is scarce.)

In India's agriculturally important Punjab state, more than 60 percent of the economy depends on agriculture. Punjab's fields grow half the grain that the government gives away to 400 million low-income people. But Punjab is heavily dependent on groundwater pumping. The state currently spends $8 to $9 billion a year subsidizing the electricity Punjabi farmers use to extract that nonrenewable resource—double what the government spends on education and half what it spends on health care.[50] The whole system of relying on groundwater is increasingly unsustainable. But by increasing water scarcity, climate change increases dependence on groundwater while holding down living standards in India and other developing nations. Moreover, because groundwater holds 10 to 100 times as much carbon dioxide as water in lakes and rivers, pumping it out on a global scale adds some 300 million metric tons of carbon dioxide to the air every year.[51]

Extreme Heat

As noted in chapter 2, the first decade of this century was the hottest on record, and 2012 was the hottest year on record for the lower 48 US states. The rate of increase of average global surface temperatures has been accelerating rapidly—twice as fast per decade during the past 50 years as during the first half of the twentieth century. Globally, the years 2005 and 2010 are tied for warmest year since official temperature records keeping began.[52] The year 2009 was especially warm in the Southern Hemisphere. New Zealand had its hottest August since its national temperature records began 155 years ago.[53] It was also Australia's hottest August and November in 60 years, and the heat contributed to an outbreak of wildfires[54] there that killed 173 people.[55]

All this is consistent with the global warming of the past century[56] and is well simulated by climate models.[57] It also fits the expected statistical tendency for extreme high temperatures to become higher and more likely as average temperatures rise, bringing hotter, longer-lasting heat waves.[58] The fact that heat waves have been lasting longer during the second half of the twentieth century than during the first is also consistent with more pervasive drought, because a decrease in moisture deprives land of the cooling effects provided by evaporation.

Years with exceptionally high average temperatures, such as 2007, are likely to include episodes of unusually extreme heat. Southern Europe was hit by temperatures as high as 113°F during the 2007 heat wave that killed

500 people. August temperatures in Japan that same year reached 105.6°F, the hottest ever recorded in that country.[59] The exceptional heat not only caused death, illness, and discomfort in Europe and Japan, but it had sudden and shocking consequences in parts of the Arctic, where the temperatures were abnormally high. The melting of frozen soil there caused dramatic increases in erosion that in turn brought significant vegetation losses and an increase in siltation, muddying waterways and choking salmon streams.

Researchers from an International Polar Year research project funded by the Canadian government were conducting studies on Melville Island, a vast, uninhabited, and normally very cold island in the northwest Canadian Arctic. But they found that during July 2007, when temperatures on the island would normally have been 41°F, air temperatures instead were 68°F.[60] The heat thawed permafrost about a yard beneath the ground surface, resulting in a slippery layer of underground meltwater. Topsoil then slid off hillsides in massive mudslides, clearing everything in its path and piling up in ridges at the valley bottom. One scientist who observed the process said, "The landscape was being torn to pieces, literally before our eyes. . . . A major river was dammed by a slide along a 200-meter length of the channel. River flow will be changed for years, if not decades to come." As pointed out in chapter 5, when this thawing occurs in populated areas, roads, pipelines, airfields, and other property may be rendered unusable.[61]

Another scientist studying the Arctic in the summer of 2007 reported, "We are in the midst of a phase of dramatic change in the Arctic. The ice cover of the North Polar Sea is dwindling, the ocean and the atmosphere are becoming steadily warmer, the ocean currents are changing."[62] Rain was observed at the North Pole for the first time that year.

Hotter, Bigger, More Frequent Fires

Because wildfires convert carbon in vegetation to carbon dioxide and—when hot enough—even roast the organic soils themselves, wildfires are a major global carbon dioxide source that release 1.7 to 4.1 billion tons of carbon to the atmosphere per year.[c] (In addition, tragically, large forest areas are still deliberately burned every year in the Amazon, Indonesia, parts of Southeast Asia, and Africa.)

---

[c] Global heating leads to an increased likelihood of the large, very hot blazes that do the most ecological damage, due to their destruction of the soil and its nutrients. For more on fire ecology, see John J. Berger, *Forests Forever: Their Ecology, Restoration, and Protection* (Chicago and SF: Center for American Place and Forests Forever Foundation, 2008).

Exactly as one would expect, global warming seems to be increasing the number of fires at high latitudes.[63] Many of these regions are now becoming warmer, drier, and therefore more susceptible to fire. The more severe storms of a hotter world will also ignite more fires, since they bring an increase in the number of lightning strikes. Touching down as they now increasingly do on dry forests and grasslands baked by higher temperatures (if not by outright drought), these lightning strikes can ignite large, sometimes uncontrollable fires. It is thus not surprising that fires had a more extensive impact on the world's northern forests in the 1980s than in any other decade recorded, and, according to some projections, the amount of carbon dioxide released to the atmosphere from fires will increase.[64] The ecological damage of vast and very hot fires is one of the many ecological impacts of global heating that serve to worsen other impacts, such as increasing the extinction of species, the topic of chapter 10.

The next few pages briefly explore what it would be like to experience extreme weather firsthand, with the help of first-person accounts.

Extreme Weather in Russia

During July and August of 2010, an unprecedented months-long heat wave took hold in much of western and central Russia, bringing drought, wildfires, and crop failure. Hundreds of fires simultaneously burned across six hundred square miles of central Russian steppe, forest, and peat bog. The roaring flames created "an infernal archipelago of 550 wildfires [that] released millions of tons of carbon dioxide and carbon monoxide into the atmosphere."[65] (See figure 9-7.)

Temperatures that exceeded 100°F in areas that rarely see temperatures above 75°F combined with a thick, 1,000-mile wide pall of suffocating smoke. Not only was this heat wave and drought the worst in the 130 years of formal weather records in Russia, but it was unprecedented in the past 1,000 years of recorded Russian history.[66]

Millions of face masks were ordered for Muscovites who had to go outdoors in the surreal and poisonous haze blanketing the city. Ordinary Russians were forced to keep their windows shut to avoid suffocation, only to stifle without air conditioning in unbearably hot apartments. One Moscow mother who suffered for days in almost intolerable heat and smoke said she felt as if she were "trapped in a burning building . . . constantly struggling to suppress the urge to grab [my] children and flee."[67] A retired Moscow school-teacher resident declared, "I woke up before dawn and thought I was going

FIGURE 9-7. A woman flees a wildfire in the area of Vyksa, Russia on July 29, 2010. The blaze was one of more than 550 wildfires that erupted within a 670-square-mile zone of Russia in the summer of 2010 during an intense drought and the most intense heat wave in Russian memory. Photograph © by Mikhail Voskresensky, Reuters.

to die of suffocation."[68] A third wrote, "With open windows, it's impossible to breathe because of the burning, and with closed windows we choke in the stifling heat."[69]

While the wildfires officially killed more than 50 people, health officials inferred that more than 300 people each day also died from the lethal combination of high temperatures and toxic smog. Sick, elderly, and susceptible individuals with respiratory problems or high blood pressure and heart disease were at greatest risk, and some people still face long-term health consequences. Russia's health ministry announced that carbon monoxide levels were six and a half times the maximum allowed in Moscow and that other poisons were as much as nine times higher.[70] Many workplaces had to shut down and, in other factories and places of business, workers were too sick with headaches and respiratory problems to continue working. Morgues were jammed with bodies, and ambulances made thousands of additional emergency runs. In one day, more than 100,000 people fled Moscow by plane.

Almost 200,000 people, including more than 10,000 firefighters backed by Russian army troops, fought the fires at one time, but were unable to extinguish all the blazes. Outside the cities, fires reportedly destroyed about a quarter of Russian grain crops,[71] and millions of acres were so badly afflicted by the drought that Russian wheat exports were banned for the rest of the year. World grain prices soared.[72] Bread prices in Russia rose steeply.

Fires threatened and may have burned areas outside the town of Ozersk in the Ural Mountains, where Russia suffered its worst nuclear disaster in 1957 prior to Chernobyl when a stainless steel tank containing high-level nuclear waste exploded.[73] The blast sent 70 to 80 tons of radioactive material into the air, contaminating hundreds of square miles of land.[d] (Fires on land contaminated with long-lived radioactivity can cause radioactive particles to become airborne, reexposing people to hazards from long-ago disasters.) Wildfires also burned contaminated forests in the Bryansk region bordering Ukraine and Belarus, where radioactive fallout from the 1986 Chernobyl nuclear reactor disaster poisoned the soil.[74] Some Russian authorities downplayed the risks, although the long-term health effects due to the fires cannot be ruled out.[75]

As may be seen from the 2010 Russian emergencies and from simultaneous extraordinary floods in Pakistan, India, and China, climate risks can suddenly become clear and present dangers and cannot safely be ignored.

Extreme Weather in Pakistan

The catastrophic floods that inundated a third of Pakistan in 2010 were the worst in that country's history and may be its biggest natural disaster ever.[76] The floods arose from severe monsoon rains in the country's mountainous northwest province of Khyber Pakhtunkhwa. The waters then traveled down the Indus River from Punjab into the deep south, spreading over at least a third of the country, affecting more than 20 million people, including at least 3.5 million children at risk of cholera and other diseases. Whole villages were swept away and homes, roads, bridges, and schools were destroyed, along with crops and livestock. Many survivors escaped with only the clothes

---

[d] The accident, known as the Khystym Disaster, exposed hundreds of thousands of people to radiation, most of whom were never evacuated. The disaster caused numerous deaths, radiation injuries, and illnesses, but was kept secret by the Soviet government for almost 30 years. According to one report, the explosion was known to the CIA, which also kept it secret. Had it become known to the American public, it might have increased opposition to the nascent US nuclear power industry. (See "Mayak," Wikipedia, http://en.wikipedia.org/wiki/Mayak for an important chronology and reference list. See also William Robert Johnston, "Chelyabinsk Nuclear Waste Accident, 1957," http://www.johnstonsarchive.net/nuclear/radevents/1957USSR2.html, updated June 27, 2004.

on their backs and were without food and drinking water or relief supplies. At least 4 million people needed emergency assistance, and vast numbers waited desperately without help for weeks. Millions were marooned by the floods, many on small, unstable islands surrounded by floodwaters. Close to 2,000 people died, and nearly 3,000 were injured. Seen from the air, much of Pakistan resembled a vast sea.[77]

The government was overwhelmed by the crisis. Many who lost their families, friends, homes, and livelihoods are bitter over the lack of government assistance. One villager said that although he survived the floods in Sindh that wrecked its economy, "the total lack of government help means dying may have been a better alternative."[78] Another villager from Shikarpur said, "It would have been better if we had died in the floods, as our current miserable life is much more painful."[79]

Even before the floods, Pakistan was a nation in crisis with severe economic troubles and ongoing insurgencies. But the floods turned a climate disaster into a full-blown economic, public health, and humanitarian crisis. They poignantly illustrate how a climate catastrophe can trigger political and social chaos capable of destroying a nation or setting its economic and social development back decades.

Unfortunately, the disaster in Pakistan is not unique and doesn't represent an end but a beginning of more frequent megafloods (see figure 9-8, next page). For example, 17 provinces in central and southern Thailand were hit by flooding in December, 2011. Fed by heavy monsoon rains and tropical storms, the flooding took more than 700 lives and affected 13 million people. It was the worst flooding in the Thai capital of Bangkok in a century.[80,81] The governments of Thailand and Pakistan were both criticized for their responses to the flooding crisis and, in the case of Thailand, for making matters worse by poor management.[82] Flood-proofing these nations will be impossible because of the magnitude of the floods to come and because of the enormous costs and management expertise required. Mass suffering from climate change is thus likely to get far worse, despite efforts at anticipating and adapting to the problem.

As climate change intensifies, situations will arise in which millions of people will be unable to escape or protect themselves from it. As the Russian, Pakistani, and Thai examples reveal, uncontrolled climate change brings chaos, not to mention the breakdown of normal life. The strain will literally be unbearable for millions of people. Some of these effects have already become unavoidable. All we can do now is to try our utmost to minimize climate change and, by slowing it, avoid the worst impacts that scientists are forecasting. If we fail to arrest its pernicious acceleration, then the extinctions projected in chapter 10 are sure to occur.

FIGURE 9-8. Children wait for food in front of their homes in Dhaka, Bangladesh, after a month of monsoon rains submerged most of the country in July 2004. Photo by Pavel Rahman, Associated Press.

CHAPTER 10

Extinction Perils

It takes hundreds of thousands to millions of years
for evolution to build diversity back up to pre-crash levels
after major extinction episodes.

—ANTHONY D. BARNOSKY ET AL.[1]

This chapter delves into the mass extinction of species that human activity is causing on the planet. Extinction rates and their implications for humanity and natural ecosystems are described. The connection to hunger and poverty is explored.

The Imminent Threat

HEATING THE WORLD by just a few degrees will cause a planetary spasm of extinction on a par with the largest five mass extinctions recorded on Earth in the past 500 million years. The current extinction rates for mammals and amphibians are already many times higher than normal, as revealed in the fossil record. (See figure 10-1.) Some authorities have estimated that the current species extinction rate may currently exceed natural background rates "by two to three orders of magnitude," i.e., by hundreds or thousands of times.[2] We have thus apparently begun the world's sixth "mega-extinction."

The causes are complex and transcend climate change. They include habitat loss and fragmentation, prodigious human population growth—four and a half billion newcomers to the planet just since 1950—widespread pollution,[a] overharvesting of species (by hunting, trapping, and fishing),[3]

---

a   The pollutants include thousands of unregulated chemicals, heavy metals (such as mercury, lead, disturbances to natural ecological processes,[4] and the degradation of many native ecosystems by invasive species. All these factors combine with climate change to increase the extinction rate.[5]

A quarter of all the Earth's terrestrial plant and animal species are thus extremely likely to die out over the next 50 years, and half of all the species on Earth are extremely likely to disappear forever by the end of the century if we continue increasing emissions along our current emissions "trajectory."[b] Unfortunately, biodiversity, when depleted, recovers even more slowly than does a disturbed climate. Earth's species are the end results of millions of years of evolution, and thus it will take Earth millions of years to replace the biodiversity that humanity appears likely to destroy through climate change and other means in a matter of decades.[6,7] If this mass extinction occurs, many surviving species will be greatly reduced in abundance and range, either because they will be unable to withstand the rising temperatures on the warmer portions of their range or, more frequently, because they will be unable to survive the ensemble of other ecological changes that climate change will bring.[8]

These adverse conditions might not only include direct climate impacts, such as abnormal precipitation and an increase in extreme events, but also ensuing changes in food supply, disease, competition, and predation. These challenges will be compounded by increasing human encroachments on the natural environment as human population heads toward 10 billion people by the end of this century.[9] With species declining so rapidly in both number and abundance, the structure and functions of ecosystems will be disrupted. Ecosystem services, such as the pollination of plants, essential to the health of the biosphere and to human welfare, will be reduced.[c,10]

---

and chromium), pesticides, herbicides, and air pollutants (such as sulfates, nitrates, ground-level ozone, particulates, black carbon), along with radiation from coal combustion and the nuclear fuel cycle (uranium mining, milling, nuclear fuel enrichment and fabrication, nuclear accidents, nuclear weapons production, nuclear fuel reprocessing, nuclear power plant routine operations, and nuclear waste management).

[b] That trajectory approximates the scenario designated as A1B by the IPCC in its *Special Report on Emissions Scenarios* and is roughly equivalent to a scenario known as Representative Concentration Pathway 8.5, which describes a path of increasing emissions sufficient to add 8.5 watts/square meter of heat energy to Earth's surface. Pursuing RCP8.5 for the twenty-first century would bring atmospheric carbon dioxide levels to more than 925 parts per million by 2100, well over the level seen on Earth for millions of years.

[c] Other ecosystem services include purification of air and water, oxygenation, carbon storage, replenishment of soil fertility, nutrient recycling, preservation of genetic resources, erosion control, and flood protection.

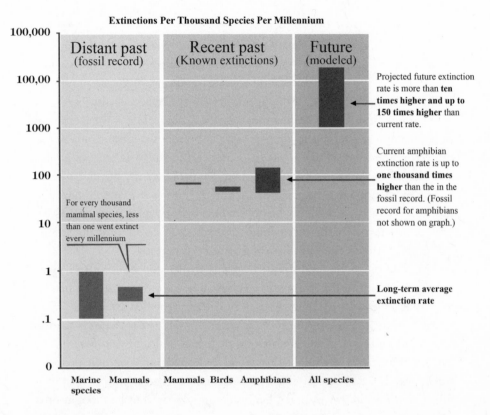

FIGURE 10-1. Current extinction rates are already more than 10 to as much as 1,000 times higher (for amphibians) than the rates found in the fossil record. Future extinction rates are conservatively projected to be more than 10 times higher than the current elevated rates. Source: The Millennium Ecosystem Assessment. UN Millennium Project, The Millennium Ecosystem Assessment (New York, NY: United Nations and Sterling, VA: Earthscan, 2005).

In a world where more than 2 billion people already live on less than two dollars a day[d] and more than 925 million already do not have enough to eat, anything that diminishes food supplies will create more hunger and poverty. A global extinction crisis and a resulting loss of ecosystem services along with ongoing environmental degradation plus climate change—all of which *independently* diminish food supplies—will likely create more hunger and poverty than the world is prepared to handle. For all these reasons, the world faces enormous peril.

---

[d] Of these 2 billion, 1.4 billion in developing countries live on $1.25 a day or less, according to the *2010 World Hunger Index*. See http://www.ifpri.org/publication/2010-global-hunger-index.

The Extinction of Populations

Populations of species are geographically and hence reproductively isolated groups belonging to the same species. Their number exceeds the number
of species by orders of magnitude; thus, while the species of the Earth number
in the millions, the number of populations is in the billions.[11] Whereas most
people think of species loss when they hear the word *extinction*, the extinction
of populations within species is even more important from the standpoint of
protecting the services that nature provides to humanity, as Hughes et al.[12]
and Ehrlich and Pringle[13] have pointed out. Yet paradoxically, the literature on
species extinction often neglects to mention the cost to the environment and
humanity of extinguishing populations whose vitality, abundance, and diversity are actually of immense importance to the perpetuation of the species
and to the performance of its natural ecological functions.

In addition to the pollination of plants mentioned earlier, these functions provide many other essential ecosystem services on which humanity
and other organisms depend. Such services are provided by nature at no
charge and include the purification of water by watersheds, the oxygenation
and decarbonization of the atmosphere by forests and other ecosystems, the
cycling of nutrients, the dispersal of seeds, the reduction of erosion, the control of pests and disease-carrying organisms, and the protection of coastal areas
from flooding and storm surges. Ecosystem goods can thus be thought of as
gifts of nature—for example, wild fish and game, natural forest products, and
medicinal as well as edible wild plants.[14] Whereas the existence of a species
doesn't necessarily generate these environmental benefits, the existence of
viable populations consisting of large numbers of individual members of the
species—and their interactions with other species—does provide a significant
flow of goods and services.[15] By contrast, a species that exists only in a zoo or
in small relict populations is unlikely to be able to deliver ecosystem goods
and services in appreciable quantity. Moreover, the genetic diversity that populations preserve due to their reproductive isolation is important for the very
survival of the species, as it confers on the species the ability to evolve more
quickly in the face of changing environmental conditions that might otherwise threaten its existence. Genetic diversity preserved in populations is also
an important resource that people use industrially and agriculturally in the
improvement of crops and pharmaceuticals.

Extinctions and Ecocide

It doesn't seem possible that—fewer than 90 years from now—we could destroy half the species on Earth. A tragedy on this scale is almost inconceivable. It is so tempting to maintain an almost clinical emotional detachment about such a devastating and alarming warning or to view it with skepticism. Yet skepticism as an excuse for inaction could be a fatal error. No doubt it also seemed inconceivable to the eighteenth-century American pioneers that we could wipe out more than a billion passenger pigeons and destroy tens of millions of bison.[16] One day, however, every last passenger pigeon was dead— and only a few bison remained standing where once the ground trembled under their thundering hooves.

Since the pioneer days, humanity has been taking giant steps along the twin paths of habitat destruction and natural resource overexploitation that jointly lead to rapid mass extinctions. We have, for example, already destroyed

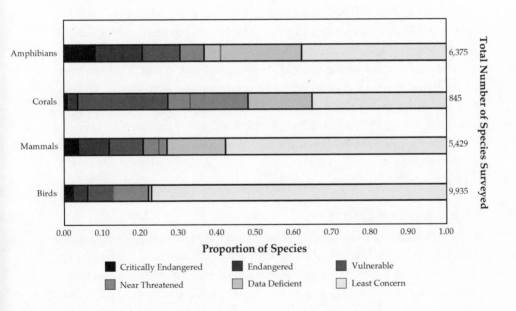

FIGURE 10-2. This bar chart shows that about 12 percent of all birds species surveyed, 25 percent of all mammals surveyed, 35 percent of all corals surveyed, and more than 40 percent of all amphibians surveyed are either critically endangered, endangered, or vulnerable to extinction. Plants (not shown) are at greatest risk as a group because the vast majority are rooted to the ground and rely on seeds for propagation. Courtesy of the International Union for the Conservation of Nature.

at least half of the world's original forests, wetlands, and grasslands. Now, at least a third of all amphibians, 12 percent of all birds, and a quarter of all mammals are already on the threatened list.[17] According to evolving estimates from the International Union for the Conservation of Nature, "28 percent of reptiles, 37 percent of freshwater fishes, 70 percent of plants, 35 percent of invertebrates assessed so far are [also] under threat."[18] (See figure 10-2.) In addition to these threatened and endangered species, certain species that once numbered in the millions, like the saiga antelope (*Saiga tatarica*), have declined steeply in number. The saiga itself is now critically endangered in its native Eurasian steppe region.

Some people may wonder why we should care about the saiga antelope and the threats it faces from humans and a changing climate. One might as well ask, who cares about the canary in the coal mine? If we commit biocide, many of us who depend on the biosphere will succumb as well. But assuming, for the sake of argument, that we could bring humanity unscathed through such a calamity, one would then need to ask, who cares about tigers, pandas, rhinos, whales, and desert tortoises? Would it be okay to bequeath a world to our children or grandchildren in which even iconic wildlife like this could only be seen in museums or zoos? Do we want a world in which the warbling of songbirds is replaced by the hum of machinery, and the chorus of frogs and insects on spring nights is silenced? Most people would say "no." Most agree it is wrong for humans to purge the world of vast numbers of vulnerable nonhuman species. Yet this is the direction in which we're heading as we expose many vulnerable species already threatened with extinction to the added stresses of climate change.[19]

Our Extinction Legacy

Modern humans are masters at converting wild ecosystems into simplified, tamer systems whose produce we harvest or whose resources we extract. We have logged lush tropical rainforests and burned them to blackened wastelands that ultimately became low-grade pasture. We turned rangelands into deserts through overgrazing. We drained productive natural wetlands for roads, housing, or farms, and paved over prime agricultural land for suburbs and cities.

Throughout our history, we hunted and killed whatever we found tasty, accessible, and slow enough to be caught. Soon after ancestral Native Americans walked over the Bering Strait into North America, for example, they began hunting to extinction most of the large mammal species they found.[e] We thus have a millennial-scale legacy of wiping out diverse and abundant wildlife populations or reducing them to relict populations. In annihilating other species, however, we destroy the matrix of life from which we emerged—and which we still need for our physical and psychological well-being. Not only do we lose the resources needed to support our current economies, but we forego precious future economic opportunities. As Harvard Professor E. O. Wilson put it, "Gone forever will be undiscovered medicines, crops, timber, fibers, soil-restoring vegetation, petroleum substitutes, and other products and amenities."[20] Granted that many impacts on the natural world are unavoidable and improve our quality of life, but much of the carnage is shortsighted and could be avoided through more enlightened resource stewardship.

Bringing Back the Dead

Extinctions—unlike some other environmental insults such as pollution— are for practical purposes irreversible.[f] Their cost is incalculable because we neither fully understand the impacts of individual extinctions on the ecosystems in which they occur nor the potential economic value of knowledge, medicines, or other products that those species might have had to offer.

While it may be possible in the foreseeable future to create species from frozen or fossil DNA or from stuffed museum specimens or even from sets of computerized instructions, that would not solve the problems of habitat destruction and climate change. Efforts to reanimate extinct creatures would be complex, very expensive, and time-consuming, and would likely have far less practical value to the environment than more conventional efforts to protect and restore habitat and species, to mitigate climate change, and to reintroduce threatened and endangered species where feasible.[21,22]

While reanimation research is likely to have scientific value and may indeed at some point make it possible to revive particular species of special

---

[e] The ancestors of most Native Americans are thought to have walked into North America across the Bering Land Bridge, starting some 11,500 years ago. Some seafaring settlers from the Pacific Islands, however, may also have arrived by boat thousands of years earlier.

[f] A team of researchers led by paleontologist Michael Archer at the University of New South Wales are trying to resurrect the extinct gastric brooding frog from the nuclei of its living tissue culture surviving in their laboratory, and to date have succeeded in producing embryos. See Richard Stone, "A Rescue Mission for Amphibians At the Brink of Extinction," *Science*, vol. 339, March 22, 2013.

interest, reanimating extinct species today remains a challenging laboratory stunt. It is emphatically no substitute for mainstream conservation and climate protection efforts.[23] We are a very, very long way from having the ability to reintroduce large numbers of currently endangered species, much less to reanimate millions of extinct species in viable populations and reintroduce them into their natural habitats, which very likely would have been radically altered and might no longer even exist. De-extinction in the context of conservation and climate change therefore represents a gross misallocation of effort and a seductive illusion that technology can compensate for even the most egregious and permanent ecological damage. It thus presents a moral hazard by fostering the erroneous belief that the advent of molecular biology and genetic engineering now render the loss of irreplaceable biodiversity and climate change less pressing concerns.[24,25] For these and other reasons, reanimation is not a sound basis for addressing climate change and the mega-extinction crisis that is unfolding.

Multispecies Extinction in a Rapidly Changing Climate

By what mechanisms does climate change cause extinctions, and how might extinctions reach epidemic proportions? How can the modification of climate by just a few degrees cause a sudden cataclysmic rise in global extinctions on a par with the five greatest extinction eras on Earth? The answer lies in the fact that what may appear to be a small shift in climate to us is not small from the perspective of the affected organisms.

Animal species must have habitat that meets their needs in terms of temperature, moisture, and nutrition if they are to complete their lifecycles successfully. They may also require a minimum territory in which to forage or a protected area in which to breed and raise young. Plants, too, require specific conditions for germination, growth, maturation, and reproduction. These conditions naturally include adequate sunlight, moisture, air, nutrients, temperatures and a space to which they're adapted, so they can grow, compete against rival species, and survive predators long enough to perpetuate themselves.

As environmental conditions approach a species' tolerance limits, individuals in the population experience stress and have a harder time completing their lifecycle; longevity and reproductive success decline. If conditions worsen and exceed tolerance limits, individuals fail in one or more of the activities vital to perpetuating their species, or they perish outright from heat, cold, exhaustion, hunger, or thirst. Climate change, for example, could directly cause an extinction by violating a species' physiological moisture or temperature requirements, or by indirectly affecting some other life-cycle requirement. For example, a changed climate may eliminate a prey or important plant species on which an animal depends. Unless the predator or herbivore can alter its diet, the loss of its preferred food supply may be fatal.

Climate change also affects species indirectly by tilting the playing field on which species interact, giving a new competitive advantage to one species that may cause the loser gradually to disappear. A recent large-scale review of empirical studies of local extinction by Cahill et al. found that heat intolerance was rarely the direct cause of extinction.[26] More commonly, extinctions occurred through decreases in food availability and climate-related losses of prey and pollinator species, as well as increases in pathogens and competitors favored by climate change. Some aquatic organisms, however, will simply suffocate from lack of oxygen in a hotter climate, as warmer water both increases oxygen demand and reduces its supply. Others will expire when heat dries up their freshwater habitat. Cahill et al. also noted that interdependent species that respond differently to climate cues by hatching or flowering at different times may perish if they no longer coexist synchronously within their habitat at critical times in their life cycles.

As climate changes, some plant and animal species will be able to adapt by moving to a new home range, while others will be unable to keep pace with the speed of the change and will die.[27] Over much of the world, "the potentially unprecedented rate of global warming" would require species to move at least 0.62 miles per year (1 kilometer per year) to keep pace with the climate to which they are adapted.[28] In some areas, they might be required to move a great deal faster.

Numerous studies have thus found, as researchers Diffenbaugh and Field have noted, that "the velocity of climate change may present daunting challenges for terrestrial organisms."[29] Some species may have the technical ability to move rapidly enough, but species with which they are interdependent may not share that ability. Even when species have the physiological ability to relocate with codependent species, human-made barriers may also block their relocation efforts.

Since temperature declines toward the poles and with increasing elevation, some extinctions are likely to occur when the imperative to relocate to a cooler habitat in effect pushes a species off the top of a mountain or beyond the poleward edge of a continent.[30]

The steepness of the landscape also plays a role in determining the success of a relocation effort. On a steep mountain slope, for example, a relatively short move might be sufficient to transport an animal or plant back into a cooler, more comfortable climate. On flat land, such as a prairie or wetland region, however, species would have to move a long way north or south (depending on which hemisphere they are in) to reach colder conditions.

In certain flat areas, scientists thus anticipate that if climate change is as rapid as projected, species might have to migrate up to six miles per year. Most plant species, however, can move only a few hundred yards a year at best, so they will be "outrun" by the worsening climate change now forecast for this century.[31] Because plants are integral to the success of all but viral and bacterial ecosystems, their failure to relocate will disrupt or destroy ecosystems on a vast scale. In addition to plants and animal species that are too slow to keep up with a fast-changing climate, some animal species move very little, if at all, as adults. Among these are clams, oysters, mussels, barnacles, and snails. Amphibians (frogs, toads, salamanders, and newts) and reptiles are also at high risk because of their slow dispersal rates.[32] The extent to which climate change has already dramatically altered the distribution of some animals is quite surprising.

Reductions in Species' Range

Recent resurveys of small-mammal populations in Yosemite National Park that were originally surveyed by the distinguished field biologist Joseph Grinnell between 1914 and 1920 reveal that the once-common Belding's ground squirrel has already disappeared from 42 percent of the sites where it had been found at the start of the twentieth century and might be extinct in its natural habitat in 50 to 75 years.[33] Contemporary researchers also found that 82 percent of the bird species Grinnell surveyed had relocated because of changes in rainfall and temperature—the park's average temperature is now 5.4°F warmer than a century ago.[34]

Much of the research on expected extinctions has focused on rare or uncommon species with restricted ranges. However, new research is showing that even those species that have large ranges or that are able to disperse quickly are likely to suffer severe reductions in range and abundance.[35] For example, a distinguished group of climate scientists led by R. Warren at the Tyndall Centre for Climate Change Research in the United Kingdom studied the likely effects of future climate changes on common and widespread species of nearly 50,000 species of plants and animals. In the absence of mitigation of the current global climate change scenario now unfolding, Warren's group concluded that more than half of these plants and about a third of these animals are likely to lose at least half of their present climatic range by the 2080s."[36] Warren et al. went on to show that early stabilization of global carbon emission rates and subsequent annual reductions in emissions of 2 to 5 percent would have major benefits in reducing extinctions and range reductions.[37]

Important Ecological Interactions

Healthy, natural, unimpaired ecosystems provide us with life-supporting services that include clean, well-oxygenated air, pure, drinkable water, edible wildlife in abundance, and fertile soil for supporting plants that in turn help prevent erosion. These ecosystems also provide many other tangible and intangible benefits—from opportunities for outdoor recreation and scientific research to contemplation and spiritual renewal. Ecological research involving hundreds of experiments just in the past 20 years has revealed that extinctions not only destroy a piece of an ecosystem, they affect its structure, its stability, its efficiency, and its most important processes, thus impairing its productivity.[38] As the loss of a species thus impairs an ecosystem's essential functions, the ecosystem's services—so vital to our welfare—are diminished or lost.[g]

The impacts of lost species on ecosystems can be huge, comparable to the impacts of drought, ultraviolet radiation, and climate change itself.[39] Bees and other insects pollinate wild plants as well as crops. If not pollinated because pollinators are absent, higher plants don't bear seed, fruits, and nuts. Unable to reproduce, wild plants eventually die, depriving useful insects, wild birds, animals, and people of food.

In chapter 6, the consequences of tropical forest destruction were discussed in the context of global climate tipping points, because of the large amounts of carbon these forests store. The loss of these ecosystems, however, also eliminates vast numbers of unique endemic species found nowhere else. The death of just a single tropical forest tree, for example, can cause the death of hundreds of beetles adapted to live only on a particular tree species. The extinction of an insect can mean the death of a fruit-bearing plant that depended on the insect for pollination. Then the extinction of that plant may leave a fruit-eating bird without food. As parts of the food web are thus eliminated, the whole web of life is weakened. Ultimately, the damaged

---

g The damage accelerates in a nonlinear manner as biodiversity loss increases, as Cardinale et al. point out in *Nature* (see reference 39).

ecosystem may no longer be able to deliver the clean air, clean water, wildlife, and nutrient-rich soil it once provided.

Future Extinctions

As previously indicated, climate change imposes new stresses atop the stresses of habitat loss, pollution, overharvesting, and invasive species. The combination of all these forces is greater than the sum of their parts. The IPCC has forecast that under rapid climate change, parts of the United States would lose roughly a third to half of its songbird species. Australia would lose up to two-thirds of its frogs, reptiles, and birds, and up to four-fifths of its mammals, along with the "eventual total extinction of all endemic species of [the] Queensland rainforest."[40] In addition, more than one-third of European birds would become extinct, and in the great game parks of Africa, up to 100 mammal species would become critically endangered or extinct.[41]

These estimates might actually be low. As pointed out by Foden et al., most previous large-scale extinction forecasts have used the "bioclimatic envelope" approach[42]—an examination of how key climate variables, such as temperature and rainfall, correlate with a species' geographic distribution. Researchers then project how future climates would affect the geographic distribution of those climate variables and thus the habitability of the species' range. From that information, they estimate the size of the species' future habitable range and its risk of extinction.[43]

These kinds of approximations, however, do not factor in the risk of extinctions caused by climate-induced disturbance to ecosystem and community relationships. Thus, although a species may be able to meet its climatic requirements in its new range, vital features of the previous range necessary for its survival—including the presence of critical plants and animals—may be absent from the new range. The bioclimatic envelope approach is also liable to neglect the effects of human ecological disruption on range habitability and the influences of unforeseen catastrophic events precipitated by rapid climate change. Therefore, the bad news this chapter brings is that most current estimates of future extinction risks are highly conservative.

For example, as described in chapter 9, when the worst heat wave and drought in at least 1,000 years struck western and central Russia in August 2010,[44] as many as 600 fires at a time burned out of control across the steppes and forests; the 1,000-mile-wide plume of smoke they created was visible from space.[45] No one knows how many species may have been reduced in range or destroyed as these fires incinerated forests and bogs.

Likewise, when the enormous floods of August 2010 submerged a large part of Pakistan, driving millions from their homes, the probability of local species extinctions was high, although the actual number of extinctions may never be accurately known. Unless the accelerated extinction of species due to climate disruption is brought under control, slowed, and then halted very soon, so many other species will follow the first wave of casualties that hundreds of millions, if not billions, of people will be endangered or die as the ecosystems on which humans depend for existence, including the oceans and forests, are degraded. (See chapter 11.)

Amphibians

Although not many extinctions have yet been directly attributed to climate change, a few harbingers of the looming extinction catastrophe are well documented. Until about 30 years ago, amphibians and lizards were seldom mentioned outside of biology texts and journals. Now these once-obscure creatures are making ominous headlines as they are pushed to extinction. The United Nations Environment Programme's World Conservation Monitoring Centre regards amphibians as the animals at greatest risk of extinction.

Amphibians evolved in the Devonian period 416 to 359 million years ago and did pretty well until the late twentieth century. But 30 percent of the world's amphibians are now threatened with extinction.[46] The tragic tale of the locally abundant golden toad (*Bufo periglenes*), native only to Costa Rica's Monteverde Cloud Forest, illustrates the sorts of predicaments in which many species increasingly find themselves.[47] As the climate in Costa Rica has warmed, the cool, moist clouds that once shrouded the forest in mist have risen above the woodland. Without cloud protection, the air and ground have grown hotter, leaving the forest floor too dry for the moisture-adapted golden toad. The shallow pools in which it laid its eggs in the wet season evaporated in the warmer climate before the eggs could hatch. Thanks to this one-two punch, the golden toad—only discovered in 1966—was last seen in 1989. Its disappearance, as biologist Tim Flannery put it, robbed the rainforest of its "brightest and most beautiful gem."[48]

The golden toad is not the only rainforest amphibian to suffer extinction or endangerment. The decline in the number of misty days at Monteverde has been associated with a decline in population of some 20 amphibian species of 50 studied.[49] Scientists from the Smithsonian Conservation Biology Institute working at amphibian rescue and research centers in Gamboa, Panama, are trying to save a few species of endangered frogs and toads there, including the

FIGURE 10-4. The brightly colored highland variety of the Panamanian golden frog (*Atelopus zeteki*) with a clutch of eggs, last seen near El Valle, Panama in 2007 and now presumed to be extinct in the wild. Copyright © Brian Gratwicke.

Panamanian golden frog (see figure 10-4). Despite their valiant local efforts, they can only protect a small fraction of the world's 7,100 amphibian species.[50] Hundreds of others are now perishing on several continents because of infection from a now-widespread amphibian chytrid fungus[h] to which certain frog embryos are made more vulnerable by increased exposure to ultraviolet light. Because the water in the ponds where the wild frogs lay their eggs is now shallower under hotter conditions, the eggs get exposed to more ultraviolent light, which slows their development, leaving them more susceptible to the fungus.[51]

The fates of the golden toad and golden frog should be recognized as a powerful warning and a symbol: Each year, thousands of less conspicuous creatures than these are dying out silently throughout the world. Even as you read these words, many amphibians, reptiles, insects, birds, and other species are vanishing in rainforests and numerous other sensitive habitats. We have no time to waste if we intend to save what can still be salvaged and avoid far worse die-offs.

---

[h] The fungus (*Batrachochytrium dendrobatidis*) was endemic to Africa but has been airlifted around the world by traffickers in wild amphibians for the pet trade.

The Extinction of Lizards

Even though scientists using climate models that project future tempera-ture and moisture conditions can forecast changes in habitable range for many creatures, the precise mechanisms that drive species to extinction in altered climates often still remain mysterious. Much important knowledge about how climate change causes animal extinctions, however, can be learned by understanding how climate change is affecting lizards. They are usually very hardy, but they have now experienced population crashes on five continents due to climate change.[i] UC Santa Cruz ecologist Barry Sinervo and his col-leagues recently resurveyed 48 spiny lizard populations on Mexico's Yucatan Peninsula that had been studied in detail from 1975 to 1995 in protected park areas. The researchers found that in less than 35 years, 12 percent of the local lizard populations in those areas had gone extinct.[52,53] (Prior to 1975, of course, additional lizard species may already have gone extinct without anyone knowing.)

Because lizards are "cold blooded," their body temperature varies with the temperature of their surroundings. They will therefore be inactive instead of hunting for food whenever the surrounding temperature is either too cold or too hot for them. Whereas many lizard species are very heat tolerant and thrive in hot, inhospitable climates, lizards must avoid temperatures that exceed their tolerance limits or they will die. To keep from getting too hot, spiny lizards in the Yucatan prefer to hunt in temperatures of about 88°F. In higher temperatures, they seek shade or underground burrows where they rest until the temperature falls. These "hours of restriction" when they can't forage vary from place to place. They are determined by air temperatures as well as the body temperature the lizard requires to become active. Because March Yucatan temperatures, which used to average 86°F, are now averaging 91°F, spiny lizards there are resting in the shade for an additional four hours a day. Since they are hunting less, they aren't getting enough to eat to success-fully reproduce.

Using such real-world data on lizard extinctions, their temperature requirements, and their hours of restriction, Sinervo and colleagues built a mathematical model capable of very accurately forecasting lizard extinctions in North and South America, Europe, Africa, and Australia, spanning tropical, temperate, rainforest, and desert habitats. The model combines climate data from the lizards' home ranges with the lizards' temperature requirements and

---

[i] Lizards belong to the class Reptilia.

tolerance. Sinervo et al. reached the alarming conclusion that globally, one in five lizard species will disappear in the next 70 years. The researchers even plotted extinction forecasts on a series of global maps showing which lizard populations are most likely to go extinct. According to this powerful analysis, lizard populations in slightly cooler habitats will follow spiny lizard populations into extinction as temperatures there continue rising.

Problems for lizard populations are also aggravated by an invasion of competing lizard species from lower elevations, whose habitats have gotten too hot for them. Not all species can migrate to more hospitable environments, however. As noted, some are on mountain slopes and can only travel so far up in search of cooler temperatures before reaching the mountaintop. Those lizard species that theoretically could migrate still generally cannot travel fast enough across the landscape to stay ahead of the rapidly changing climate. Nor will evolution be rapid enough to enable lizard populations to adapt successfully to the changing climate. Because it will take decades of effort to substantially change global energy use patterns and slow the rise in global temperature, the forecasts Sinervo made for lizards of the world are unlikely to change before midcentury, and possibly not by 2080. Thus, the last simple and tragic words of Sinervo's paper read: "Lizards have already crossed a threshold for extinction."[54]

Should we disregard the fate of lizards and other lowly creatures in remote or obscure parts of the world? Recall here that the loss of one species can have important effects on entire ecosystems. Lizards, for example, help keep insect populations in check. Insects can carry disease, such as typhus, plague, sleeping sickness, Lyme disease, tick-borne encephalitis and leishmaniasis, as well as malaria and yellow fever. And when lizards go extinct, bird and snake populations that prey on lizards will have less to eat. Snakes prey on rodents, including disease-carrying mice and rats. Similarly, the loss of amphibians will reduce food supplies for fish. And so on. Even if one had no moral scruples about wiping out supposedly unimportant species, it is obviously a bad idea to condone the loss of any part of nature's intricately interwoven, multifaceted architecture to which we ourselves belong.

The Tragedy of the Corals

Coral reefs are one of the best known and most significant ecosystems at risk from climate change. More than 1,000 species of coral inhabit the world's oceans, and their reefs are of extraordinary ecological and commercial importance. They are home to an estimated nine million species of marine life,[55] and they protect developed coastal areas as well as valuable marine and coastal ecosystems, such as seagrass beds and mangrove swamps.

A billion people worldwide depend to some extent on fish from coral reefs, according to the National Oceanic and Atmospheric Administration. In the developing world, a quarter of the fish caught by 30 million small fishermen are from coral reefs.[56] But all too many reefs are already sick or dying from the combined effects of higher water temperatures, pollution, acidification, overfishing, and dismemberment by souvenir hunters. By some estimates, a third of the world's coral reefs have already been destroyed just in the past 20 years.

Tropical coral reefs provide food and shelter to countless specialized organisms that have adapted to live in and among coral. The reef's brilliantly colored inhabitants include coral species, algae, sponges, anemone, conch, scallops, oysters, clams, shrimp, seahorses, jellyfish, and more, in a densely populated landscape. The reef's main structure is created by coral polyps—tiny colonial animals. Each polyp is a simple organism with stinging tentacles and a hard outer shell of calcium carbonate. The coral polyp lives in symbiosis with single-celled algae known as zooxanthellae. The coral provide the algae with a firm mooring, along with carbon dioxide and nourishment from waste products eaten by the coral. In return, the algae provide the polyp with oxygen and nutrients that they produce by photosynthesis and that can account for up to 98 percent of the polyp's nutrition. Thus, without the functioning zooxanthellae, the coral eventually die. (See "Adding Insult to Coral Injury," p. 201, for more on this symbiotic relationship.)

Using dissolved calcium carbonate, which they extract from seawater, coral polyps—some no larger than a pinhead—slowly fashion the vast, durable, and intricate colonial superstructures that form the foundation of each reefs. The reef surface is often encrusted with a vast array of shapes, from staghorn-like stalks to delicate lacy fans, often in brilliant luminescent colors, thanks to the cohabiting algae. As coral colonies grow, the reef rises upon the shells of countless dead polyps. The process can continue for thousands of years and can form four basic characteristic kinds of large reef structures.[57] Originating in shallow near-shore waters and growing seaward, a reef at first buttresses the coast. At times, however, the near-shore seafloor areas subside, leaving a lagoon behind an offshore barrier reef. Reefs also extend directly outward into the sea from the coast to form peninsular barrier reefs. Reefs in the near-shore waters abutting islands develop into circular coral atolls if the island in their center subsides. No matter how large the reef—and the

Great Barrier Reef off Australia is more than 1,200 miles in length—only a few millimeters of the surface are alive, supported by the shells of long-dead coral polyps.[58] Those who snorkel over a healthy reef, however, will be able to observe an intricately wrought undersea garden pulsating with life. Its nooks and crannies offer a diverse variety of habitats for millions of organisms, including stunningly beautiful creatures with some of the oddest shapes and brightest colors in the ocean.[59]

They have existed for about 200 million years and are some of the world's oldest living ecosystems, often 5,000 to 10,000 years old. Despite their size and antiquity, coral reefs are now highly vulnerable to the combined effects of direct human abuse, ocean acidification (see chapter 11) and to the warmer seawater that accompanies climate change. All of these factors combine to imperil corals. Their prognosis is still uncertain; the direct damage from humans and local polluted runoff is easier to assess and control than the future effects of climate change and ocean acidification.

Sophisticated mathematical models of coral species' heat tolerance under various emission scenarios suggest that coral ecosystems might completely collapse by 2050 or, alternatively, might be able to maintain themselves beyond 2100.[60] Their fate depends both on corals' resilience and adaptability to greater ocean acidity, and on the success of emission-reduction efforts.[61,62] To some extent, their fate also depends on how carefully reefs are managed and protected from the conventional degradation they are currently experiencing. In a very insightful, far-ranging review of threats to corals in *Science*, John M. Pandolfi et al. wrote, "The non-climate-related threats already confronting coral reefs are likely to reduce the capacity of coral reefs to cope with climate change."

Other Threats to Coral

The existential threats to coral reefs have emerged in what is but a flash on a geological time scale. Beginning mostly in the twentieth century, people began destroying or damaging coral reefs by dynamiting them, or by poisoning the waters around them with cyanide, or by trawling over them for fish with dragnets, or by chiseling pieces off for souvenirs and aquaria. Many ancient reefs that could otherwise have provided a sustainable harvest of fish indefinitely if fished lightly were sacrificed so a lot of fish could be caught quickly and easily—for a short time. More than 80 percent of all shallow coral reefs are currently being overfished by conventional means.[63] The overfishing not only affects the fish but often disrupts or destroys the predator-prey relationships that keep algae in check and prevent it from growing over the reef and destroying it.[64]

Reefs are also quietly being poisoned with pollution from farms and cities, and choked with sediment from soils eroding from farming, construction projects, and lands scorched by forest fires. Fine particles or other pollution from all these sources can make seawater too opaque for photosynthesis by the zooxanthellae and may coat the reef with sediment, killing the polyps. Sediment may also transport nutrients that create red tides—potentially enormous growths of suspended algae that produce toxins capable of devastating reefs or dying off suddenly, leaving rotting masses of algae on the ocean floor where they can suffocate coral.

About 15 percent of the world's coral reefs were already dead or dying in 2006, according to Professor E. O. Wilson of Harvard, who projects that nearly half of all corals could be lost just in the next 30 years. Even this, however, might be an understatement, depending on future global emissions and coastal management practices.

Adding Insult to Coral Injury

Coral reefs are not only extremely sensitive to increasing ocean acidity but also to warmer water and pollution. Temperature increases of only a degree or two Fahrenheit can harm coral colonies and cause them to lose their zooxanthellae. Without these colorful partners, coral bleaches to a stark white.[65] If water temperatures do not return to an acceptable range within about a month after bleaching, the coral will starve to death, killing the reef. Some species are more resistant to bleaching than others; however, even if the coral survives a mass bleaching event, it is likely to suffer high mortality and depressed rates of growth and reproduction.[66]

Not only does warmer water cause bleaching, it also makes the corals more susceptible to acidity.[j] (When carbon dioxide diffuses into the ocean and dissolves, it forms carbonic acid.) At 450 to 500 ppm of atmospheric carbon dioxide, ocean water will become so acidic it will be difficult or impossible for some corals to form shells or grow. Currently the atmosphere has about 400 ppm, but at current rates of increase, it will reach 450 within 20 years. An important new report from the International Geosphere-Biosphere

---

[j] In addition, as discussed in Pandolfi et al. (see reference 61), a warmer ocean may be more stratified with a lower mixed-water layer in the tropics, leading to a reduction in ocean productivity. That in turn would likely reduce the nutrients available to coral polyps. The coral would then have less energy to invest in calcification, a process already made more difficult by the ocean's increasing acidity.

Programme (IGBP) has concluded that once carbon dioxide levels reach 560 ppm, most coral reefs will be breaking down faster than they can accrete.[67] The same report warns that if we continue with a business-as-usual emissions scenario, conditions by 2100 will be unfavorable for coral reef growth throughout the entire tropical surface waters of the ocean. The water by then will be corrosive to the shells of organisms that depend on aragonite (a more soluble form of calcium carbonate) over 60 percent of the tropical oceans. The IGBP researchers also estimate that by 2100, 70 percent of cold-water corals will be in corrosive waters.[68]

Already, however, "Coral reefs, the biologically rich 'rainforests of the sea,' are retreating worldwide. . . . " warns Professor Wilson. "Those around Jamaica and some other Caribbean islands have largely disappeared. Even the Great Barrier Reef of Australia, the largest and best protected in the world, declined 50 percent in cover between 1960 and 2000."[69] Biologist Tim Flannery warns that when the oceans warm by a single additional Celsius degree (1.8°F), four-fifths of the Great Barrier Reef will be bleached, and at 2°C, only 3 percent of the reef would still remain viable.[70] The corals off the coasts of Sri Lanka, Tanzania, Kenya, the Maldives, and the Seychelles are also at high risk.[71]

We now know, however, that much of the heat currently accumulating in the atmosphere will be absorbed by the acidifying oceans in a few decades. Rising water temperatures will then wreck many coral ecosystems.[72] Should this trend continue, coral reefs will become pale ghostly mausoleums, shadows of their former dazzling selves.

What are the implications of damaging or destroying coral reefs that support 25 to 33 percent of the creatures of the ocean? How will the decline in fish affect marine mammals and birds that depend on fish, and how will this impact the vast numbers of people all over the world who depend on marine life for their food or livelihood?[73]

Marine scientist Ove Hoegh-Guldberg of the Centre for Marine Studies at the University of Queensland, Australia, declares, "We are witnessing the end of corals as a major feature in the oceans." The only hope he sees for saving corals is if we recognize that climate change is a "Code Red" emergency, slash carbon emissions to zero as quickly as possible, and begin drawing down the concentration of carbon dioxide in the atmosphere.

Time will tell if Hoegh-Guldberg's forecast is correct, but like the desperate situation of amphibians and lizards, coral reef endangerment ought to be taken as one of nature's most alarming warning signs that environmental stress and climate change in tandem are destroying and jeopardizing even the Earth's largest ecosystem, the ocean.

Acquiescing to Extinction

By 1992, many types of ecosystems around the world with vast storehouses of biodiversity at risk had for decades been in a steep decline or collapse, as extensively documented by the UN and other global organizations. Aware of the problem, representatives of 178 nations, including 100 heads of state, attended the 1992 United Nations Earth Summit in Rio de Janeiro. There they signed an international Convention on Biological Diversity along with a Framework Convention on Climate Change. But the powerful and steadfast resolve to effectively implement these conventions was lacking. The intervening two decades have instead been marked by halfhearted measures.

In support of its Framework Conventions, the Earth Summit did generate an action program called Agenda 21, accompanied by a statement of voluntary principles on sustainable development known as the Rio Declaration and an accompanying Forest Principles statement. As admirable as many of these goals are, no specific responsibilities for protecting species were assigned to any of the signing nations, nor was any enforcement mechanism included.[74] The principles were explicitly declared to be "non-legally binding." So they therefore could be comfortably ignored.

Most importantly, in the absence of political will, strong leadership, and widespread popular pressure, little funding was subsequently provided by the countries that signed the conventions to ensure that their commitments to save biodiversity were met. Instead, more talking and meetings ensued. Meanwhile, trillions of dollars were spent on wars by some of the signatory nations. Another decade passed. In 2002, world leaders at the Johannesburg World Summit on Sustainable Development pledged to significantly reduce or reverse the losses of biodiversity in their countries by 2010. The UN then declared 2010 to be the International Year of Biodiversity. However, not a single country has met Johannesburg targets, according to the UN Convention on Biological Diversity's general secretary.[75]

At a 2010 meeting of 193 nations in Nagoya, Japan, additional admirable goals were set, most notably to halve the rate of natural habitat loss during the next decade.[76] The jury is out on whether the Nagoya goals will be attained.

The Rio+20 United Nations Conference on Sustainable Development was convened in June 2012 on the twentieth anniversary of the first conference. More than 50,000 people from government, business, and the environmental community attended, and hopes were high. The global extinction crisis, of course, was by this time even more grave than in 1992. However, according to reports provided by Greenpeace, major corporations like Shell lobbied hard to ensure that no binding commitments to protect the Earth were imposed on participating nations. The lobbyists apparently were successful. Few tangible agreements or significant funding commitments emerged from the meeting. The convention's final declaration was a lukewarm statement calling yet again for a further refinement of goals at a new international meeting in 2015.[77] Time will tell whether anything substantive comes out of the new meeting. But without lots of inspired political organizing and highly visible public pressure, the odds are not good.

Meanwhile, as chronicled in this chapter, biodiversity is disappearing fast, with heavy losses across a range of species, from simple to complex, and probably from every ecosystem and geographic region. Not only mammals and flowering plants, but insects, crustaceans, amphibians, bacteria, and fungi are being destroyed. Data from the International Union for the Conservation of Nature show that from 1980 to 2008, 52 species a year moved closer to extinction.[78] The problem may appear to be financial, but it is not. The world has no lack of capital to protect endangered species. It found trillions of dollars to shore up shaky banks in the Great Recession of 2008[k] and, since 1992, has invested additional trillions in military hardware and fossil fuel subsidies combined. As of late 2013, the Federal Reserve Bank of the United States was spending $85 billion a month to prop up the bond market and stimulate the economy, but no commensurate actions were being taken by the US government or other nations to protect nature, the world's more important and irreplaceable natural asset.

Taking Action

To save those species that can still be saved, nations of the world collectively need to adopt a global climate protection plan as quickly as possible. Under its provisions, each nation would have an obligatory and steadily declining annual greenhouse gas emission limit. However, even with emission caps in place, nations everywhere, rich and poor alike, must also be motivated and, where necessary, financially assisted to invest adequate funds in habitat and wildlife protection.[1] The International Convention on Biological Diver-

---

[k] See Nomi Prins, *It Takes a Pillage: An Epic Tale of Power, Deceit, and Untold Trillions* (New York: John Wiley & Sons, 2010).

[1] The tragic but preventable scourge of poaching, still disgracefully widespread and tolerated in many places, needs to be brought under control.

sity[m] (CBD)[79] outlines measures for the conservation and sustainable use of biological diversity and the sharing of its benefits. The Conference of the Parties to the CBD has prompted the creation of more than 100 national action plans for biodiversity protection and has raised public awareness about the issue.

Global Biodiversity Outlook 3 (GBO3)[80] presents a number of shorter-term actions that can be taken to protect specific ecosystems on a stop-gap basis while longer-term solutions are implemented. The document summarizes national biodiversity reports as well as the latest scientific research and projections of future impacts on biodiversity.[n] *GBO3* demonstrates that the measures needed to save biodiversity are widely known and understood by scientists and resource managers. Yet according to the report's sponsors, "None of the twenty-one subsidiary targets accompanying the overall 2010 biodiversity target can be said definitively to have been achieved globally. . . ."[81] Thus, year after year, despite an avalanche of reports and untold meetings, the international community has failed to fully fund or enforce the modest protective measures to which it has agreed. Governments of the world clearly have not made protection of global biodiversity a high enough priority, despite the practical and moral hazards of sending other creatures to their doom. Global biodiversity is being lost not because people have no idea how to save it, but because sufficient political will, political pressure, and commitment have not been generated. One simple way to begin would be for the parties to the CBD and all its action plans to implement the agreements speedily and diligently.

The next and final chapter focuses in additional detail on other dangers that climate change presents to the world's oceans.

---

[m] The CBD was opened for signature at the 1992 UN Conference on Environment and Development (commonly known as the Rio De Janeiro Earth Summit) and went into effect in December 1993.

[n] *Global Biodiversity Outlook 3* also discusses the need to address large systemic issues affecting biodiversity.

Oceanic Perils

The ocean is the life support system for our planet.

—DOROTHÉE HERR AND GRANTLY R. GALLAND

To provide a more comprehensive and coherent impression of the ominous and immense changes occurring within the oceans, this chapter both revisits and explains in greater detail several oceanic phenomena touched on earlier: sea-level rise, sea-ice loss, ocean acidity, and the rarely publicized decline in the abundance of the tiny floating plants (phytoplankton) that form the base of the ocean food web.

Ocean Degradation

CLIMATE CHANGE IS ALTERING the temperature and chemistry of seawater as well as ocean currents and other features. The ocean is thus becoming more carbonated, more acidic, more heat sensitive, lower in oxygen, and fresher (less salty). As the oceans warm, sea level is rising fast, dead zones are spreading, and coastal wetlands are shrinking. Some of these changes alter critical oceanic processes in ways that over the long term could permanently tip the climate into a new state, with devastating consequences for human welfare and survival. For example, climate change may be slowing the ocean's "conveyor belt" circulation that transfers heat from the equator to the poles and could eventually stop it completely.

It could also potentially disrupt El Niño and La Niña cycles (ENSO), leading to the drying out and destruction of the Amazon rainforest, as discussed briefly in chapter 6. Finally, over the long term, climate change could release billions of tons of frozen methane hydrates from the seabed, leading to an uncontrollable planetary warming.[a,1] (This and other tipping point topics were discussed in chapter 6.)

To explain these and other climate-related ocean changes in more depth now, I will outline the basic biological, physical, chemical, and geological processes that go on in the ocean, unseen by the casual observer. (Further details can be found in the appendix to this chapter, page 252.)

The Protective Ocean

The oceans are, of course, a vital part of Earth's life support system. They stabilize the climate by absorbing a quarter of all the carbon dioxide that people release to the atmosphere, protecting the climate from additional heating. The oceans' carbon management work is done through a complex web of biological, physical-chemical, and microbial processes that ultimately locks carbon away in geologic storage where it remains for eons. But vast as they are, the oceans' capacity to remove and store carbon dioxide can be compromised. The warmer the oceans get, the less of it they can absorb from the atmosphere. Thus, as time goes on and the world warms, more of the carbon humanity puts into the air stays there, further warming the planet—another example of a positive climate feedback.

Ocean Structure and Currents

Climate change not only alters ocean water chemistry and temperature, but higher temperatures cause polar ice to melt, freshening the sea surface, reducing seawater salinity and density, and hence changing ocean currents that depend on salinity and density gradients and play an important part in the planet's heat transport system.

As explained in chapter 6, the progressive freshening of the sea surface could slow or stop cold salty water in the polar region from plunging down into the ocean depths to drive the ocean's thermohaline circulation, which transports heat from the equatorial regions toward the poles. The surface water's density is also affected by temperature. Heating of the ocean surface leads to increased layering of the ocean water by temperature, with warmer, less dense waters floating like a cap above cooler, denser waters. Because this stratification reduces the vertical mixing of seawater, it further limits the ocean surface's ability to absorb atmospheric carbon dioxide.[2]

---

[a] In addition to ocean warming, seabed mining, drilling, and dredging can also disturb these deposits and release significant amounts of the gas.

Reduced vertical mixing also interferes with the natural upwelling of colder, nutrient-rich ocean waters. This deprives plankton of the nutrients they need for growth and survival. In turn, this affects the operation of the ocean's biological carbon pump (see the appendix to this chapter, page 253). Plankton production then falls and, consequently, less of their cellular carbon drifts down into the deep ocean where it can either become incorporated into the ocean sediments or circulate at great depths without returning to the surface for long periods of time. Not only is less carbon safely stored over long periods of time, but the entire ocean food pyramid that depends on this planktonic "fuel" at its base then suffers due to a lack of nutrients. The ocean then produces fewer and smaller fish and shellfish. Coastal regions whose economies depend heavily on commercial fishing are hurt. Globally, the market value of marine fish as they leave the boat is $80 to $85 billion a year. But when the impact on related industries is included, like shipyards, boat sales, canning, marinas, markets, and seafood restaurants, the total economic value of the commercial marine fish catch is at least triple that amount.[3] As for small or subsistence fishermen, lack of fish can be devastating for them.[b]

Swelling Seas

Earlier chapters focused briefly on how sea-level rise from the melting of ice caps and glaciers and from the expansion of warmed seawater threatens to disrupt the lives of hundreds of millions of people living near the ocean. Let's now consider a few hypothetical future sea levels to get deeper insight into the enormous consequences of rising seas.

The oceans have on the average already risen by about eight inches during the twentieth century due to climate change. Sea level is now rising at a rate of about 13 inches per century—double the pace in the past decade compared to that of the previous 50 years.[4] Yet this is only a hint of the much larger changes likely to come in time. For at times in the Earth's distant geologic past, sea level has been as much as 393 feet lower than at present and as much as 229 feet higher.[5] During the last interglacial period about 130,000 to 118,000 years ago, sea levels were 13 to 20 feet higher than at present.[6] Some evidence exists that during that interglacial period, the ocean gained almost 10 feet in height within only a 50- to 100-year period![7]

Imagine what would happen nowadays in the unlikely event that the ocean someday was rising 10 feet every 100 years. Given all the coastal cities and vast shoreline development we have, a 10-foot increase would threaten

---

[b] Currently, however, their fortunes are still more at the mercy of overfishing than climate change.

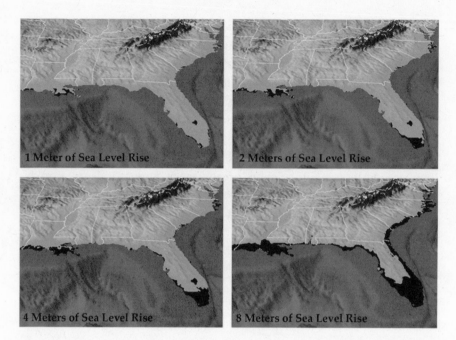

FIGURE 11-1. Moving clockwise from top left, the darkened areas on the map of a portion of the United States show the progressive effects of 3.3, 6. 6, and 9.8 feet of sea-level rise (one, two, and three meters). Some projections suggest that sea level might rise by roughly 6.6 feet by the end of this century if the world remains on its high carbon emissions path, and it would go on rising for centuries after carbon emissions were halted.

the lives and livelihoods of the billions of people living near the coast, and would flood valuable coastal infrastructure. Moreover, the damage would be irreversible. We couldn't just refreeze Greenland or the polar ice caps after they had melted in order to lower the seas again.

Whereas the IPCC's *Fourth Assessment Report* projected that sea level would likely rise by less than two feet by the end of this century, more current research indicates that sea-level rise could possibly reach six feet by then.[8] A hotter world then will produce more intense storms, flinging them shoreward from a higher base, producing larger, more damaging surges.[9] Even under the conservative IPCC projections about rising seas, however, hundreds of millions people will be exposed to storm-surge flooding just by 2080.

Threats to Islands, Deltas, and Coasts

Obviously, sea-level rise will adversely affect all seafront nations, and people in large, low-lying, densely populated river deltas, such as the Nile, the Ganges-Brahmaputra, and the Mekong will be particularly severely impacted.[10,c] At a rise of three feet, "Cairo, Bangkok, London, Shanghai and Venice (among other cities) would be critically endangered."[11] Flood events that once happened once or twice in a century would then be likely every few years. In the United States, some 5,000 square miles of dry land and 15,000 square miles of wetlands would be exposed to the risk of permanent flooding from the sea unless protected by barriers of some sort.[12] Cities and towns worldwide are increasingly likely to be faced with the choice of relinquishing large portions of valuable, developed territory; or of creating expensive offshore defense systems, with movable barriers to keep out the sea; or of diking their lands. Levees of some kind would also need to wrap back inland, roughly perpendicular to the sea from the dikes at added expense to prevent the seas from flowing around them.

Dikes would not be a panacea, however. Sea-level rise threatens the existence of small, low-lying island states. Many islands of the Pacific and Indian Oceans, as well as those of the Caribbean Sea (like Antigua, Barbados, and the Bahamas) or of the South China Sea will be entirely swamped or will lose much of their land. Large parts of the Republic of Kiribati in the tropical Pacific, and the Republic of Seychelles and the Republic of Maldives, both in the Indian Ocean, are less than 40 inches above sea level. The average elevation of more than 1,000 islands comprising the Maldives is 3 to 5 feet. Storm surges from hurricanes can be 20 feet (6 meters) high and will tower over islands that are left.[13] Sea-level rise also is of great concern to the Marshall Islands, where a rise of about three feet would cost the country 60 percent of its arable land. Small island states typically lack the financial resources to build expensive defenses against the sea. Just protecting the shoreline of the Caribbean islands would cost more than $11 billion, which the small islands can ill afford.[14] Those that manage to remain above sea level 50 to 100 years from now will still be especially vulnerable to storm surges, hurricanes, and tsunamis. To make matters worse, as seas rise and coasts erode, saltwater will seep into the islands' shallow coastal aquifers, reducing freshwater supplies for drinking water and farming. Even where engineered defenses are feasible, over time they can fail, as they did in New Orleans during Hurricane Katrina. They may simply be too low or in disrepair when needed. Another issue is land subsidence, commonly caused by extensive pumping of water from the ground. Land may also subside for geological reasons as it is

---

[c] As discussed in chapter 3, in the United States, parts of Washington, DC, and New York City, along with large parts of South Florida will be permanently submerged; so will North Carolina's Outer Banks and areas around Pamlico and Albemarle Sounds, if current trends continue.

famously doing in the now partially submerged historic village of Dunwich in Sussex, England.

On time scales of tens of thousands of years, a countervailing elevation of land also occurs in certain areas once buried in ice, produced by a slow phenomenon known as glacial rebound. It occurs when the melting of Ice Age Ice Sheets lifts a great weight from the land. The land mass gradually rebounds and rises. Land also gradually rises over geologic time when enormous plates of the Earth's crust collide and plunge beneath the continents in subduction zones. Ocean currents and gravitational forces, too, play a role in local and regional sea levels. All these complicating factors operating on very different time scales can combine to worsen or alleviate the effects of sea-level rise at local and regional scales, making sea-level rise nonuniform around the world.

Long-Term Sea-Level Rise

The consequences of sea-level rise in the twenty-first century are but a small foretaste of the vastly more serious oceanic flooding to come. The oceans will continue to rise for at least the next 500 years due to the future effects of the long-lived greenhouse gases already aloft. With more than a billion people already living in coastal areas—a number that could swell to 5.2 billion by 2080 (according to some IPCC scenarios)[15]—sea-level rise will have enormous worldwide effects and is one of the most consequential but most often underestimated impacts of global warming.

While the current pace of sea-level rise is *very* rapid on a geological time scale, during the last interglacial period, when the Greenland Ice Cap and possibly parts of Antarctica were melting, seas may have risen much faster—well over three feet per century, a pace similar to the projected sea-level rise for 2100.[16] It may not sound like much, but had the seas been rising that fast over the 400 years since the Pilgrims landed in Massachusetts, the ocean by now would have risen 13 feet, flooding thousands of square miles of coastal land, not to mention Plymouth Rock itself.

Adapting to Sea-Level Rise. People commonly build homes and businesses in harm's way on oceanfront property, sometimes right on the beach itself. Public officials have barely begun to educate the public about the interrelated economic and environmental impacts of sea-level rise. But programs to protect wetland areas, other shoreline buffers, and potentially defensible economic assets (for example, utility corridors and wastewater treatment facilities), are still in their infancy—if they exist at all.[17] Visionary public officials already know, however, that they will need to begin acquiring the property rights necessary for a gradual phased retreat from the shore ahead of expected sea-level rise as part of comprehensive coastal management planning. Orderly anticipatory action will help prevent loss of property and life during storm surges and other extreme weather events. It will also enable natural habitat to migrate landward with rising water. That in turn will help wetlands reestablish themselves farther inland, where they will again be able to buffer coastal land against storm surges while preserving essential ecosystem services and wildlife.[18]

Without advance planning and shoreline preparation, however, including the eventual closing of some facilities, industrial sites (past and present) and port facilities commonly found near the shore will likely be damaged and may well release their contaminants to bays and estuaries. For example, some 80 percent of 11,000 acres that would be lost to a five-foot sea-level rise in San Francisco are heavily contaminated.[19]

Wetland Losses and Other Coastal Impacts

Sea-level rise will cover large tracts of remaining coastal saltwater wetlands. These natural areas are already very badly degraded by the discharge of pollutants and by wetland destruction for residential, commercial, agricultural, or industrial use,[20] including oil and gas drilling and the construction of roads and pipelines. The United States is currently losing 80,000 acres of coastal wetlands every year, an alarming rate according to the US Fish and Wildlife Service and the National Oceanic and Atmospheric Administration, which jointly conducted a recent study on the worsening problem.[21]

Healthy coastal marshes provide many important benefits. They protect biological diversity, buffer the coast against storms, reduce shoreline erosion, preserve fish and wildlife habitat, and store large amounts of carbon in plant roots and sediments,[d] and support the ocean food web. But because development so often crowds the shoreline, the remaining narrow wetlands perched precariously between development and the sea cannot migrate inland in response to rising seas as they would naturally. Sea-level rise therefore will destroy much of this valuable remaining fringe of coastal marshes.

Half of Europe's coastal wetlands are expected to be lost by 2020 because of sea-level rise.[22] According to one global study, a rise in sea level of only

---

[d] Carbon dioxide is naturally held in coastal marshes, mangrove swamps, and seagrass beds as plant roots and floating particles of organic matter are sedimented out of the water column. The beauty of these natural carbon sinks is that the carbon is immobilized in newly created marsh peat or other sediments for long periods of time.

2.4 feet by 2080 will result in the loss of 44 percent of the world's coastal marshes, excluding sea grasses.[23] That will leave valuable fish, shrimp, and other shellfish with reduced spawning, feeding, and nursery areas, and will harm global fisheries and reduce the ocean's biodiversity. Sea grass beds, like coastal marshes, also provide important fish and wildlife habitat and reduce erosion. Half of all sea grass beds have already been lost, and only a third to one-half of all the world's valuable mangrove swamps remain.[24] An anticipated sea-level rise of more than three feet over the next century will cannibalize much of the remnants. Losses of wetland plants and animals will then send ripple effects through the global food web, reducing its productivity.

Saltwater Intrusion and Coastal Erosion

In addition to outright flooding, when sea level rises along shores, natural estuaries will become saltier as seas flow farther inland. This will affect public water supplies and also will force brackish water organisms to move through the estuary higher into the watershed. Freshwater coastal marshes will tend to become brackish and brackish marshes will become saltier, impacting their plants and animals.

Rising seas will also threaten groundwater supplies with saltwater contamination. The situation varies from region to region because if climate change causes a local increase in rainfall and raises the local water table, that could help stave off intruding saltwater, protecting freshwater supplies in that area.[25] Although water tables in general are expected to rise, that can create drainage problems. The dangerous combination of higher water levels and more intense storms will cause more flood damage and more coastal erosion over larger areas.[26]

Many beaches, wetlands, levees, and protective offshore barrier islands will be damaged, washed away, or submerged. Combined with the loss of coastal marshes, the loss of barrier islands and reefs is already starting to make many mainland coastlines more vulnerable to hurricanes and erosion. Over the past 30 years, Mississippi and Texas have lost up to 10 feet per year to the advancing sea, while Louisiana lost about 39 feet per year.[27] Erosion is also a serious problem along 20 percent of the European Union's coastline.[28]

Climate change also alters the amount and timing of runoff to the ocean. This changes the inflow of sediment to coastal waters near river and creek mouths. The circulation of the suspended sediment within the ocean along the coasts is also changed because of the increase in sea level. Sediment then may fill in estuaries and lagoons, which are costly to dredge.

Impacts of Ocean Acidification on the Marine Food Web

Like a blotter that soaks up carbon, the ocean has taken up about half of all the carbon dioxide released by the burning of fossil fuels since preindustrial times. But the ocean's ability to absorb additional carbon is gradually becoming saturated.[e] By the end of this century, the ocean will only be able to absorb a third as much atmospheric carbon dioxide released each year as it did before industrialization.[29] A continuation of this trend could ultimately transform the oceans from a carbon sink that removes some of the carbon dioxide society emits to a carbon source that not only doesn't remove carbon but on balance releases it, strongly amplifying global heating.

The ocean's ability to neutralize acidity caused by dissolved carbon dioxide is also being saturated. (When dissolved in water, carbon dioxide produces carbonic acid. See the appendix to this chapter, page 257 for more details. Normally, seawater (which is alkaline) tends to neutralize dissolved acid through a process known as buffering. However, the absorption of additional carbon dioxide gradually depletes the ocean's buffering capacity. The oceans are then less efficient at neutralizing acid. Each additional unit of absorbed carbon dioxide thus makes the ocean slightly more acidic than did the previous unit.

Ocean acidity has been gradually rising since the start of the Industrial Revolution and has already increased by 26 percent. But according to the UN Convention on Biological Diversity, "by 2050, ocean acidity could increase by 150 per cent, 100 times faster than any change in acidity experienced in the marine environment over the last 20 million years...."[30]

By 2100, the ocean may be three times more acidic than before industrialization, according to the Royal Society of London,[31] and 170 percent more acidic, according to a study by the International Geosphere Biosphere Programme and others. That would make the oceans more acidic than at any time within hundreds of thousands and possibly millions of years. Commenting on the ominous trend in ocean water chemistry, the IPCC stated in its *Fourth Assessment Report*: "The expected continued decrease [in pH and consequent rise in acidity] may lead within a few centuries to an ocean pH estimated to have occurred most recently a few hundred million years before present."[32]

---

[e] The ocean is actually capable of taking up 85 percent of all the carbon dioxide released from the burning of fossil fuels, but this requires very long oceanic turnover time scales. See reference 18, page 291.

The more acidic the ocean gets, the lower the carbonate concentration of seawater. Because many sea organisms need carbonate molecules to build their shells or skeletons (see chapter 10), higher acidity thus reduces the growth, health, reproduction, abundance, and shell-forming ability of a vast number of marine organisms.[33] The most obviously affected are mussels, oysters, and clams as well as crabs, lobsters, shrimp, and coral. The larvae and juveniles of these organisms tend to be most sensitive to acidification.[34] If carbonate levels get low enough, these animals eventually die. A lack of carbonate adversely affects even smaller planktonic organisms. This holds true for tiny organisms like foraminifera, a class of single-celled protozoa with shells, and for pteropods (marine snails). Their shells are already dissolving in parts of the Southern Ocean near Antarctica where pteropods had been a key food source for pink salmon.[35] Because these and even much smaller organisms are at the base of the ocean food web, harming or destroying them can cause damage of unknown magnitude to cascade through the entire marine food chain. Their responses could potentially cause major ecological disruptions leading to large reductions in fish production and ocean biodiversity, and to significant changes in the cycling of carbon to and from the atmosphere. Climate change thus presents humanity with the risk of an ocean food chain catastrophe.

Destroying the Ocean Food Web

As previously noted, the entire complex food web of the ocean, including the fish and seafood that we eat, all depends on phytoplankton. They operate an elegant energy conversion and transmission system that ultimately feeds almost all life forms of the sea. Phytoplankton do this by floating at or near the sea surface, where they are able to capture solar energy from sunlight and use it in photosynthesis to combine carbon dioxide and water to produce edible plant tissue and a significant amount of the Earth's oxygen. The phytoplankton are then grazed on by floating animals of many types and sizes known as zooplankton. These range in size from tiny protozoa unseen by the naked eye to juvenile fish, jellyfish, crustaceans, worms, and the one- to two-inch-long shrimp-like krill that provide excellent nourishment to larger fish, seabirds, and whales.

Less energy flowing into the base of the food web in the form of a reduction in phytoplankton will thus inevitably lead to a reduction in the amount of consumable seafood for organisms like us at the top of the web. Whatever harms the ocean's phytoplankton could therefore wreak havoc on the whole ocean food web and shut down the minute oxygen-producing factories in the sea. Phytoplankton are also very sensitive to pollution, ocean water temperature, and increases in ultraviolet radiation.[f] Recently, a group of scientists at Canada's Dalhousie University in Halifax, Nova Scotia, reached the shocking conclusion, after a three-year study published in the prestigious journal *Nature*, that marine phytoplankton has declined by 40 percent over the past century.[36] They believe these results cannot be explained by natural variations and are caused by rising sea-surface temperatures associated with climate change. As pointed out earlier, higher temperatures can cause more pronounced layering of the waters near the equator, thereby depriving phytoplankton of nutrients needed from cooler, deeper waters. Increasing acidity also may be taking its toll. The drop in plankton abundance—unknown to most people—indicates that a colossal degradation of ocean conditions is in progress.

Rot in the Ocean

The average surface temperature of the world's ocean reached a 130-year record high in July 2009—62.6°F—driven not only by global warming but also by the concurrent onset of a natural El Niño system.[37] In August 2009, the normal frigid waters off the coast of Maine—which used to be endurable by swimmers for only a few minutes—hit 72°F near Scarborough, Maine, a temperature more typical of Mid-Atlantic coastal waters. Reflecting the widespread ocean warming, summer coastal water temperatures off the Mid-Atlantic states resembled those usually found off Florida's coast.

When the phytoplankton in ocean surface water produce oxygen, they at the same time "fix" carbon, incorporating it in their tissue. This then helps to maintain the ocean's capacity for soaking up additional carbon from the atmosphere. An oversupply of plant nutrients, however, stimulates a rapid, dense overgrowth of phytoplankton known as an "algal bloom." Warm water enhances the algal growth, making algal blooms larger and more common.

Often these blooms are so dense that they color surface waters green, red, golden, or brown, and will block sunlight to lower layers of the water column. Later, if the plankton bloom's growth rate outpaces the ability of zooplankton to consume it, the phytoplankton will eventually exhaust their nutrient supply. The bloom then ends when the phytoplankton begin dying, sinking toward the bottom in a large mass of decaying vegetation. As the algae rot at

---

[f] Stratospheric ozone absorbs ultraviolet light and protects life from its damaging effects. However, ozone levels over the oceans in the Arctic and Antarctic have been reduced by the release of now-banned synthetic chlorofluorocarbons.

depths, the bacteria that break them down consume available oxygen near the bottom. This creates a watery graveyard, often visible as a cemetery of dead creatures—crabs, fish, shrimp, worms—all lying on the seafloor shrouded in a white bacterial film.

In recent years, such offshore dead zones have stretched off the Oregon coast from Washington all the way south to the California border.[38] Out of sight and therefore out of mind for most people, hundreds of square miles of once healthy, once productive ocean bottom now lies in ecological ruin. This dead ocean bottom area in 2006 was roughly the size of Rhode Island. Another well-known dead zone is expanding in the Gulf of Mexico off the coast of Louisiana, fueled mainly by nutrient-rich farm runoff.[g] Much of the pollution comes from the hundreds of millions of pounds of nitrogen fertilizer applied to Mississippi Valley cornfields, a third of which are devoted to the production of ethanol from corn, made possible by expensive federal subsidies.[h]

The Gulf Coast dead zone is roughly the size of the state of New Jersey; in 2010, the dead zone was projected to extend for 8,500 square miles, reaching from Alabama to Texas.[39] As a dead zone expands in size, it becomes more and more difficult for sea life to escape, and slow or immobile bottom-dwelling creatures like crabs and oysters are among the first casualties engulfed by the anoxic conditions.

To digress for a moment: What occurs on land affects what happens in the nearby water. Ironically, the federal ethanol subsidy program that contributes to this ecological destruction of the Gulf has also led farmers to withdraw land from the Department of Agriculture's Conservation Reserve Program (CRP), under which farmers have received an annual federal payment to take marginal cropland out of cultivation and revegetate it.[40] Some of that CRP land had been planted with native prairie grasses capable of trapping large amounts of carbon in the soil.[41] Now that the land is tilled, deteriorating, and eroding, it is probably releasing stored carbon to the air. This further undermines any climate protection that corn ethanol fuel additives were supposed to foster. (The Environmental Working Group, a renewable energy advocacy group, estimated that ethanol subsidies from 2005 to 2009 cost taxpayers

g Soil erosion, sewage, and industrial pollution also contribute to the problem.

h This is an economic boondoggle and environmental burden since corn ethanol production and combustion when analyzed systematically have long been known to actually release only marginally less carbon to the atmosphere than gasoline and diesel fuels. The harm is thus done for negligible benefits.

$17 billion that could have been better spent supporting carbon-free energy sources like wind and solar.)[42]

Returning now to the story of algal blooms: Accelerated by global heating, algal blooms not only make pristine seawater murky and create dead zones, but some produce toxins, as mentioned in chapter 8.[i] Some 60,000 cases of algal poisoning occur annually worldwide, and about 900 people a year die from it.[43] In addition to algae and coral, pathogens, too, are sensitive to sea surface temperatures, so as ocean temperatures rise, the warmth promotes the growth and reproduction of disease-causing waterborne organisms, including bacteria, viruses, and fungi that kill corals and other marine life, or cause diseases in humans, including seafood poisoning. Research conducted in the late 1990s by James W. Porter, an ocean studies specialist at the University of Georgia, and Joan B. Rose, a scientist at the University of South Florida, found a nearly 500 percent increase in disease at 160 coral reef sites in Florida. They suspect that the increase in pathogens found in ocean water and marine life might be due to an increase of only 1.8°F in sea-surface temperature.[44]

Another research group has suggested that the outbreaks of pathogens and other organisms seen both in disturbed and relatively undisturbed parts of the ocean may be a consequence of an increase in the movement of iron-rich dust from the desiccated Sahel region of Africa, where climate-related drought has become more prevalent. It's yet another "who knew?" unexpected negative consequence of climate change.

Polar Ice Loss Revisited

During the decade ending in 2008, average global temperatures increased 0.9°F, but average temperatures in the Arctic rose nearly four times as much,[45] so the Arctic is now warmer than at any time in the past 2,000 years.[46] Even these increases are minor, however, compared to what lies ahead. Average temperatures in northeast Greenland are expected to rise by close to 11°F.[47] Researchers James Screen and Ian Simmonds have concluded that most of the added heating in these regions is due to the effects of sea ice melting in response to the rise in air temperature. Prior to the Screen and Simmonds study, the relative roles of changes in cloud cover, humidity, and ocean current

---

[i] If algal toxins are absorbed by shellfish or other marine life, the algae can poison people who consume that contaminated seafood or who come in contact with contaminated water. Certain algal blooms thus cause deaths, illnesses, fish kills, and loss of tourism and fishing revenue.

circulation were thought to be more important than the extent of sea ice.[48] To recap the sequence of events as discussed in chapter 6, when sea ice melts, heat that was formerly reflected by the ice is absorbed by the dark ocean. This creates a positive feedback effect that results in a cycle of more warming, which causes more ice melting, which in turn triggers more warming.

The Arctic is now warming a great deal faster than the IPCC projected in its 2007 *Fourth Assessment Report.* According to a 2009 account in the *Geophysical Research Letters*, summer temperatures in the Arctic are up to 9°F warmer than the IPCC forecast.[49] (They were even higher—9 to 13°F warmer than normal—over the Chukchi Sea in 2008.)[50] These high temperatures and warmer ocean water will very likely result in the complete disappearance of Arctic summer sea ice in less than 30 years.[51] Ice in the Arctic today is not only shrinking in area, but is thinning and breaking up earlier and freezing later than normal. Over the past 30 years, average sea ice cover has declined by 20,700 square miles each year.[52]

Multiyear Arctic sea ice used to be 30 feet or more thick in some places and once covered 90 percent of the Arctic basin.[53] The thicker the ice, the more likely it is to endure seasonal warming and temperature changes. Prior to the current onset of Arctic warming, more than 60 percent of its sea ice was tough, perennial ice more than six years old.[54] By 2009, 90 percent of the ice was under two years old.[55] The remaining older multiyear ice is now only six feet thick at most.[56] After the summer melting in 2009, more than 70 percent of the remaining ice was young and thin—under a year old.

In Summary: The Oceans Are in Big Trouble

Whereas the oceans appear imperturbable, they are highly vulnerable. We cannot expect them to provide the vital ecosystem services to which we're accustomed if we allow them to become acidic, saturated with carbon, and overheated. Signs of serious disturbances within the oceans abound. Seas are rising, sea ice is melting, ice shelves are breaking up, the Greenland Ice Cap is melting, numbers of tiny organisms, including plankton, are decreasing overall, while in some areas, algal blooms and dead zones are spreading. Marine mammals are suffering and dying, wetlands are shrinking, acidity is rising, and significant volumes of frozen methane are bubbling out of the shallow seabeds of the East Siberian Sea in powerful plumes as much as a half a mile wide, foreshadowing a possible eventual massive methane release to the atmosphere over the long term.[57] As a global society that depends heavily on the oceans for sustenance, livelihood, and well-being, we cannot afford to continue abusing the oceans. Unfortunately, no simple, easy, practical solutions exist for protecting them. They can only be protected in the context of protecting the Earth's climate system, of which they are a crucial part.j Hopefully, this will happen before it is too late to save the climate and the oceans that depend on it.

---

j A detailed discussion of recommended actions and policies for protecting the climate may be found in my forthcoming book, *Solving the Climate Crisis: Turning Global Peril Into Jobs, Prosperity, and a Sustainable Future.* (See www.JohnJBerger.com for further availability information.)

Conclusions

There is still time to avoid the worst impacts of climate change,
if we take strong action now.

—Sir Nicholas Stern[1]

The issue of climate change is one that we ignore at our own
peril. . . . Unless we free ourselves from a dependence on these
fossil fuels and chart a new course on energy in this country,
we are condemning future generations to global catastrophe.

—President Barack Obama[2]

I WILL NOW BRIEFLY RECAP the basic facts about the climate crisis that show why action is urgently needed, what the general nature of that action ought to be, and what obstacles stand in the way.

The Indisputable Facts

Since the start of the Industrial Revolution, humans have released 545 metric tons of carbon to the atmosphere.[a] It now contains the highest levels of heat-trapping gases—carbon dioxide, methane, and nitrous oxide—of any time in the past 800,000 years. Annual emissions of carbon dioxide—from fossil fuel burning and cement production plus land-use changes[b]—have surged 54 percent just from 1990 to 2011 to 10.5 billion metric tons of carbon a year.[3] The Earth has responded to all these heat-trapping gases by getting warmer: its average land and sea temperature has thus risen by about one and a half degrees F since the mid-nineteenth century. The Greenland and the Antarctic Ice Sheets are melting at increasing rates, as are the world's

---

[a] That is equivalent to two trillion metric tons of carbon dioxide.

[b] The land-use changes include deforestation and emissions from agricultural activities, such as animal husbandry, rice cultivation, and fertilizer application.

glaciers. Positive climate system feedbacks, such as the warming, melting, and thinning of Arctic permafrost, are appearing. Arctic sea ice is also melting very fast, adding more positive feedback. Meanwhile, sea level is rising at an accelerating rate; ocean temperature, currents, and salinity are changing; and the oceans are growing dangerously more acidic. Because global temperature has risen, heat waves and other weather extremes have become more common, the onset of seasons has altered, and the global water cycle and atmospheric circulation have been affected.

These trends are likely to continue and accelerate for the foreseeable future. But even after emissions stop, adverse climate effects will continue for millennia. Eventually, society will have to cease its discharge of heat-trapping gases. Delay only allows the atmospheric burden of heat-trapping gases to swell. The greater these cumulative emissions, the higher the Earth's final temperature, and the more severe the consequences—longer-lasting droughts, more insufferable heat, larger deserts, scarcer food and water, higher oceans, more corrosive seawater, more fetid ocean bottoms, and a dreadful paroxysm of species extinctions.

The Earth cannot indefinitely withstand the ravages of habitat destruction, the strain of an exploding human population, *and* abrupt climate change. The outcome is predictable.

Healthy natural ecosystems will lose their diversity or collapse outright. As their productivity declines, so will the Earth's life-support capacity. People will suffer and populations will contract. These impacts are now so imminent and devastating that it is time to declare that the planet is in a climate emergency.

Every emergency has two basic aspects: (1) a grave threat to life, liberty, property, or the environment, and (2) a need for immediate action. Millions of people have already died from disease and malnutrition brought on by climate change. Even more harm is likely, according to hundreds of authoritative scientific studies. So climate change presents a grave threat to life, liberty, and property as well as to the environment. And it is irreversible for the foreseeable future, so immediate action is necessary before further avoidable harm is done. Even if heat-trapping gas emissions miraculously fell to zero tomorrow, the atmosphere would get another 1 to 2°F degrees hotter, just from excess heat already absorbed by the oceans.

In a matter of decades, billions of people will lack adequate food and water if society continues on its current emissions trajectory. Governments and relief organizations today are already struggling to care for millions of refugees. In an overheated world, tens of millions *more* environmental refugees will be on the move—hungry, sick, and desperate. This is a recipe for conflict and chaos. This problem and that of climate change in general are greatly compounded by rapid global population growth, most of it in developing countries, exacerbated by child marriage and inadequate access to family planning resources.[c]

Time Is Running Out

Because of the cumulative nature of carbon emissions and the decades required to convert global economies from fossil fuels to clean energy, the chance to protect the Earth from horrific consequences is slipping away. Merely to have a two-thirds chance of avoiding a global temperature increase of more than 3.6°F means we cannot add more than another 270 billion metric tons of carbon to the atmosphere, according to the latest assessment report of the Intergovernmental Panel on Climate Change (IPCC). Thus, even if the world held its current emissions constant at 10.5 billion metric tons of carbon a year, instead of increasing them rapidly, the world would have only 26 years to avoid crossing the 270 billion metric ton carbon threshold. Future emissions would then need miraculously to fall to zero in 2039 to avoid overshooting 3.6°F, the nominal boundary between safe and unsafe climate change.

We are therefore now clearly on the precipice of extremely dangerous changes: by between 2080 and 2100 we are on track to increase global average temperatures by 6 to 10°F, as compared with preindustrial times, according to the scientifically conservative IPCC. Some experts are projecting that 7°F could be reached by 2060. Such temperatures haven't been seen on this planet in five million years. Moreover, those average temperatures would be roughly doubled in continental interiors.

In the overheated world of a few decades from now, up to 30 percent of the world would be in drought at any time, up from 1 percent today. Fifty percent of land where crops now grow would become unsuitable for farming. A 7°F temperature increase could cause most of the world's old trees to die from a combination of drier conditions, heat, and climate-related diseases.[4] Even a temperature increase of 3.6°F could eventually drive the Earth's climate past

---

[c] One in three girls in developing countries (excluding China) is married before age 18 and one in nine is married before age 15. This not only raises birth rates but violates girls' human rights and jeopardizes their health, often curtailing their education and vocational choices. (See United Nations Population Fund, *Marrying Too Young: End Child Marriage* (New York, UNFPA, 2012).

various "tipping points" at which the climate system itself begins to multiply the effects of human greenhouse gas releases. Such feedbacks could defy all conceivable human control.

Making Remedial Action a Top Priority

If our current emissions trajectory continues, a quarter of all land plant and animal species will likely be gone within just 50 years—far less than a human life span. Then by 2100, half of all the species on Earth would likely disappear—a catastrophe unprecedented in human history.[d] For all these reasons, the climate emergency, too long neglected, must become a top financial as well as political priority. It is even more threatening to our long-term security than terrorism and conventional military threats—on which the United States spends hundreds of billions a year—and the financial crisis of 2008 and beyond, when the Federal government committed trillions to bail out troubled banks and insurers. Economies have recovered from financial crises. But once a critical climate tipping point is passed, no financial manipulation will un-tip it. Whereas a healthy climate is essential for economic prosperity, a runaway global climate catastrophe would devastate rich financiers along with poor subsistence farmers, dwarfing the 2008 financial crisis.

Fortunately, many global studies confirm that we have the technology, financial capability, and renewable energy resources to successfully transition to an energy economy largely free of fossil fuels. But this will require some hard technological and political choices. Very large global programmatic investments in energy efficiency, renewable energy technology, agriculture, forestry, as well as carbon capture and storage will be needed to protect the climate. Yet affordability is not the main impediment.[5] The United States currently has a gross domestic product of about $16 trillion, but like most other nations, it prioritizes military and other spending over climate protection. The United States thus spends $1.0 to $1.4 trillion a year of its $3.45 trillion federal budget on defense[6]—which amounts to about 40 percent of the entire world's military spending.[7] All the nations of the world together, however, spent only $145 billion for all renewable energy technologies and systems in 2009.[8,9,10]

---

[d] That trajectory has been designated as A1B by the IPCC in its *Special Report on Emissions Scenarios* and is roughly equivalent to a scenario known as Representative Concentration Pathways 8.5—a path of increasing emissions sufficient to add 8.5 watts/square meter of Earth's surface to the planet. Pursuing RCP 8.5 for the twenty-first century would bring atmospheric carbon dioxide levels to more than 925 parts per million by 2100, well over the level seen on Earth for millions of years.

Redirecting Energy Investments

The International Energy Agency estimates that the world needs $38 trillion in energy infrastructure investment between 2010 and 2035—an average of over $1.5 trillion a year.[11] If past is prologue, most would be spent on gas, oil, and coal energy infrastructure. Yet if those dollars were redirected from fossil fuel infrastructure into efficient and renewable energy systems, they would make more energy available more cleanly and with vastly more new employment than business-as-usual fossil fuel investments. It is therefore hard to escape the conclusion that socially irrational energy decisions are being made due to the political and economic influence of fossil fuel producers. A relatively small number are responsible for a disproportionate share of the world's carbon emissions. From 1854 to 2010, nearly two-thirds of all human-induced carbon dioxide and methane were attributable to just 90 major commercial and state entities, according to Richard Heede of the Climate Accountability Institute.[12]

A redirection of capital would partially de-fund these entities. The new funding for renewable and efficiency could come from many sources. Using federal, state, and local financial incentives, governments could leverage public money to encourage private investment. Loan guarantees, revolving credit, public-private cost-sharing, accelerated depreciation, tax exemptions, tax credits, and "feed-in tariffs,"[e] can all tilt markets in favor of climate-safe energy sources. Funds could also be made available by ending direct fossil fuel subsidies that totaled $500 billion worldwide in 2010. As resource policy expert Lester R. Brown wrote, "All together, governments are shelling out nearly $1.4 billion per day to further destabilize the earth's climate."[13] Additional funds to supercharge a global transition to climate-safe energy sources can also come from fees on carbon-based fuels. In short, if intelligent clean energy and transportation programs are interwoven with enlightened agricultural and forestry policies, humanity can avoid aggravating the climate crisis. "Our progress here will be measured . . . ," said President Obama in his 2013 Georgetown University climate speech, "in crises averted, in a planet preserved."

Sensible Steps Toward Climate Protection

A comprehensive national energy plan for each nation on Earth is needed—aimed at nothing less than a total transformation of its national energy system is necessary. The plan needs to provide for a steadily increasing

---

[e] A "feed-in tariff" is a guaranteed price for renewable power set by a regulatory body, such as a public utility commission or a legislature.

national renewable energy requirement, the electrification of the transportation system, energy storage technologies, and modernization of the electric transmission grid. Such plans could also aim at achieving full employment and economic revitalization, so ordinary people would both benefit from, and support, the plan. Jobs would be created in energy efficiency services as well as in manufacturing, installing, transporting, financing, and maintaining new renewable energy equipment. Millions of people could be put to work restoring and enhancing damaged natural resources that naturally remove carbon from the atmosphere, including forests, agricultural lands, grasslands, and wetlands. My forthcoming book, *Climate Solutions: Turning Climate Crisis Into Jobs, Prosperity, and a Sustainable Future,* provides an in-depth look at exciting opportunities to simultaneously reduce emissions while creating enormous economic opportunities and environmental benefits that would accompany a massive clean-energy infrastructure program. Nonetheless, momentous political and logistical challenges currently stand in the way of implementing such a solution.

Barriers to Climate Protection

Efforts to pass sweeping climate protection legislation in the US Congress have been stymied over the past two decades by the alliance of powerful fossil fuel interests and wealthy corporations described in my earlier book, *Climate Myths: The Campaign Against Climate Science* (2013). America's Climate Security Act of 2007, introduced by Democratic Senator Joseph Lieberman and Republican Senator John Warner, would have capped US carbon dioxide emissions at 2005 levels and reduced them 63 percent by 2050. The fossil fuel industry and the US Chamber of Commerce opposed the bill, however, claiming it would damage the economy. Senators seeking to delay or derail the bill insisted on having all 491 pages read aloud in the Senate, and two Republican senators offered over 150 amendments. Although supported by a majority of senators, the bill then fell short of the 60 votes to override a filibuster and so was killed by Senate Republicans before a final vote could be taken.

Another major cap-and-trade climate bill, the American Clean Energy and Security Act of 2009, introduced by Representative Edward Markey (D-MA) and Representative Henry Waxman (D-CA), would have cut carbon dioxide emissions by 83 percent by 2050. It also would have required US electric utilities to get 20 percent of their power from renewable energy or energy efficiency by 2020. After more than 400 amendments were introduced by House Republicans hoping to delay the bill, and after provisions generous to industry were included—85 percent of the emissions allowances were to be given away free—the weakened and more industry-friendly bill passed the House, only to be defeated in the Senate. The bill's opponents included the US Chamber of Commerce by the National Association of Manufacturers, the American Petroleum Institute, the Heritage Foundation, and the Competitive Enterprise Institute. Although the Congressional Budget Office had found the bill to be deficit-neutral for its first decade, these groups claimed the bill would cause egregious harm to the economy.

Congress's failure to assertively respond to the climate emergency reflects a deeper crisis in American democracy, which is under assault from powerful interests that have consolidated their political power, thanks to a growing concentration of wealth and income since the 1980s.[14] Former US Secretary of Labor Robert Reich and others have documented how Wall Street financiers, wealthy corporations, and superwealthy individuals have garnered virtually all the income gains since the financial crisis of 2008 while ordinary Americans have been left behind. Forty-seven million of those ordinary Americans now depend on food stamps. Others have been left with stagnant wages, high unemployment, and decreased social mobility. They are discontented, and many have grown disillusioned and cynical about government. They are therefore more likely to be receptive to demagogic, antigovernment rhetoric than to government climate protection programs.

The experience with climate legislation in Congress has needlessly delayed measures to protect the climate. The executive branch has tried to take action through "jaw boning" and executive orders but has been largely unable to compensate for Congress's inaction.[f] The climate in fact cannot be effectively protected until the excessive political influence of fossil fuel interests and other large corporations is reduced. The conventional energy industry—oil, coal, natural gas, and electric utility companies—vastly outspends and out-lobbies environmental advocates. In the 2008 elections, it spent about 20 times what environmental advocates spent to influence elections, according to Common Cause. By funneling millions of dollars of campaign contributions to tractable legislators and their Political Action Committees (PACs), the industry gains access to these lawmakers and on occasion even helps draft legislation. All this not only stalls progress on climate protection but also undermines American democracy. PACs offer wealthy individuals

---

[f] Its own energy policies have sent mixed messages. The administration embraced a fossil fuel–friendly "all of the above" energy policy that included an expansion of domestic oil and gas production while it also advocated climate protection and modestly increased support for both renewable energy and energy efficiency.

and corporations convenient conduits for channeling large amounts of cash to parties, candidates, and their media campaigns, often while keeping donors' identities secret. Later, prominent lawmakers who "voted correctly" or others in government who did the lobbyists' bidding are rewarded with plush jobs or consulting contracts.

Removing Obstacles to Climate Protection

Far-reaching campaign finance reform is needed in the United States. Campaigns for public office should be publicly funded to fend off the corrupting influence of large donations. A government untainted by de facto institutionalized bribery would in time induce more people of merit, distinction, and knowledge back into Congress and political life. Were the inordinate influence of wealth reduced, a nascent climate-protection movement could wield more influence in electoral politics, not only supporting candidates for office but fielding its own. More than 40 Green Parties worldwide do so.

A major step toward fairer elections in the United States would be passage of a constitutional amendment to overturn the Supreme Court's 2010 Citizens United v. Federal Election Commission decision in which corporations, associations, and unions were accorded the same rights as individuals to spend as much money as they want on TV commercials and other ads to frame political issues and elections. The court concluded that the contributions in were an exercise of free speech.

To quickly build a broad and knowledgeable constituency for climate protection, the public needs to hear the truth about climate change, and climate science denial needs to be vigorously rebutted. Restoring the Federal Communications Commission's Fairness Doctrine (abolished in 1987) would be a step in this direction.g The doctrine requires broadcasters to provide contrasting views on political issues and also requires that people subject to on-air political attack be given advance notice, when possible, and an opportunity to respond.

What Is to Be Done—Taking Action

What can concerned individuals do now to improve the quality of government and bring climate protection to center stage? "Those with the privilege to know, have the duty to act," Albert Einstein declared. Studying the

---

g   The doctrine was abolished by the FCC in 1987 on the executive order of President Ronald Reagan. Its abolition was followed in the late 1980s and 1990s by an explosion of strident right-wing talk shows that often mock environmental concerns and deride environmental advocates.

issues and talking about them with family and friends as well as with the media and government officials is a good first step. A few people of high moral purpose, however, will not be sufficient to defeat entrenched conventional energy interests. In the twenty-first century, an unaided voice crying in the wilderness—however passionate—is not as powerful as a voice raised on CNN, Fox News, or the BBC.

Unfortunately, the movement for climate protection and clean energy does not have 50 years to put clean energy proponents in high offices and end business-as-usual energy policies. Fortunately, grassroots political activism plus new technology sometimes produces faster results. President Obama himself used grassroots organizing and social media to gain the White House in 2008. Social media was also indispensable to the Arab Spring revolutions that began in Tunisia in 2010. But whereas many climate organizations are already active on the web, their initiatives are often lost in internet cacophony, much of it created by powerful commercial interests.

The mass marketing of forceful climate protection messages to hundreds of millions of people ultimately requires a powerful mass-media network devoted to the planetary environmental emergency. Thought leaders, articulate scientists, plus entertainment and sports celebrities with large followings need to speak out for climate protection. Large radio and TV networks need to provide daily, in-depth coverage of climate and energy news and analysis, along with relevant scientific, political, and economic developments. More of this programming could be available if organized groups demanded it and helped broadcasters recruit audiences for it. Advertisers follow audiences and pay for programming. What if the climate-protection and safe-energy movement had the same radio and TV clout as Rush Limbaugh, Sean Hannity, Bill O'Reilly, or others who cast doubt on climate science, muses environmental broadcaster Betsy Rosenberg. So long as the movement lacks an effective mass-media platform, is it any surprise that the masses are not mobilizing around climate issues? she wonders.[15]

Powerful Reasons for Hope

Even though governments tend to be captured by special interests and resist needed change that challenges the global fossil fuel industry, it is possible to overcome even very powerful minority interest groups and to force bad governments from power. Through years of struggle, Nelson Mandela and the global campaign to end apartheid demonstrated the power of a well-organized and coordinated international boycott to bring down the racist government of South Africa.

Protecting the climate is still possible, but it is something, like apartheid, that needs to be fought for with steadfast determination and implacable will. Wendell Berry said, you can't ask if you're going to win, but if it is right. While it is not a lost cause, there is no magic bullet, no formula for protecting the climate—just long, hard work and a great deal of political organizing by many committed people to generate the pressure that will create change. Thankfully, the 7.2 billion people on Earth do not all need to be convinced of the policies needed to protect the climate. Only a small fraction need to be mobilized to create intense pressure on the people at the apex of power resisting climate protection. Life must become less comfortable and less profitable for them. People can vote against fossil fuels by supporting the right candidates for office and with their dollars, by curtailing reliance on fossil fuel products and by urging others not to invest in fossil fuel companies.

The argument that climate catastrophe can't be prevented or that creating a clean-energy economy is too expensive or will take forever is a disempowering myth fostered by fossil fuel interests. When the world's power brokers, heads of states, and oligarchs feel enough pressure and therefore finally decide the climate should be protected, things can start to happen very quickly.

FIGURE C-1. During a major flood in Mexico in 2009, members of Greenpeace protest the failure of politicians to act to reduce the magnitude and risks of climate change. The yellow sign reads, "Politicians: You Failed. Solve Your Climate Disaster!" Photo courtesy of Marco Ugarte, Associated Press.

FIGURE C-2. Fifty years after 200,000 people turned out for the historic August 28, 1963, March on Washington for Jobs and Freedom at which Martin Luther King Jr. delivered his inspiring "I Have a Dream" speech, crowds once again gathered in 2013 (above) at the Lincoln Memorial to commemorate the 1963 event. They were addressed by President Barack Obama, America's first black president.

Radical change comes from the bottom up. Millions of people do care about the Earth, their children, the future, and the climate. Ordinary people are powerful when deeply committed to a cause. They defeated slavery, guaranteed women the right to vote, fought for civil rights in the United States and South Africa, ended colonialism in India and Africa, and brought down governments in the Middle East and elsewhere.

There is no time to lament the climate predicament or make excuses for inaction. The movement for climate protection needs active, wholehearted support. Millions of people believe we are in a climate crisis. They will stand up if inspired or asked to do so. So take action. Inspire them. And ask their help. Chances are you have more power and influence than you think!

APPENDIXES

Appendix to Chapter 1

ECOLOGICAL EFFECTS OF ICE LOSS AND TEMPERATURE INCREASES ON THE ARCTIC

Higher Arctic temperatures already have had profound ecological effects in the Arctic and can be seen on ice-dependent species like polar bears, Pacific walrus, hooded seals, ringed seals, narwhals, and ivory gulls. According to the Fourth International Polar Year—a $1.5 billion program involving 10,000 scientists from 63 countries—these species are declining in numbers, condition, and birth weight.[1]

Thin, new ice is unable to support the weight of a polar bear, restricting the bears' range to the vanishing fringe of multiyear ice. Without ice on which to hunt seals, polar bears are starving, and a number of them have been observed attacking and eating cubs due to hunger.[2] As the multiyear ice becomes scarcer and farther from shore, more and more polar bears are drowning as they attempt to swim back to land. Scientists from the US Geological Survey have forecast that two-thirds of the world's polar bears and all of those in Alaska will be dead by 2050.[3]

Walruses also depend on ice floes for rest and as safe places to keep their pups while the adults feed in near-shore shallows. As distances between the shore and the thick ice increase, walruses are forced to abandon their young.[4] The walrus population of Alaska, some quarter million animals, was already showing signs of stress a decade ago, before the extreme ice losses of more recent years.[5]

Meanwhile, other climate-related changes are occurring on land in the Arctic. The Arctic fox, for example, is in decline because of range expansion by the competing red fox, and lemming populations critical to the Arctic food chain appear to be in a climate-driven depression that could lead to a population collapse.[6]

Arctic snow cover is also decreasing, bringing profound ecological changes that affect the balance between shrub, tree, and grassland communities, wildlife distribution and numbers, and the cycling of nutrients and CO_2 in Arctic soils.

Caribou are another barometer of Arctic ecological health. A staple food supply of the Inuit peoples, caribou are in Arctic-wide decline. They appear to be suffering both from an increase in biting insect populations as well as from a mismatch in timing between the availability of forage and the needs of pregnant females, due to earlier seasonal plant growth in the Arctic.[7] Clearly many important indicators of Arctic well-being and health are pointing downward.

Appendix A to Chapter 4

THE ROLE OF WATER VAPOR AND OTHER NONCARBON HEAT-TRAPPING GASES IN CLIMATE CHANGE

Although noncarbon heat-trapping gases are eventually destroyed in the atmosphere, while they remain there, they have important warming effects. Nitrogen, roughly four-fifths of the air we breathe, enters the atmosphere from natural sources and human activities (such as fertilizer use and fossil fuel burning) and is cycled through the web of life by plants, animals, microorganisms, and geochemical forces.

From the climate perspective, we need only be concerned with the additional nitrous oxide (N_2O) burden placed on the natural nitrogen cycle by human activity. (High natural background levels of nitrogen, however, may amplify these effects.)[1]

The atmospheric concentration of nitrous oxide is roughly 1,000 times less than carbon dioxide and is therefore measured in parts per billion.[2] However, over a period of 100 years, its per-molecule potential for warming the planet is more than 310 times greater than that of carbon dioxide.[3] Nitrous oxide remains in the atmosphere for an average of 114 years. Fortunately, we are adding far less of it to the atmosphere than carbon dioxide, so it adds a little less than a tenth as much global warming as carbon dioxide.

Methane is produced naturally by the decomposition of organic matter in anaerobic soil and wetlands and by human activities, such as fossil fuel extraction and transportation, landfill use, and livestock raising. Methane stays in the atmosphere for only a dozen years but has a per-molecule warming impact about 84 times greater than carbon dioxide's over a 20-year period and 28 times greater than carbon dioxide's, when averaged over a 100-year period, in part due to methane's longer-term effects on atmospheric chemistry in contributing to the creation of ozone, another heat-trapping gas.[4]

The good news about methane is that, if captured and burned, it is converted to the far less powerful carbon dioxide. The importance of controlling methane tends to be overshadowed by the global concern about carbon dioxide, but University of California researcher Stacey Jackson has pointed out that medium-term pollutants like methane, plus short-lived atmospheric pollutants like soot, which stay airborne for days or weeks, together account for half the human-induced warming over a 20-year period.[5] Reductions in these climate forcing agents can therefore be of great value in bringing about a quick decrease in the aggregate global warming potential of all climate-disturbing emissions over the short-term.

The chlorofluorocarbon class of now-banned, synthetic refrigerants known as halocarbons destroys vast quantities of ozone in the upper atmosphere and exerts a powerful long-lasting heating effect on the Earth. Chloropentafluoroethane (CFC 115), for example, has an atmospheric lifetime of 1,700 years, while carbon tetrafluoride (CF_4) has a lifetime of 50,000 years.

Sulfur hexafluoride, an insulating fluid for heavy electrical equipment, has a long atmospheric lifetime and thousands of times the per-molecule warming effect of carbon dioxide. Fortunately, the quantities released are minuscule compared with those of carbon dioxide.

Direct increases in the amount of water vapor in the atmosphere due to human activities like irrigation are not a major source of climate change. Water molecules have a very short residence time in the atmosphere. Climate alterations, however, do greatly affect the Earth's water cycle, triggering important climate feedbacks.

Water is constantly being cycled between the oceans, freshwater reserves, polar ice caps and glaciers, soil and other subsurface zones, plants, and the atmosphere. The continual cycling is due to natural processes, such as evaporation, condensation, and runoff. Average global temperature provides a measure of the energy driving this water cycle.

In a warmer world, more water evaporates from the oceans and freshwater bodies into the atmosphere, trapping heat and thereby causing a positive (amplifying) climate feedback. It also increases humidity and precipitation.

Another result is an increase in snowfall in polar regions. This produces a negative climate forcing because it tends to slow global warming by increasing reflection from the land's surface, reducing the Earth's absorption of heat. However, the dominant effect of increasing evaporation is amplified warming.

Appendix B to Chapter 4

RESTORING THE ATMOSPHERE

Living plants pull carbon out of the air during photosynthesis and then incorporate the carbon into carbohydrates that form plant tissue, also referred to as *biomass*. When those tissues are burned, they are converted back to carbon dioxide and water, releasing exactly as much carbon dioxide to the atmosphere as the plant initially extracted from the air to make the carbohydrates.

Biomass can thus be combusted in an energy cycle that adds no net carbon to the air, but only if: (a) no fossil fuels were used in its preparation or transportation; or (b) a greater or equal compensatory amount of biomass is grown but is deliberately not burned and also prevented from oxidation (combining with oxygen to form carbon dioxide).

Carbon-containing biomass might be stored in the soil as a relatively stabilized char,[a] so it doesn't burn or reenter the atmosphere. This whole process, if done on a large scale, could somewhat reduce the concentration of atmospheric carbon but would not alone be sufficient to stabilize the climate. Large-scale biomass cultivation nonetheless could be part of a global strategy for scavenging carbon from the atmosphere.

Unfortunately, serious consideration of this "biosequestration" option—which basically exploits natural processes to protect the climate—is not yet a prominent part of the national climate dialogue. It would be useful if policymakers requested (and provided incentives for) scientific research organizations to intensively explore these options, and if legislators proposed specific legislation to encourage the adoption of the most cost-effective ways of enhancing the areal extent and performance of natural carbon-removal systems.

---

[a] Char is a charcoal-like substance produced when biomass is heated in a low-oxygen atmosphere. It can be coproduced when electricity is produced from biomass by gasification and pyrolysis. When incorporated into the soil, biomass increases soil carbon content, and improves soil fertility, tilth, and nutrient retention while reducing nitrous oxide emissions. (See Robert Crowe, "Could Biomass Technology Help Commercialize Biochar?" October 31, 2011, renewableenergyworld.com.)

Fast-growing algae, aquatic weeds, and cover crops can be grown specifically to capture carbon dioxide from the air for incorporation into the soil. Perennial prairie grasses can also be planted with the same goal; they build large, deep, carbon-rich root systems that, upon the plants' death, gradually become assimilated into the soil. During the plants' lifetime, their aboveground structures provide habitat for hundreds of native prairie species.[b]

Biomass comprised of pruned tree branches and orchard and logging waste could also be partially burned to charcoal and then mixed with soil to enrich it and slow the carbon's return to the atmosphere. (Carbon in tree trunks, wood furniture, or buildings can also remain apart from the atmosphere for long periods of time if protected from combustion.)

The use of carbon-free energy technologies (like solar, wind, geothermal, hydropower, tidal, and wave energy) is far preferable to any process, including biomass combustion, in which a carbon-based fuel must be burnt.[c] Biomass, however, still has a valuable role to play in today's energy economy, and is widely used, especially in developing countries, for fuel and fertilizer.

Even in developed economies, biomass is especially valuable as a lower-carbon, and vastly preferable alternative to fossil fuel for producing liquid fuels for aircraft and other machinery for which electrical drive motors will likely be unavailable.

---

[b] We cannot assume, however, that a forest will be impervious to fire or to an insect attack that might kill it and thereby release much of its carbon to the atmosphere again.

[c] Biomass energy sources are sometimes represented as zero net carbon emissions technologies. Zero net carbon emissions means that the emissions released in combustion are equal to the carbon extracted from the atmosphere by the plant during its lifetime. Often what may appear on superficial analysis to be a zero net carbon balance, however, turns out to be a positive source, introducing carbon into the air because of fossil fuel inputs consumed during some phase of the biomass' agricultural production or transportation.

Appendix C to Chapter 4

A BRIEF NOTE ON THE DISPLACEMENT OF ECOSYSTEMS

Biomes are ecosystems dominated by characteristic vegetation—forest, wetland, prairie, savanna, tundra, and so forth. The boundaries of these ecosystems shift in response to new climatic conditions and their side effects: changes in interspecies competition and predation. Some biomes will be able to adapt to these changes, especially if climate change is mild to moderate; others will be displaced. But biomes do not move as a unit. Individual species within biomes adapt or move at their own rates, creating new and sometimes novel assemblages of species.

The current rapid pace of climate change will require that many biomes migrate ten times faster than their historical migration rates. Few will be able to meet this challenge. Biomes unable to adapt will be changed or damaged and will eventually die, releasing stored carbon to the atmosphere. This amplifying feedback will warm the atmosphere and worsen global warming. True, lost ecosystems, such as northern forests, would eventually be replaced by a new biome, such as tundra, better suited to the new climatic conditions. Before that occurs, however, the atmosphere would have received another large dose of carbon dioxide as the previous biome died and decomposed.

Appendix to Chapter 8

ECONOMIC OPTIMISM

Some economists and agricultural experts believe that future economic activity, slower population growth, or a hoped-for steep reduction in future global emissions (enforced by a comprehensive global climate treaty) will mute or slow the effects of global climate change. I hope they are correct, but am skeptical that economic progress will neutralize the impact of climate change or that a climate treaty is about to trigger a steep decline in emissions.

Those who are still not too deeply worried about climate change today despite all the warning signs include some scientists and economists who place great stock in the presumed benefits of climate change and the notion that even if these benefits won't actually outweigh the costs, they will at least mitigate them into relative insignificance. Their arguments will now be summarized.

Climate Change and Agriculture

Regarding the impact of climate change on agriculture, optimists correctly note that the growing season will increase in some temperate lands, such as Russia. Indeed, moderate warming will expand the area of land suitable for cropping in some temperate areas. And meanwhile, all else being equal, higher levels of atmospheric CO_2 will increase the yields of certain crops (such as rice, soybeans, and wheat)—provided that a number of "big if's" happen:

1. enough water and nutrients are available;
2. the warming remains conveniently moderate;
3. heat waves, hailstorms, and other forms of extreme weather, including droughts, don't destroy the crops.

Nature, however, provides no such guarantees, as in the fierce 2010 Russian drought when 55,000 people died as a result of the abnormally hot weather and ensuing firestorms. Moreover, not only crop yields but crop quality will fall steeply once the mercury exceeds a crop's optimum temperature.[1]

Grape yields, for example, would decline in California's wine country, along with five of the state's other major perennial crops by 2050 under climate change scenarios now forecast.[2]

Some analysts, however, remain doggedly optimistic about our ability to adapt to climate change, because they believe that socioeconomic progress will be so great over the next 100 years that it will basically outweigh climate change's ill effects. They rely on projections of relentless, robust economic growth and generally rising incomes in a world of swelling population. They posit, for example, that although rice and sugar prices may increase by as much as 80 percent,[3] most of the world's people will be better off and therefore will find ways to cope with a less stable, less secure food supply.[4] They ignore projections that future climate change may cost the world tens of trillions of dollars.[5]

They further speculate that freer world trade and improved transportation systems will make food more available to countries with projected food shortages. They are unfazed that world population might approach 14 billion by 2100, even though today's world of 7 billion is already unable or unwilling to care for poor nations' needs.

The optimists also assume that Middle Eastern economies built on oil and gas will flourish indefinitely. They ignore the coming day of reckoning when the inevitable long-term global trend away from fossil fuel—and the eventual depletion of Middle Eastern oil fields—dampens the region's economic fortunes.

Another optimistic scenario goes like this. First, population growth will begin to gradually level off midcentury as more prosperous people decide to have fewer children. Secondly, investments in irrigation will make it possible to increase water supply in arid lands to feed nine billion people. Then, perhaps some genetically modified crops, inexpensive fertilizers, and more powerful pesticides will insure food security.

The rosy projections, however, neglect the effects of sea-level rise on agricultural production along with the impacts of extreme climate events, and the possibility of a crash in ocean fish production. Notably, they also don't factor in possible competition between food production and the large-scale production of bioenergy crops. Nor do they consider the possible loss of crop pollinators from multiple causes,[6] nor the continuing widespread desertification of land that in some countries is already substantial enough to negate economic growth even today.[7]

Essentially, these scholars are projecting that economic growth can go on compensating for abuse of natural resource systems, apparently indefinitely, and that except for the very poor of sub-Saharan Africa and South Asia, the developed world can expect to enjoy rising living standards and excellent food despite chronic environmental abuse.[8] They believe that even in an increasingly globalized world, the prosperous will somehow be able to continue to live in glorious detachment, untouched by the coming suffering. That scenario is not very likely.

Irrevocable Damage

It is more likely that if we do permit 9 or 10°F of warming by 2100—a consequence of the emissions trajectory we're now on—then something very like the year 2100 global scenario described in chapter 1 will indeed occur. If that happens, our children and grandchildren will find themselves in a hungrier, thirstier, harsher, unhealthier, more crowded, and far more unstable world.

Appendix to Chapter 10

EXTINCTIONS

The Paradoxical Antiquity and Fragility of Life

Extinctions in nature, apart from those provoked by humans, generally tend to occur slowly; that is, at rates on the order of one to perhaps ten a year. Of course, mass extinctions have occurred at times in the planet's history in response to violent natural events, such as earthquakes and volcanic eruptions, and in response to abrupt natural climate change. A mass extinction was even brought on by the collision of a large asteroid with Earth 65 million years ago at the end of the Cretaceous period. Although the causes of past mass extinctions were natural, the Earth paid a penalty in terms of its species richness and required millions of years to recover from these cataclysmic events.

The differences between the normal slow pace of extinctions for most of the past 600 million years and modern extinctions are twofold. First, today's extinction rate is so high—and climbing all the time—that the pattern of species loss resembles a disastrous "ecospasm" rather than the slow disappearance of species whose loss nature replenishes through the constant occurrence of new mutations, a small number of which survive to become new species.

The second major difference between the normal extinctions of the past and the cascade of extinctions today is that the bulk of current extinctions are triggered by human activity. If the elevated pace of extinction is allowed to continue, the Earth will be left a less habitable, more ecologically impoverished, less comfortable place. The multimillion-year recovery cycle would be so slow that humans as a species might not even be around to see it.

Every species has an ancient lineage extending back in time to the earliest life. Thus when we cause a species to go extinct, we are discarding the high point of 3.6 billion years of evolution, which transformed the first species of archaebacteria into the enormous variety of species on Earth today. As we now drive more and more species faster and faster to extinction, we are destroying complex organisms far faster than evolutionary biologists, geneticists, ecologists, and biochemists can study them. This could be called "death without documentation."

Biologists using rigorous mathematical analyses tell us that most of the Earth's species—up to 86 percent of all land species and 91 percent of all sea species—are still undiscovered.[1] We are doubtless destroying species of whose existence we are not even aware as we destroy their habitats. This could be called "death before discovery."

Of the species that are known, only a small fraction have ever been thoroughly studied by scientists. Who knows what wonderful secrets of nature we aren't learning when a species of life is erased? When thirty amphibian species were wiped out recently in a Panama forest, five of the species lost were new to science.[2]

As E. O. Wilson eloquently writes, even obscure life forms are "a masterpiece of evolution, exquisitely well adapted to the niches of the natural environment in which [they] occur. The surviving species around us are thousands to millions of years old. Their genes, having been tested each generation in the crucible of natural selection, are codes written by countless episodes of birth and death. Their careless erasure is a tragedy that will haunt human memory forever."[3]

How rapidly are we driving these fellow creatures to extinction, and why should an increase in this rate due to climate change be cause for alarm? Let's look at this question in the context of past extinctions and the natural rate at which extinctions have occurred.

Current Extinction Rate

As noted earlier, the natural background rate of extinction averaged over time is thought to be quite low—on the order of one species extinction a year for every 10 million species on Earth, although some scientists believe it to be up to 10 times that number. The losses that are occurring today, however, may be 100 times or so that natural background rate. (Some scientists believe it may be 1,000 to 10,000 times the natural rate.) By any of these estimates, the destruction of species is so rapid and so extensive that we can now refer to it as one of only six known "megaextinction events" that occurred over a period of time spanning hundreds of millions of years.[4] While the estimates of extinction rates vary from 0.001 percent to 0.1 percent extinctions per year, we are still uncertain how many species of life exist on Earth. If there are indeed 10 million species, then we would be losing 100 to 10,000 species a year.

In 2011, researchers at Dalhousie University predicted that there are likely to be 9.7 million species on Earth, including microorganisms.[5] (Some scientists have proposed that the number of species discovered is only 10 percent of all existing species.)ᵃ There are more species of insects on Earth than any other class of organisms, since some 900,000 have been documented.[6] E. O. Wilson predicts that there may be as many as 10 million species of insects alone.[7] There are at least 1.5 to 1.8 million species of life on Earth, and are likely to be as many as 100 million species or more that have not yet been discovered if the millions of species of bacteria, viruses, fungi, and slimes are included, he writes.[8]

In contrast to the projections based on ecological theory that 0.001 percent to 0.1 percent of all species are going extinct each year, extinction estimates based on known rates of tropical forest destruction and tropical forest biodiversity suggest that perhaps 35 species a day are going extinct from tropical deforestation. This would put the number of species extinctions from that cause alone at more than 12,000 per year. Some current estimates suggest that the loss may be 20,000 to 30,000 species per year. Based on the assumption that tropical deforestation has been roughly as severe for the past 40 years as in recent years, this would imply that a million species have already been driven to extinction in the tropics.

Extensive as this damage to nature is, it understates the severity of its impact because of the losses of populations that don't result in the extinction of the species, and because for every species that is eliminated, the abundance of countless other species is reduced to remnants barely able to perform their ecosystem functions. For example, the United Nations Environment Programme's World Conservation Monitoring Centre and the Secretariat of the International Convention on Biodiversity noted recently, "The abundance of vertebrate species, based on assessed populations, fell by nearly one-third on average between 1970 and 2006, and continues to fall globally, with especially severe declines in the tropics and among freshwater species." Such a dramatic drop in population of vertebrate species in less than 40 years should be considered sufficient warning and cause for extreme global alarm—front-page news, to say the least. It has not, however, been sufficient to capture much public attention, nor to dispel global complacency at the highest levels of government on biodiversity and climate issues. (Sports, Hollywood movies, TV, politics, cultural issues, and the slightest ups and downs of the economy get vastly more attention.)[9]

ᵃ To infer the number of species actually on Earth, E. O. Wilson assumes that perhaps fewer than 10 percent of all species are known to science. Thus, if we have 1.65 million known species, he would infer that there might be some 16 million species on Earth.

Some might wish to trivialize the extinction threat by downplaying the importance of tiny, even microscopic organisms. If the majority of extinctions prove to be tiny organisms, are they less significant? It turns out that "undersized" does not mean of less importance. While insects are small, they are huge in number, populating the Earth by the trillions. They and other small and microscopic life are the real caretakers of the natural world, breaking down dead matter, recycling nutrients, forming soil, and creating conditions that support higher forms of life, including humans. Without these caretakers, life on the planet would stumble, since flowering plants would lack pollinators and the soil would lack ants and termites to bring up nutrients and aerate the earth.[b] Insects even work directly for us: managed colonies of honeybees are trucked around the US from Maine to California to pollinate commercial crops. "About one mouthful in three in our diet directly or indirectly benefits from honey bee pollination," according to the US Agricultural Research Service.[10]

The Costs and Benefits of Destroying Biodiversity

A project known as The Economics of Ecosystems and Biodiversity works to estimate the value of services that nature provides us. The group has valued the annual loss of forests at $2 to 5 trillion.[11] The value of the services rendered by the totality of all healthy ecosystems has been estimated at $30 trillion.[12] Do even such enormous estimates make sense? It seems more plausible that the value of all healthy ecosystems would be infinite, since we would eventually die without them. However, if the number is a meaningful estimate of marketable services from those ecosystems, it adds another powerful argument in favor of preventing further climate deterioration.

E. O. Wilson contrasts the enormous benefits of biodiversity protection with much more modest costs of spending a few tens of billions of dollars to protect Earth's biodiversity "hot spots" where large concentrations of the Earth's species are found. "Conserving biodiversity," Wilson wrote, "is the best economic deal humanity has ever had placed before it since the invention of agriculture."[13] If climate change continues, however, biodiversity will be lost even if protected areas are set aside, for climate change knows no boundaries.

---

b Ants and termites aerate the soil with their tunnels and permit better water infiltration. Termites and ants also bring up mineral-rich particles of subsoil to the surface in their mounds, fertilizing the soil and causing plants and organisms that depend on them in their vicinity to flourish. They process dead and decaying organisms, breaking down complex molecules into simpler compounds more easily assimilated by plants. Colonization of soil by termites has such a dramatic effect on ecosystem productivity that termite colonies can be spotted from space by satellite, according to E. O. Wilson in *The Creation* (see endnotes to this chapter).

Human Population Growth and Resource Demands

From a historical perspective, the speed with which humans have multiplied is truly astonishing and has implications for future extinctions. Anatomically modern humans only evolved 150,000 to 200,000 years ago. Then roughly 74,000 years ago, during a period of rapid climate change, we almost went extinct; the world's human population fell to only about 10,000 adults of reproductive age.[14] At the dawn of agriculture about 12,000 years ago, global population was still sparse. Outside of the equatorial zone, in what is now all of Europe, the human population numbered only about 30,000 individuals.[15] Even when modern industrial technology began 250 years ago,[16] the world still had fewer than 800 million people.

If the age of the Earth were represented as a 24-hour timeline, all recorded human history would be less than a second long, and the industrial age would be a vanishingly small fraction of a second. But in that blink of an eye on an evolutionary time scale, the world's population exploded, increasing almost ninefold. Just during the twentieth century, human population tripled,[c] and since 1987, we have been adding a billion more people every dozen years. To imagine that these staggering growth rates in population, resource use, and economic activity are sustainable or compatible with a stable climate is mistaken. Nothing on Earth can continue increasing exponentially without eventually collapsing. As humans continue degrading the Earth's natural resource systems while changing the climate, the number of species extinctions will ultimately soar into the millions. Thus, if present land- and energy-use practices and population growth rates are not radically changed in time, we will produce an extinction catastrophe beyond anything ever seen by human beings.

Recommended Resources on Biodiversity Protection

For additional information on what needs to be done to protect biodiversity, see *Agenda 21*, "The Conservation of Biological Diversity,"[17] and other publications of the United Nations Environment Programme, which has a vast list of resources available, and the studies of the International Union for the Conservation of Nature.[18] As the public announcement of *Global Biodiversity Outlook 3* points out, "For a fraction of the money summoned up instantly by the world's governments in 2008–2009 to avoid economic meltdown, we can avoid a much more serious and fundamental breakdown in the Earth's life support systems."[19]

---

[c] While it took humans about 12,000 years to reach an initial global population of one billion people, most recently, a billion people were added to the Earth in only 12 years of the twenty-first century, from 2000 to 2012.

Appendix to Chapter 11

OCEANIC CARBON STORAGE AND OTHER PROCESSES

How the Oceans Store Carbon

Overview. The solubility of carbon dioxide (CO_2) and related compounds known as carbonates in the ocean is affected by temperature and other factors. As the ocean gets warmer, for example, CO_2 solubility diminishes. So global warming progressively reduces the ocean's ability to absorb and retain carbon. That capacity is actually a net result of three separate mechanisms known as *the carbon solubility pump, the biological carbon pump*, and *the carbonate counter-pump*. These three pumps all cycle carbon within the ocean and between the ocean, the atmosphere, and the ocean's bottom sediments. Their operation determines the ocean's ability to absorb excess carbon discharged into the atmosphere by human activity.

In reality, both the biological carbon pump and the carbonate counter-pump are intrinsically biological in nature, with the former cycling organic carbon and the later cycling inorganic carbon as calcium carbonate, which is present in seawater and is incorporated in the shells of marine organisms. When these organisms die and drift down into the deep ocean, they dissolve and their carbonate upwells. It then becomes available again to both buffer CO_2 dissolved in the ocean and to be reformed into the shells of other marine organisms.

The Carbon Solubility Pump. About a third of the CO_2 dissolved in the ocean gets into the ocean across the sea surface–atmosphere boundary. This carbon solubility pathway is therefore an important aspect of the oceanic carbon sink.[1] However, when global warming raises sea surface temperatures and reduces CO_2 solubility, the fraction of newly released atmospheric CO_2 "pumped" into the ocean across this boundary declines.[a] Direct surface

---

[a] This decrease in CO_2 solubility is accentuated by the increased acidity just described, caused by an increase in dissolved fossil fuel CO_2.

warming of the ocean will reduce this CO_2 uptake mechanism by an estimated 9 to 15 percent by the year 2100. Simultaneous changes in ocean
circulation induced by global heating may reduce the effectiveness of this
carbon solubility pump by another 3 to 20 percent.[2]

Most of the remaining two-thirds of the CO_2 dissolved in the ocean that
does not diffuse across the sea surface–atmosphere boundary enters the water
via the organic biological carbon pump from plant tissue and other dissolved
substances produced by plankton in ocean surface waters,[3] including CO_2
emitted by plants during respiration.

The Biological Carbon Pump. The oceans not only cover 70 percent
of the planet, but are a living primordial soup. Their sunlit surface waters
(known as the photic zone) are filled with tiny, floating phytoplankton—
microscopic plants that are responsible for half the oxygen produced on the
planet at any given time.[4] This is *gross* oxygen production. The phytolankton
are also responsible for consuming almost an identical amount of oxygen.
That's because plankton, like other plants, breathe 24 hours a day and, at
night, when the sun isn't shining and no photosynthesis occurs, they are net
consumers of oxygen. Photosynthesis and respiration are approximately in
balance. Therefore, the *net* oxygen production of the marine phytoplankton
(photosynthetic production minus oxygen respired) is less than half a percent
of the total phytoplankton oxygen production. The net imbalance occurs
because, when phytoplankton die, although most are decomposed by bacteria
that consume oxygen in the process and convert the carbon in the phytoplankton back to CO_2 by combining the carbon with oxygen, a small fraction
of the phytoplankton reach the bottom of the ocean and are incorporated
into sediments where they and their carbon stay for a very long time. This
sedimented carbon does not recombine with oxygen again for eons. Eventually it finds its way into terrestrial rock where it is ultimately oxidized during
rock weathering. It thus returns on time scales of millions of years to the stock
of atmospheric carbon dioxide.

The Carbon Cycle. Although the rate of oxygen that accumulates in
the atmosphere from net photosynthetic production on land and the ocean is
small, over geologic time, it has created all the oxygen in the atmosphere. The
oceans, just like terrestrial vegetation, are part of the planetary carbon cycle
and are also an integral part of our planet's long-term oxygen production and
life support system. So whereas we often hear the tropical rainforest described

as "the lungs of the planet," the oceans are even more like "a planet-sized set of lungs that inhale and exhale CO_2"[5] than are the Earth's forests.

Ocean waters store vast amounts of dissolved carbon in their surface waters (900 billion metric tons) and 37,000 billion metric tons in the intermediate and deep ocean.[b] Even the surface ocean thus stores more carbon than all land vegetation (300 to 500 billion metric tons). The surface ocean even has about 70 metric tons more carbon than does today's industrial atmosphere (829 billion metric tons). Prior to industrialization, the atmosphere had only about 590 metric tons of carbon. Thus, the surface ocean in preindustrial times had an even greater relative mass of carbon compared with the atmosphere than today, since during the industrial era, humans have added 240 billion metric tons (± 10 metric tons) to the atmosphere. If someday we are able to safely and cost-effectively enhance the rate at which carbon is taken up and sequestered in the oceans while we curtail new atmosphere emissions, we might be able to gradually draw down the amount of carbon dioxide in the atmosphere, slowing global heating.

One process researchers are in early stages of exploring to enhance oceanic uptake of carbon is the microbial conversion of ordinary dissolved carbon in the ocean to refractory carbon. Ordinary carbon is readily consumed by bacteria and returned to the atmosphere as respired carbon dioxide. But some carbon is rendered relatively inedible to the microbes that normally break it down. These refractory forms of carbon—thousands of compounds whose life cycles are not well understood—are able to circulate in the ocean virtually unmolested by bacteria for thousands of years.[6]

The bidirectional cycling of carbon between the atmosphere and the ocean occurs across the boundary layer between the atmosphere and the ocean. Carbon flows out of the ocean as CO_2, and simultaneously, CO_2 from the atmosphere dissolves in the ocean surface. There it is taken up by plankton and used as a nutrient during photosynthesis. In this process, plants and some bacteria absorb solar energy and, through a series of chemical reactions that combine CO_2 and water, store this energy in their tissue as carbohydrate and release oxygen as a reaction product.[7] These organisms thus act as a biological pump, removing carbon from the sea while building new tissue. Other carbon flows occur in the ocean between the surface ocean and the intermediate and deep ocean, and between the deep ocean and the ocean's bottom sediments.

---

[b] These estimates and the rest of the data in this paragraph are from the draft version of the IPCC's *Fifth Assessment Report*.

Today there is a net positive flow of carbon from the atmosphere to the ocean of about 2.3 billion metric tons per year.

Just as we assimilate food and exhale CO_2, most phytoplankton are eaten, and their carbon is soon returned to the water or atmosphere as CO_2 by their tiny predators. On balance, the oceans thus only retain about 2 percent of the carbon that enters into the sea across the air-ocean boundary. (A small amount of carbon [0.9 metric tons] also flows into the ocean from the land through surface runoff.) If the current imbalance in carbon flows between the atmosphere and the oceans continues for a long period of time, however, this small annual carbon removal process will eventually add up to a vast oceanic carbon hoard. How is it all stored and kept from immediately returning to the atmosphere? As noted, a crucial part of the story is that dead phytoplankton sink and their tiny calcium carbonate exoskeletons drift downward through the ocean water column toward the ocean floor. The calcium carbonate that makes it all the way to the bottom without being dissolved or consumed becomes incorporated in the deep ocean bottom.[8] This phase of the carbon cycle is augmented by the action of the microbial pump and by the carbonate counter-pump.

The Microbial Pump. A very important microbial carbon pump also operates in the ocean. Through a strange sequence of microbial events, the microbial pump transforms bioavailable organic dissolved carbon into the much more stable and complex refractory dissolved carbon compounds mentioned earlier that can circulate in the ocean for thousands of years without reemerging in the atmosphere as carbon dioxide. Fully 95 percent of all the dissolved carbon in the ocean is now thought to be in this relatively inert form that is hard for bacteria to digest.[9]

A fascinating "who knew" part of the story concerns a relatively newly discovered class of photosynthetic bacteria known as AAPBs (aerobic, anoxygenic, phototropic bacteria) that live in the surface ocean.[10] When these bacteria absorb CO_2, they produce the refractory carbon compounds. They are then attacked by viruses known as bacteriophages that rip open the bacteria, releasing the refractory carbon to the open water. Perhaps if researchers ever discover a means to increase the rate at which this process occurs, that might reduce the amount of carbon released through biological pathways from the ocean back to the atmosphere, or increase the rate at which additional carbon can be dissolved in the surface ocean by reducing the amount of chemically active dissolved carbon in circulation, thus enhancing the ocean as a carbon sink.

An unwanted consequence of ocean warming could be a shift in the ocean's phytoplankton composition from large diatoms to smaller plankton and bacteria because of differing responses by ocean phytoplankton to warming, acidity, light, vertical circulation, and a diminished nutrient supply. Reducing plankton size is a bit like adding an additional lower level to the ocean's food pyramid. Because nutritional energy is lost at every step in a biological food pyramid, the addition of another step would result in a large decrease in the efficiency of energy transfer throughout the ocean's food web and would cause a drop in fish production at the top of the food chain.[c]

The Carbonate Counter-Pump. Working against the biological pump is the carbonate counter-pump. This carbonate pump removes CO_2 from the water in a roundabout way. When marine organisms form their calcium carbonate shells, they combine two bicarbonate ions. The reaction byproducts are a carbonate ion and a molecule of dissolved CO_2. The dissolved CO_2 raises the partial pressure of CO_2 in the seawater and thus forces more CO_2 into the atmosphere.[d] But there is more to the carbonate counter-pump story. When the calcium carbonate–containing organisms die, they drift downward in the water column. As the water becomes colder and the pressure rises, much of the carbon in these calcium carbonate shells is redissolved in the deep ocean where it is likely to stay for hundreds of years. Eventually, albeit on these very long time scales, it recirculates from the deep ocean back to the ocean's upper layers. As noted previously, a small but significant fraction of the carbonaceous detritus does escape the remineralization process and is eventually incorporated in seafloor sediments.

The Solubility Pump. As mentioned in the overview, CO_2 is less soluble in warm water than in cool water, so as the ocean's surface waters heat up, less CO_2 from the atmosphere dissolves. More CO_2 is then left in the atmosphere to block the transmission of heat from the Earth back to outer space and thus, the Earth is warmed. The more we overload the atmosphere with CO_2, however, the greater is CO_2's partial pressure—the pressure it would exert if alone in a similar volume of air. (A gas will flow from a region of higher partial

---

[c] As energy is transmitted from the primary producers (photosynthesizing plants at the base of a food chain) to higher levels of the system, there is a significant energy loss at each stage of the process. So if additional steps are introduced by the downsizing of organisms at the bottom of the web, less energy is transmitted through the entire system to support higher-level commercially harvested predatory organisms that we would recognize as dinner.

[d] See the next section, "The Solubility Pump," for more on partial pressures.

pressure to one of lower partial pressure.) Therefore, all else being equal, the higher the partial pressure of CO_2 in the atmosphere, the more CO_2 is forced by the solubility pump process across the atmosphere–ocean boundary to dissolve in the ocean surface water. The higher the concentration of CO_2 dissolved in the ocean, however, the more the ocean resists and the greater the partial pressure gradient needed to drive each additional molecule of CO_2 across the ocean surface boundary and into solution. Thus, the ocean's carbon absorption capacity is progressively lessened. Over the long term, this reduces the ocean's capacity as a carbon sink.

Even more important than the rise in the partial pressure of CO_2 within the ocean is the depletion of the carbonate supply in the surface waters of the ocean. When CO_2 dissolves in the ocean, a series of reactions occur that result in the production of hydrogen ions in the water, causing the ocean to become more acidic. Marine life is harmed in various ways at the same time as the ocean's ability to absorb additional carbon dioxide is depleted. However, carbonate, a chemically active ion of carbon (CO_3^{-2}), neutralizes the acidifying effects of dissolved CO_2 in a process known as buffering that not only reduces acidification but enables more CO_2 to dissolve in the ocean. The removal of carbonate by reactions with CO_2—in effect a saturation of the carbonate buffering capability of the ocean—then reduces the solubility of CO_2 in seawater. (See "The Ocean Acidification Process" box, below.)

=====

THE OCEAN ACIDIFICATION PROCESS

The detailed discussion of ocean acidification in this box is based on the work of Ulf Riebesell and his colleagues[11] and The Royal Society of London's 2005 ocean acidification report.[12]

Acidity is usually measured in pH units, a logarithmic scale from 0 to 14 based on the number of hydrogen ions present in solution. According to a recent National Research Council study, the acidity has risen by 0.1 pH units from 8.2 to 8.1, representing a change of 26 percent in the concentration of hydrogen ions in seawater.[13] This increase, the council noted, is expected to double or triple by the end of the century, reflecting a rate of change unprecedented for 800,000 years.[14]

When the ocean takes up CO_2, the CO_2 forms carbonic acid (H_2CO_3), which ultimately dissociates (breaks apart) into hydrogen ions (H^+), bicarbonate (HCO_3^-) ions, and carbonate (CO_3^{-2}) ions. Ions are simply molecules or atoms with a net electrical charge. The first step of the process that occurs when CO_2 dissolves in the ocean to produce carbonic acid is represented as:

$[H_2O] + [CO_2] => [H_2CO_3]$ (Equation 1)

The carbonic acid itself dissociates into hydrogen and bicarbonate:

$[H_2CO_3] => [H^+] + [HCO_3^-]$ (Equation 2)

Some of the newly formed free hydrogen ion now combines with carbonate:

$[H^+] + [CO_3^{-2}] => [HCO_3^-]$ (Equation 3)

This chain of chemical reactions culminates both in an increase in the acidity of seawater and a decrease in its ability to assimilate additional CO_2.[e] Because there is interaction among all the reaction products as they seek chemical equilibrium (and reactions can proceed in both directions), it is the final net reaction that is of practical interest.

Seawater is naturally alkaline—carbonate ions are already present in it in addition to the bicarbonate indirectly produced when CO_2 dissolves. An end result of all this is that the dissolution of more and more CO_2 in seawater liberates hydrogen ions that combine with available carbonate to form additional bicarbonate. The carbonate thus acts as a buffer, taking hydrogen ions out of solution and thereby reducing the ocean's acidity.

However, as more and more carbonate is, on balance, used up in this process, the ocean's finite buffering capacity is gradually reduced along with the solubility of additional CO_2 in the ocean. The loss of carbonate and its buffering ability then allows the ocean to become increasingly more acidic for each unit of CO_2 that does dissolve, putting the calcium-based shells and the skeletons of marine organisms at risk of dissolving.

Saturating the Ocean's Carbon Sink Capacity. Although the *rate* of CO_2 absorption in the ocean slows as the ocean surface waters become more and more heavily saturated with CO_2, the total *amount* of CO_2 dissolved in the ocean continues increasing as human CO_2 emissions continue rising. Meanwhile, the ocean's reduced ability to absorb CO_2 creates a positive feedback effect.

Now a higher proportion of any CO_2 added to the atmosphere remains there instead of being stored in the ocean. That means that each additional

---

e Whereas acidity is a universal measure of the presence of hydrogen ions, alkalinity is used in ocean chemistry to mean the amount of carbonate and bicarbonate ions present. In the context of the carbonate chemistry of seawater, alkalinity is thus not the opposite of acidity but means the presence of the two basic ions, carbonate (CO_3^{-2}) and bicarbonate (HCO_3^-).

release of CO_2 causes ever greater planetary warming. In addition to soaking up carbon, the ocean also absorbs 90 percent of all excess heat collecting in the Earth's climate system.[15] Thus, the ocean's surface temperature rises in response to the higher atmospheric temperatures, and the ocean's ability to take up carbon is thus not only reduced by saturation of its buffering ability but also by the increase in temperature. (See "The Ocean Acidification Process" box, page 257.) Moreover, these surface temperature gains are eventually transmitted over long periods of time to the ocean depths. Ultimately, however, all the stored heat within the ocean will become more uniformly distributed and will subsequently reach a new equilibrium with the atmosphere, guaranteeing that the rise in global surface temperature now occurring will affect the planet's climate for thousands of years to come.

The Rising Seas

With more than a billion people already living in coastal areas—a number that could grow to 5.2 billion by 2080 (according to some IPCC scenarios)[16]—sea-level rise will have pervasive global effects and is one of the most consequential but most often underestimated impacts of global warming. At greatest imminent risk from the higher tides and higher storm surges that sea-level rise will bring are countries in Asia and Africa; residents of the world's large river deltas; and the small, low-lying Pacific island states that are likely to be flooded later in this century. The higher storm surges will also be occurring in conjunction with the more intense storms expected in a hotter world.

According to a suite of economic growth projections made by the IPCC using very modest (and increasingly questionable) assumptions for sea-level rise (8 to 12 inches), more than 300 million people will be exposed to storm-surge flooding by 2080. "Increased storm intensity would exacerbate these impacts, as would larger rises in sea level, including [rises] due to human-induced subsidence," wrote the IPCC.[17] The IPCC projects that without flood defense upgrades, more than 100 million people worldwide will be subjected to coastal flooding from storm surges every year, a 1,000 percent increase just since the start of the millennium.[18] As we pointed out earlier, however, seas will continue rising for at least hundreds of years beyond this century.

In addition to the melting of land ice and the thermal expansion of seawater, some additional factors also affect the height of the land relative to the oceans. Extensive water withdrawal from the ground, for example, causes land to sink. This process occurs on a fairly rapid time scale. By contrast, the melting of glacial ice over very long periods of time causes land to rebound from the lifting of the great weight of ice, and thus to rise. Another long-term process, the tectonic plunging of crustal plates of the Earth beneath the continents in certain locations, also causes the land slowly to rise. Ocean currents and gravitational forces, too, play a role in local and regional sea levels.

Average global sea level is currently rising at about 13 inches per century, and the rate of the rise has doubled in the past decade compared to the rate of the previous half century.[19] Whereas the *Fourth Assessment Report* of the IPCC projected that sea level would likely rise by less than two feet by the end of this century (the blink of an eye on a geological time scale), current research indicates that sea-level rise could easily exceed 3.28 feet by then, with even higher levels of up to two yards possible.[20]

Evidence collected by coring coastal Irish Sea sediments reveals that when seas began rising at the end of the last glacial maximum of 19,000 years ago, they rose about 33 to 49 feet over 100 to 500 years.[21] The elevated sea-level rise estimates of the *Fourth Assessment Report* of the IPCC for 2100 are a result of new knowledge about polar ice sheet dynamics and the mechanisms that lead to their loss. Whereas earlier climate models assumed fairly simple relationships between rising global average temperatures and ice sheet melting, more recent studies (informed by actual observations of melting ice sheets) have shown that ice sheet melting can be much accelerated by the infiltration of meltwater to the base of glaciers through vertical cracks known as moulins. When the water reaches the base of the ice, it lubricates the land-ice interface and speeds up the movement of the ice mass toward the sea. (This is a problem in Greenland, but not in Antarctica because air temperatures there are too cold.) As recently as 40 years ago, moulins were rare in Greenland. Now, according to American polar expert Robert Correll, "They are like rivers 10 or 15 meters in diameter and there are thousands of them."[22]

Ice Loss on Greenland. For a variety of reasons, many of Greenland's largest glaciers are accelerating their descent into the ocean, some at two or three times their 1990 speeds.[23] The massive Ilulissat Glacier now moves at six and a half feet per hour, a speed readily seen by the naked eye. As gigantic slabs of ice lubricated by meltwater slip against their rocky underpinnings, they have even begun triggering earthquakes that measure up to 3 on the Richter scale, a phenomenon previously unheard of in northwest Greenland.[24] Greenland air temperatures have increased, and deep subtropical water originating far to the south seems to be melting the undersea toes of the glaciers, allowing them to flow far more freely.[25] During 2012, very rapid glacial melting was observed in Greenland. The huge northern Greenland Petermann Glacier, one of Greenland's largest, cracked in July 2012 and a 46 square mile iceberg floated away. Two years earlier, an iceberg double that size and four times the size of Manhattan broke away from the same glacier.[26] Even more shocking was the discovery by NASA's ice research program during the summer of 2012 that Greenland experienced an enormous sudden thaw in which 97 percent of the 683,000 square mile ice sheet began to melt.[27] Just four days earlier, 60 percent of the land had still been frozen. Temperatures have risen about 4°F in just the past 30 years in northern Greenland and Canada, a warming rate about five times faster than the global average.[28] Scientists have estimated, however, that it will probably only take a global temperature increase of from 1.4 to 5.8°F (best estimate: 2.9°F) to completely melt the Greenland Ice Sheet.[29] Yet because of Greenland's size and the amount of ice there, completing the entire melting process would take an estimated 2,000 years. That would eventually raise sea level by about 23 feet.[30]

Submarine Anchorages. One mechanism that accelerates the destruction of ice shelves and glaciers in both the Arctic and Antarctic is the filling of cracks in the glacial surface by meltwater on relatively warm days. Then when temperatures drop and the water freezes, it expands in the cracks, widening them and causing the edges of the glacier nearest the sea to break off. When this happens with floating ice shelves like the Larsen B Ice Shelf in Antarctica, which famously disintegrated in 2002, land-bound ice—previously held back behind the ice shelves—can begin moving seaward. In the case of the Larsen B, when it broke up, the ice behind it began flowing eight times faster into the sea.[31]

Some ice sheets on the margins of land masses are grounded below sea level. An important mechanism thought likely to accelerate the movement of ice sheets into the sea—particularly in Antarctica—is erosion of these underwater anchorages. When subjected to warm ocean currents, these anchorages gradually melt, releasing the ice sheets above from their seafloor moorings. Climate change, aided by ozone depletion over Antarctica, has sped up polar winds that drive surface water offshore and cause the upwelling of deep warmer waters. (Unlike the temperature profile of warm tropical seas, the surface waters in polar regions tend to be colder than the deeper waters.) As a result, contact between underwater ice and the warm currents has likely increased. Thus, on the West Antarctic Ice Sheet, huge masses of ice able to raise sea level by more than 11 feet are now increasingly being made vulnerable to the attack of warmer waters upwelling from the deep.[32,33] Were deep ocean currents to shift in the future and attack the underpinnings of the gigantic Ross or Ronne Ice Shelves in Antarctica as well, far greater rise in sea level would be possible.[34]

Ice streams are another important mechanism by which both the West and East Antarctic Ice Sheets can quickly move ice to the ocean.[f,35] They are fast-moving rivers of ice flowing within a much slower-moving surrounding glacier. They can be 30 to 60 miles wide and can move 10 to 100 times as fast as the glacier.[36]

The Contribution of Melting Ice to Sea-Level Rise. Whereas the relative contribution to sea-level rise caused by the heat-driven expansion of seawater and by meltwater from land ice changes greatly over time with changes in sea surface temperature, most recent estimates indicate that ice mass losses are accounting for about 60 to 70 percent of the rise in sea level, whereas expansion is producing about 30 to 40 percent.[37,38] If the polar ice sheets continue their breakup at current rates, or if the rate accelerates, major portions of the polar ice will be gone in a few centuries. To know where all this leads, one need look no farther than the last interglacial period 130,000 to 118,000 years ago when large portions of the Greenland Ice Sheet apparently melted, and seas were one to two stories higher. Sophisticated coupled atmosphere-ocean climate computer model indicate that the high northern latitude temperatures required to melt the Greenland Ice Sheet were then likely only 6.3°F warmer than today's temperatures![39] As noted, parts of the Arctic have already experienced temperature increases of 4 to 5°F .

On balance, Greenland is losing a couple of hundred cubic kilometers of ice per year, which is a small fraction of the 2.5 million cubic kilometers of ice in the Greenland Ice Cap. While the rate of loss (melting minus snowfall accumulation) is still relatively small, it is still 150 billion tons of ice per year, and we have mentioned some of the danger signals that are appearing as the rate of melting continues its rapid increase.[40,41]

---

f The West Antarctic Ice Sheet loses large volumes of ice to the ocean through ice streams that work their way through glacial interiors many times faster than the surrounding ice. When the first radar images were made of the entire Antarctic continent by a Canadian satellite known as Radarsat in the late 1990s, scientists discovered that the even larger East Antarctic Ice Sheet also contained extensive ice streams, some reaching the center of what had previously been thought to be the solid center of the Antarctic Ice Cap.

The melting of the Greenland Ice Cap might go even faster than in the last interglacial period because of airborne soot particles from fossil fuel burning falling on its surface. These fine, dark particles, consisting of industrial carbon and impurities, absorb heat and speed the disintegration of snow and ice.[42]

Not All Seas Rise Equally. Sea-level rise is nonuniform and varies greatly around the world from region to region because of irregularities in the Earth's rotation, winds, differences in ocean circulation, salinity, prior glaciation history, and (as mentioned earlier) local crustal settling or rebound along with local inputs of freshwater and meltwater.[43] Therefore, the effects of sea-level rise will depend greatly on the geography and topography of specific coastal land areas, as well as on the melting rates of ice in Greenland and Antarctica. The Pacific and Atlantic coasts of the US will experience an additional 31 inches of sea-level rise above the global average because of such factors.[44]

Research reported in 2009 in *Geophysical Research Letters* warned that continued acceleration in Greenland's rate of melting would eventually send so much freshwater into the North Atlantic that dominant ocean currents (part of the transoceanic conveyor belt circulation) would change and cause sea level along the northeastern US to rise 12 to 20 inches *above the global average*.[45]

In areas that previously were heavily glaciated, land is likely to already be rising due to a glacial rebound effect, an uplifting that tends to counteract the effects of sea-level rise. But in the many locales where the land is sinking, the effects of sea-level rise will be increased. The extent of sea-level changes also varies a lot from one locality to another because of the effects of human activities, such as groundwater withdrawal, which can cause ground sinking. Reduction or diversion of river sediment flows into coastal areas can also lead to coastal erosion, which accentuates the effects of sea-level rise. These latter two types of human activities are more likely to occur and be most evident in densely populated urban areas.

One reason for hope that sea-level rise will not require abandoning all low-lying populated areas is the fact that a number of major metropolitan centers have already sunk by more than three and a quarter feet and yet have managed to protect themselves against higher flooding risks by engineered flood-control structures and systems. But whereas flood protection can be cost-effective for highly urbanized, densely populated areas, more numerous less-developed areas will typically not be protected because of the high costs and reduced economic benefits, or a lack of political power. Thus, they will be permanently susceptible to flooding.

Despite selective moderation of the effects of sea-level rise in some relatively prosperous areas, its effects on coastal ecosystems as a whole are likely to be extremely negative. They will include shoreline retreat, the loss or degradation of valuable real estate, erosion, ecosystem destruction, saltwater intrusion into surface water, and groundwater, and greater vulnerability to storms and storm surges. As pointed out in early chapters of the book, sea-level rise will also require the very expensive and complex relocation of transportation, energy, sewage, other waste, and communications infrastructure, as well as the relocation or reconstruction of shipping infrastructure, such as ports, docks, and loading facilities.

Risks of Ocean Acidity

As previously explained, organisms with calcium-based shells or body structures require carbonate ions from seawater to build their shells and skeletons. The greater the acidity of seawater, however, the lower the concentration of carbonate, and the harder it is for these organisms to form shells.

Moreover, once formed, calcium-rich shells, reefs, and bones tend to dissolve in acidic environments. This is bad news for corals as well as for the world's oyster reefs, all but 15 percent of which have already been destroyed by human activities. The remaining reefs are now vulnerable to destruction from climate-induced acidification.[46] Oyster larvae being commercially raised at hatcheries in Washington state have reportedly already been unable to form shells.[47]

The ability of coral to absorb calcium and then use it to build shells is expected to decline "in the near future."[48] Using data on 9,000 coral reefs worldwide, an international research team projected that by the time atmospheric CO_2 concentrations double (compared with the preindustrial era concentration of 280 ppm), stony coral reefs will no longer be able to grow and will begin to dissolve.[49]

Not only does it become more difficult for mollusks and coral to acquire carbonate in usable form at higher acidity levels, but the acidity also subjects the carbonate compounds in their skeletons or external skeletons to corrosion. Acidification of ocean water ultimately will therefore erode coral reefs faster than they can grow, making it impossible for them to survive. "With unabated CO_2 emissions, 70 percent of the presently known reef locations will be in corrosive waters by the end of this century."[50] This will eliminate thousands of species and the breeding grounds and nurseries for commercially important fish species.

Organisms with calcium-based shells like pteropods are sure to be affected by higher acidity. Pterods are little swimming molusks with a single soft foot that propels them through the water. One pteropod, *Limacina helicina*, serves as an important food source for North Pacific salmon and also nourishes mackerel, herring, and cod.[51] If other alternative food sources that these commercial species would normally rely on are also similarly eliminated, it's not hard to imagine a drastic drop in the population of these valuable food fish. Increased ocean acidity can therefore trigger the collapse of entire marine food webs. Because of the complexity of interrelated marine food webs, we don't know precisely how resilient they are to species losses and to reductions in species abundance.

Along with coral, oysters and microscopic organisms at the base of the food web called foraminifera are particularly susceptible to acidification because of their calcium-containing shells. By examining the fossilized shells that foraminifera left behind in ocean sediments over a 50,000 year period, Australian scientists have tracked changes in shell weight and compared them with modern foraminifera. They found that the shells of modern foraminifera have already shrunk by 30 to 35 percent in weight since the Industrial Revolution![52] This is very disturbing indeed because it reveals how greatly acidification is already affecting important marine organisms at the base of the food web that are intimately involved with the cycling of carbon through the ocean: Australian researchers cited previously estimate that foraminifera account for a quarter to a half of the total open-ocean marine carbonate "cycling."

ENDNOTES

Chapter 1

1. K. Anderson and Alice Bows, "Cumulative Carbon Emissions, Emissions Floors and Short-term Rates of Warming: Implications for Policy," *Philosophical Transactions of the Royal Society A (Mathematical, Physical & Engineering Sciences)* 369, no. 1934 (January 13, 2011).

2. H. Damon Matthews and Ken Caldeira, "Stabilizing Climate Requires Near-Zero-Emissions," *Geophysical Research Letters* 35, no. 4 (February 2008).

3. Taroh Matsunos et al., "Stabilization of Atmospheric Carbon Dioxide via Zero Emissions—An Alternative Way to a Stable Global Environment. Part 1: Examination of the Traditional Stabilization Concept," *Proceedings Japan Academy of Sciences, Series B Physical and Biological Sciences* 88, no. 7 (July 25, 2012).

4. Mark Lynas, *Six Degrees: Our Future on a Hotter Planet* (Washington, DC: National Geographic, 2008). See references and assumptions in chapter 6.

5. "Quick Facts on Ice Sheets," National Snow and Ice Data Center, Boulder, CO, http://nsidc.org/cryosphere/quickfacts/icesheets.html.

6. "Eemian," Wikipedia, accessed June 4, 2013, http://en.wikipedia.org/wiki/Eemian_interglacial.

7. Alister Doyle, "Greenland Ice Loss Accelerating: Study," Reuters, November 12, 2009, citing research by Michiel van den Broeke of Utrecht University in the Netherlands.

8. Stefan Rahmstorf, "A Semi-Empirical Approach to Projecting Future Sea-Level Rise," *Science* 315, no. 5810 (January 19, 2007).

9. Anil Ananthaswamy, "Projections of Sea Level Rise Are Vast Underestimates," *New Scientist* (November 29, 2012).

10. E. J. Rohling et al., "High Rates of Sea-level Rise During the Last Interglacial Period" (Letter), *Nature Geoscience* 1 (2008): 38–42.

11. Ananthaswamy, "Projections."

12. Aiguo Dai, "Increasing Drought Under Global Warming in Observations and Models" (Letter), *Nature Climate Change* 3, no. 171 (August 5, 2012).

13. Garrett Hardin, "The Tragedy of the Commons," *Science* 162, no. 3859 (December 13, 1968):1243–1248.

14. Mark Fischetti, "Storm of the Century Every Two Years," *Scientific American* 308, no. 6 (June 2013).

15–19. Ibid.

20. Laura Tom, "Climate Adaptation and Sea-Level Rise in the San Francisco Bay Area," Delta Urbanism Series, American Planning Association (January 2012).

21–22. Ibid.

23. Intergovernmental Panel on Climate Change, *Climate Change 2013: The Physical Science Basis, Summary for Policymakers*, WGI (September 27, 2013).

Chapter 2

1. Joe Romm, "Very Warm 2008 Makes This the Hottest Decade In Recorded History By Far★," *Climate Progress* (blog), December 7, 2008, http://climateprogress.org/2008/12/07/very-warm-2008-makes-this-hottest-decade-in-recorded-history-by-far/.

2. "Year Seen as Hottest in Lower 48 States," *Bloomberg News*, November 16, 2012.

3. Joe Romm, "Sorry Deniers, Hockey Stick Gets Longer, Stronger: Earth Hotter Now Than in Past 2,000 Years," *Climate Progress* (blog), September 3, 2008, http://climateprogress.org/2008/09/03/sorry-deniers-hockey-stick-gets-longer-stronger-earth-hotter-now-than-in-past-2000-years.

4. "Heat Wave Devastates the South of India, Killing Hundreds," *The New York Times,* May 18, 2002. See "Southern India—Heat Wave, May 2002," *Sri Aman Eco Rangers* (blog) (Item 90), http://sa-er.blogspot.com/2010_04_01_archive.html.

5. "Malaria in Southern Africa," http://www.malaria.org.za/Malaria_Risk/Update/update.html.

6. Fred Guterl, "Climate Armageddon: How the World's Weather Could Quickly Run Amok [Excerpt]," *Scientific American* (May 25, 2012). Accessed at http://www.scientificamerican.com/article.cfm?id=how-worlds-weather-could-quickly-run-amok.

7. IPCC, 2007: S. Solomon, D. Qin, M. Manning, Z. Chen, M. Marquis, K. B. Averyt, M. Tignor, and H. L. Miller, eds., *Climate Change 2007: The Physical Science Basis. Contribution of Working Group I to the Fourth Assessment Report of the Intergovernmental Panel on Climate Change* (Cambridge and New York: Cambridge University Press, 2007). Hereafter: IPCC, *Climate Change 2007*, vol 1; as well as the earlier major multivolume IPCC assessments published in 2001 and 1995.

8. Stephane Hallegatte, "Future Flood Losses in Major Coastal Cities," *Nature Climate Change*, published online August 18, 2013.

9. Rhett A. Butler, "Coral Reefs Decimated by 2050, Great Barrier Reef's Coral 95% Dead," *Mongabay, Conservation and Environmental Science News* (November 17, 2005; published online March 16, 2012), http://news.mongabay.com/2005/1117-corals.html.

Chapter 3

1. Anthony D. Barnosky et al., "Approaching a State Shift in Earth's Biosphere," *Nature* 486, no. 7401 (June 7, 2012): 52–58.

2. Richard A. Kerr, "How to Make a Great Ice Age, Again and Again and Again," *Science* 341, no. 6146 (August 9, 2013).

3. Ibid.

4. Inez Fung, "A Hyperventilating Biosphere," *Science* 341, no. 6150 (September 6, 2013).

5. James Hansen, *Storms of My Grandchildren: The Truth About the Coming Climate Catastrophe and Our Last Chance to Save Humanity* (New York: Bloomsbury USA, 2009), 101.

6. "Timeline of Glaciation," *Wikipedia*, http://en.wikipedia.org/wiki/Timeline _of_glaciation.

7. Hansen, *Storms of My Grandchildren*.

8. Ibid.

9. NOAA Satellite and Information Service, National Climatic Data Center, NOAA Paleoclimatology, "Astronomical Theory of Climate Change," accessed February 7, 2011, http://www.ncdc.noaa.gov/paleo/milankovitch.html.

10. Ibid.

11. "MilutinMilanković,"*Wikipedia*,http://en.wikipedia.org/wiki/Milutin_Milan kovi%C4%87.

12. National Snow and Ice Data Center, "All About Glaciers: How Are Glaciers Formed," accessed September 18, 2013, http://nsidc.org/cryosphere/glaciers/questions/formed.html.

13. "Milutin Milanković," *Wikipedia*.

14. "The Carbon Cycle and Earth's Climate," http://www.columbia.edu/~vjd1/carbon.htm.

15. NOAA Satellite and Information Service, National Climatic Data Center, "Greenhouse Gases, Frequently Asked Questions, Water Vapor," https://www.ncdc.noaa.gov/monitoring-references/faq/greenhouse-gases.php.

16. Windows to the Universe, "The Water Cycle—A Climate Change Perspective" (Boulder, CO: The National Earth Science Teachers Association, 2011), accessed May 16, 2012, http://www.windows2universe.org/earth/Water/water_cycle_climate_change.html.

17. National Air and Space Administration, "Tracking Earth's Most Abundant Greenhouse Gas," *Climate Change: Vital Signs of the Planet*, October 30, 2008, http://climate.nasa.gov/news/index.cfm?FuseAction=ShowNews.

18. Dieter Lüthi et al., "High-Resolution Carbon Dioxide Concentration Records from the EPICA Dome C Ice Core (EDC99)" (2009), Supplement to: Dieter Lüthi et al., "High-Resolution Carbon Dioxide Concentration Record 650,000–800,000 Years Before Present," *Nature* 453 (2008): 379-382.

Chapter 4

1. Senator Barack Obama, "The Coming Storm: Energy Independence and the Safety of Our Planet," April 3, 2006.

2. Tom Boden, Gregg Marland, and Bob Andres, "Global CO_2 Emissions from Fossil-Fuel Burning, Cement Manufacture, and Gas Flaring: 1751-2008" (Oak Ridge, TN: Carbon Dioxide Information Analysis Center, 2011), accessed via US Department of Energy Carbon Dioxide Information Analysis Center, http://cdiac.ornl.gov/ftp/ndp030/global.1751_2008.ems.

3. OECD iLibrary, *OECD Factbook 2013: Economic, Environmental and Social Statistics*, www.oecd-ilibrary.org.

4. Environmental Protection Agency, "Nitrous Oxide: Sources and Emissions" (Washington, DC: Environmental Protection Agency, 2010).

5. P. Friedlingstein et al., "Update on CO_2 Emissions," *Nature Geoscience* 3, (2010): 811-12.

6. Kevin Trenberth, personal communication, August 2, 2013, and Magdalena A. Balmaseda, Kevin E. Trenberth, and Erland Källén, "Distinctive Climate Signals in Reanalysis of Global Ocean Heat Content," *Geophysical Research Letters* 40, no. 9 (May 10, 2013): 1754-1759.

7. IPPC, *Climate Change 2007*, vol. 1.

8. J. R. Petit et al. "Climate and Atmospheric History of the Past 420,000 Years From the Vostok Ice Core, Antarctica," *Nature* 399 (1999): 429-436.

9. See http://www.esrl.noaa.gov/gmd/ccgg/trends.

10. http://www.ncdc.noaa.gov/paleo/icecore/antarctica/law/law.html.

11. Associated Press, "Greenhouse Gas Level Highest in Two Million Years, NOAA Reports" (Update 2), May 10, 2013, accessed at http://www.phys.org/news/2013-05-carbon-dioxide-atmosphere-historic-high.html, July 18, 2013.

12. Ibid., and Benjamin P. Flower. "Relationships between CO2 and Temperature in Glacial-Interglacial Transitions of the Past 800,000 Years," Search and Discovery Article #110116 (2009), posted September 8, 2009, based on an oral presentation at AAPG Annual Convention, Denver, Colorado, June 7–10, 2009.

13. IPPC, *Climate Change 2007*, vol. 1: 25.

14. IPCC, *Climate Change 2007*, vol. 1: 66.

15. Caroline Ash et al., "Once and Future Climate Change," *Science* 341, no. 6145 (August 2, 2013): 472–473.

16. Ibid., 25.

17. IPCC, *Climate Change 2007*, vol. 1: 37.

18. Joe Romm, "Science: CO2 levels haven't been this high for 15 million years . . . ," *Climate Progress*, October 18, 2009, http://climateprogress.org/2009/10/18/science-co2-levels-havent-been-this-high-for-15-million-years-when-it-was-5%C2%B0-to-10%C2%B0f-warmer-and-seas-were-75-to-120-feet-higher-we-have-shown-that-this-dramatic-rise-in-sea-level-i.

19. James Hansen, *Storms of My Grandchildren: The Truth About the Coming Climate Catastrophe and Our Last Chance to Save Humanity* (New York: Bloomsbury USA, 2009), 51.

20. IPCC, *Climate Change 2007*, vol. 1: 69, and Sir Nicholas Stern, *The Economics of Climate Change: The Stern Review* (Cabinet Office–HM Treasury, 2006).

21. Trenberth, *op. cit.*, and http://www.climate.gov/news-features/understanding-climate/climate-change-ocean-heat-content; http://www.skepticalscience.comlevitus-2012-global-warming-heating-oceans.html; http://www.epa.gov/climatechange/science/indicators/oceans/ocean-heat.html; and D. M. Murphy et al., "An Observationally Based Heat Balance for the Earth," *Geophysical Research-Atmospheres* 114, no. D17 (September 16, 2009), which graphically shows the quantity of heat stored in the ocean since 1950 relative to the heat stored over the same period in the soil and atmosphere.

22. "Global Ocean Heat and Salt Content," National Oceanographic Data Center, National Oceanic and Atmospheric Administration, www.nodc.noaa.gov/OC5/3M_HEAT_CONTENT/index.html.

23. Hansen, *Storms of My Grandchildren*, 65-67.

24. Ibid., 99.

25. IPCC, *Climate Change 2007*, vol. 1: 69

26. IPCC, *Climate Change 2007*, vol. 1: 12.

27. Caroline Ash et al., "Once and Future Climate Change," *Science* 341, no. 6145 (August 2, 2013): 472–473.

28. James Hansen, "Game Over for the Climate," *The New York Times*, May 9, 2012.

29. "Growth of Global Greenhouse Gas Emissions Accelerating," Environmental News Service, November 29, 2006 (based on research Dr. Mike Raupach, co-chair, Global Carbon Project, Commonwealth Scientific and Industrial Research Organization, CSIRO). See US Department of Energy, Oak Ridge National Laboratory at http://cdiac.ornl.gov/trends/emis/prelim_2009_2010_estimates.html for additional sources.

30. Hansen, *Storms of My Grandchildren*, 86 and 255–256.

31. Jerry X. Mitrovica, Natalya Gomez, and Peter Clark, "The Sea Level Fingerprint of West Antarctic Collapse," *Science* 23, no. 5915 (Feb 6, 2009).

32. IPCC, *Climate Change 2007*, vol 1: 777 and 814.

33. Richard B. Alley, Peter U. Clark, Philippe Huybrechts, Ian Joughin, "Ice-Sheet and Sea-Level Changes," *Science* 310, no. 5747 (October, 21, 2005): 456–460.

34. IPCC, 2001: J. T. Houghton, Y. Ding, D. J. Grggs, M. Noguer, P. J. van der Linden, X. Dai, K. Maskell, C. A. Johnson eds., *Climate Change 2001: The Scientific Basis. Contribution of Working Group I to the Third Assessment Report of the Intergovernmental Panel on Climate Change* (Cambridge and New York: Cambridge University Press, 2007), 639–694.

35. Hansen, *Storms of My Grandchildren*, 76 and 81–88.

36. Hansen, *Storms of My Grandchildren*, 84.

Chapter 5

1. Thomas R. Karl, Jerry M. Melillo, and Thomas C. Peterson, eds., *Global Climate Change Impacts in the United States* (Cambridge: Cambridge University Press, 2009).

2. Ibid., 80.

3. Ibid., 85–87.

4. Ibid., 81.

5. Ibid., 156.

6. Ibid., 151.

7. Ibid., 149.

8. Philip W. Mote et al., "Climate Change in Oregon's Land and Marine Environments," *Oregon Climate Assessment Report*, Oregon Climate Change Research Institute, December 2010, http://occri.net/wp-content/uploads/2011/04/cover.pdf.

9. Karl, *Global Climate Change Impacts*, 78.

10. Ibid., 52.

11. Ibid., 60.

12. Ibid., 56.

13. Ibid., 150.

14. Ibid., 26.

15. Ibid., 148–149.

16. Yereth Rosen, "Fish Die as Alaska Temperatures Continue to Break Records," Reuters, August 2, 2013.

17. Karl, *Global Climate Change Impacts*, 140–141.

18. Ibid., 141.

19. Ibid., 143.

20. Ibid., 137.

21. J. Battin et al., "Projected Impacts of Climate Change on Salmon Habitat Restoration," *Proceedings of the National Academy of Sciences* 104, no. 16 (2007): 6720–6725, http://www.pnas.org/content/104/16/6720.

22. Karl, *Global Climate Change Impacts*, 132.

23. Ibid., 124

24. Ibid., 113.

25. Ibid., 114.

26. Ibid., 107.

Chapter 6

1. Anthony D. Barnosky et al., "Approaching a State Shift in Earth's Biosphere," *Nature* 486, no. 7401 (June 7, 2012): 52–58.

2. Daniel A. Lashof et al., "Terrestrial Ecosystem Feedbacks to Global Climate Change," *Annual Review of Energy and the Environment* 22 (1997): 75–118.

3. Ibid.

4. Robert Sanders, "Trees Invading Warming Arctic Will Cause Warming Over Entire Region, Study Says," *UC Berkeley News*, January 12, 2010.

5–6. José A. Rial et al., "Nonlinearities, Feedbacks and Critical Thresholds Within the Earth's Climate System," *Climate Change* 65 (2004): 11–38.

7–8. Ibid. (See the article for internally cited authorities.)

9. Douglas J. Kennett et al., "Development and Disintegration of Maya Political Systems in Response to Climate Change," *Science* 338, no. 6108 (November 9, 2012): 788–791.

10. Mario Molina et al., "Reducing Abrupt Climate Change Risk Using the Montreal Protocol and Other Regulatory Actions to Complement Cuts in CO_2 Emissions," *Proceedings of the National Academy of Sciences* 106, no. 49 (December 8, 2009): 20616–20621.

11. Hans Joachim Schellnhuber, "Tipping Elements in the Earth System," *Proceedings of the National Academy of Sciences* 106, no. 49 (December 8, 2009): 20561–20563.

12. Robert H. Frank, "A Small Price For a Large Benefit," *The New York Times*, February 21, 2010.

13. Molina, "Reducing Abrupt Climate Change," 20616–20621.

14. Joeri Rogelj et al., "2020 Emissions Levels Required to Limit Warming to Below 2°C," *Nature Climate Change*, published online December 16, 2012, http://www.nature.com/natureclimatechange.

15. Barnosky, "Approaching a State Shift," 52–58.

16. Matthias Hofmann and Stefan Rahmstorf, "On the Stability of the Atlantic Meridional Overturning Circulation," *Proceedings of the National Academy of Sciences* 106, no. 49 (December 8, 2009): 20584–20589.

17. Anna Steynor and Craig Wallace, "The Gulf Stream—Atlantic Meridional Overturning Circulation," UK Climate Impacts Programme (UKCIP), Environmental Change Institute, University of Oxford, http:www.ukcip.org.uk/resources/gulf-stream.

18. Dorothée Herr and Grantly R. Galland, *The Ocean and Climate Change: Tools and Guidelines for Action* (Gland, Switzerland: International Union for the Conservation of Nature, IUCN US Multilateral Office, Gland, Switzerland, 2009), 36.

19. Natalie Hoare, "Is Time Running Out for the Tundra?," *Geographical* (March 2010).

20. Ibid.

21. Katey W. Anthony, "Methane: A Menace Surfaces," *Scientific American* 301, no. 6 (December 7, 2009).

22. D. F. McGinnis et al., "Fate of Rising Methane Bubbles in Stratified Waters: How Much Methane Reaches the Atmosphere?" *Journal of Geophysical Research: Oceans* 111, no. C9 (September, 2006), http://onlinelibrary.wiley.com/doi/10.1029/2005JC003183/abstract.

23. Marianne Lavelle, "Good Gas, Bad Gas," *National Geographic* (December 2012).

24. Sidney Draggan, ed., "Potential for Abrupt changes in Atmospheric Methane," in Cutler.J. Cleveland, ed., *Encyclopedia of Earth* (Washington, DC: Environmental Information Coalition, National Council for Science and the Environment, March 3, 2010; revised May 7, 2012), http://www.eoearth.org/article/Potential_for_Abrupt_Change_in_Atmospheric_Methane?topic=49491.

25. David Archer et al., "Ocean Methane Hydrates as a Slow Tipping Point in the Global Carbon Cycle," *Proceedings of the National Academy of Sciences* 106, no. 49 (December 8, 2009): 20596–20601.

26. James Hansen, "A New Age of Risk," presentation, Low Memorial Library, Columbia University, New York, NY, September 22, 2012.

27. Hoare, "Is Time Running Out?"

28. F. Liggins et al., "Projected Future Climate Changes in the Context of Geological and Geomorphological Hazards," *Philosophical Transactions, Royal Society A* 368, no. 1919 (2010): 2347–2367, cited in Mark Maslin et al., "Gas Hydrates: Past and Future Geohazard?" *Philosophical Transactions, Royal Society A* 368, no. 1919 (2010): 2369–2393.

29. Eli Kintisch, "Ticking Arctic Carbon Bomb May Be Bigger Than Thought," *ScienceNOW* (December 7, 2012).

30. Hoare, "Is Time Running Out?"

31. Michael Slezak, "Arctic Permafrost is Melting Faster Than Predicted," *New Scientist* (November 28, 2012).

32. Natalia Shakhova et al., "Extensive Methane Venting to the Atmosphere from Sediments of the East Siberian Arctic Shelf," *Science* 327, no. 5970 (March 5, 2010): 1246–1250.

33. BBC News.com, January 6, 2009.

34–38. Natalia Shakhova, "Extensive Methane Venting."

39. David Archer, "Ocean Methane Hydrates."

40. BBC News.com, January 6, 2009.

41. "Arctic Permafrost Leaking Methane at Record Levels, Figures Show," *The Guardian*, January 14, 2010.

42. Miguel Llanos, "Climate-Changing Methane 'Rapidly Destabilizing' off East Coast, Study Finds," NBC News, October 24, 2012.

43. Benjamin J. Phrampus and Matthew J. Hornback, "Recent Changes to the Gulf Stream Causing Widespread Gas Hydrate Destabilization," *Nature* 490, no. 7421 (October 25, 2012): 527–530.

44. Robert M. DeConto et al., "Past Extreme Warming Events Linked to Massive Carbon Release From Thawing Permafrost," *Nature* 484, no. 7392 (April 5, 2012): 87–91.

45. Al Gore, *An Inconvenient Truth* (Emmaus, PA.: Rodale Press, 2006).

46. Anders Levermann et al., "Basic mechanism for abrupt monsoon transitions," *Proceedings of the National Academy of Sciences* 106, no. 49 (December 8, 2009): 20572–20577.

47. Ibid.

48. Scott Powers et al., "Robust Twenty-First-Century Projections of El Niño and Related Precipitation Variability," *Nature* (published online October 13, 2013).

49. M. Latif and N. S. Keenlyside, "El Niño/Southern Oscillation Response to Global Warming," *Proceedings of the National Academy of Sciences* 106, no. 49 (December 8, 2009): 20578–20583.

50. S. McGregor et al., "Inferred Changes in El Niño-Southern Oscillation Variance Over the Past Six Centuries," *Climate Past* 9 (2013): 2269-2284, as cited in "El Niño Southern Oscillation Activity and Intensity Increasing With Global Warming" (posted online October 31, 2013), http://www.iIndymediaClimate.org.

51. Ibid.

52. Wenhong Li et al., " Impact of Two Different Types of El Niño Events on the Amazon Climate and Ecosystem Productivity," *Journal of Plant Ecology* 4, no. 1–2 (2010).

53–54. M. Latif, "El Niño/Southern Oscillation."

55. Yadvinder Malhi et al., "Exploring the Likelihood and Mechanism of a Climate-Change-Induced Dieback of The Amazon Rainforest," *Proceedings of the National Academy of Sciences* 106, no. 49 (December 8, 2009): 20610–20615.

56. C. B. Field et al., "Primary Production of the Biosphere: Integrating Terrestrial and Oceanic Components," *Science* 281, no. 5374 (1998): 237–40. Cited in Wenhong Li et al., " Impact of Two Different Types of El Niño Events on the Amazon Climate and Ecosystem Productivity, *Journal of Plant Ecology* 4, no. 1–2 (2010).

57. P. M. Cox et al., "Amazonian Forest Dieback under Climate-Carbon Cycle Projections for the 21st Century," *Theoretical and Applied Climatology* 78, no. 1–3 (June 2004): 137–156.

58–63. Malhi, "Exploring the Likelihood."

64. Scott Powers et al., "Robust Twenty-First-Century Projections of El Niño and Related Precipitation Variability," *Nature* (published online October 13, 2013); Tong Lee and Michael J. McPhaden, "Increasing Intensity of El Niño in the Central-Equatorial Pacific," *Geophysical Research Letters* 37, no.14 (July 2010); S. MacGregor et al., "Inferred Changes in El Niño-Southern Oscillation Variance Over the Past Six Centuries," *Climate Past* 9 (2013): 2269–2284.

65–68. "Ecuador in New Probe to See if Climate Change Worsens El Niño," Agence France-Presse, November 9, 2013.

69–71. Richard Washington et al., "Dust as a Tipping Element: The Bodélé Depression, Chad," *Proceedings of the National Academy of Sciences* 106, no. 49 (December 8, 2009): 20561–20563.

72–75. Dirk Notz, "The Future of Ice Sheets and Sea Ice: Between Reversible Retreat and Unstoppable Loss," *Proceedings of the National Academy of Sciences* 106, no. 49 (December 8, 2009): 20590–20595.

76. James Hansen, *Storms of My Grandchildren: The Truth About the Coming Climate Catastrophe and Our Last Chance to Save Humanity* (New York: Bloomsbury USA, 2009), 82–85.

Chapter 7

1. IPCC, 2007: S. Solomon, M. L. Parry, O. F. Canziani, J. P. Palutikof, P. J. van der Linden, and C. E. Hanson, eds., Climate Change 2007: *Impacts, Adaptation and Vulnerability. Contribution of Working Group II to the Fourth Assessment Report of the Intergovernmental Panel on Climate Change* (Cambridge and New York: Cambridge University Press, 2007), 77. Hereafter: IPCC, *Climate Change 2007*, vol 2.

2. Jonathan Koomey, "Moving Beyond Benefit-Cost Analysis of Climate Change," *Environmental Research Letters* 8, no. 041005 (December 2, 2013), http://iopscience .iop.org/1748-9326/8/4/041005.

3. Heather Stewart and Larry Elliott, "Nicholas Stern: 'I Got It Wrong on Climate Change—It's Far, Far Worse,'" *Observer*, January 26, 2013.

4. National Research Council, *Hidden Costs of Energy: Unpriced Consequences of Energy Production and Use* (Washington, DC: National Academies Press, 2010).

5–6. Koomey, *Moving Beyond*.

7. IEA, *World Energy Outlook* 2010 (Paris: International Energy Agency, Organization for Economic Cooperation and Development, November 9, 2010).

8. Surender Kumar and Amsalu W. Yalew, "Economic Impacts of Climate Change on Secondary Activities: A Literature Review," *Low Carbon Economy* 3, no. 2 (2012): 39–48.

9. John Hassler and Per Krusell, "Economics and Climate Change: Integrated Assessment in a Multi-Region World," *Journal of the European Economic Association* 10, no. 5 (October, 2012): 974–1000.

10. IPCC, *Climate Change 2007*, vol. 2: 37.

11. Ibid., 45.

12. Surender Kumar and Amsalu W. Yalew, "Economics and Climate Chage."

13. Ibid.

14. Ibid., 48.

15. Ibid., 49.

16. Lester R. Brown, *Plan B 4.0: Mobilizing to Save Civilization* (New York and London: W.W. Norton & Co., 2009).

17. Ibid.

18. Sayed Mohammad Alavi-Moghaddam, "Overview of Drought & Climate Change Adverse Effects on Iranian Water Resources Management," paper presented at the International Conference on Dams and Hydropower, Tehran, Iran, February 9, 2012.

19. Joel Brinkley, "Environmental Crisis Gives Iran, US a Common Enemy," *San Francisco Chronicle* and sfgate.com, November 17, 2013.

20. Brown, *Plan B 4.0*.

21. Bureau de la Coordination des Affaires Humanitaires des Nations Unies, "Climate Change: Himalayan Glaciers Melting More Rapidly," Irin, Nouvelles et Analyses Humanitaires, July 20, 2012, http://www.irinnews.org/fr/report/95917/climate-change-himalayan-glaciers-melting-more-rapidly.

22–23. Brown, *Plan B 4.0*.

24. Palash Ghosh, "Western Investors Buying up African Farming Properties in 'Land Grab,'" *International Business Times*, June 8, 2011.

25. Ibid., and *Wikipedia* "Land Grabbing," http://en.wikipedia.org/wiki/Land_grabbing.

26. Brown, *Plan B 4.0*.

27. IPCC, *Climate Change 2007*, vol. 2: 74.

28. California Department of Agriculture, "California Agricultural Production Statistics," accessed February 2014, http://www.cdfa.ca.gov/statistics.

29. Associated Press, "Great Lakes Drying Up Tourism," November 27, 2012.

30. Mike Pearson, "New Water Lows for Great lakes Could Drain Local Economies," CNN.com, January 17, 2013.

31. Wyatt Buchanan, "Climate Change: Worst-Case Planning," *San Francisco Chronicle*, December 3, 2009.

32. Wilfried Thuiller et al., "Climate Change Threats to Plant Diversity in Europe," *Proceedings of the National Academy of Sciences* 102, no. 23 (June 7, 2005): 8245–8250.

33. IPCC, *Climate Change 2007*, vol. 2: 52.

34. Hayden Washington and John Cook, *Climate Change Denial: Heads in the Sand* (London and Washington, DC: Earthscan, 2011).

35. NASA Earth Observatory, "2013 Arctic Sea Ice Minimum," September 21, 2013.

36. Brown, *Plan B 4.0*.

37. IPCC, *Climate Change 2007*, vol. 2: 631.

38. Solomon M. Hsiang et al., "Quantifying the Impact of Climate on Human Conflict," *Science* 341, no. 6151 (September 2013).

39. IPCC, *Climate Change 2007*, vol. 2.

40. IPCC, *Climate Change 2007*, vol. 2: 40.

41. IPCC, *Climate Change 2007*, vol. 2: 69.

42. IPCC, *Climate Change 2007*, vol. 2: 71.

43. Georgetown Climate Center, Georgetown Law, *State and Local Adaptation Plans*, published online March 15, 2012, http://www.georgetownclimate.org/node/3324?page=2.

44. Ben Chou et al., *Ready or Not: An Evaluation of State Climate and Water Preparedness Planning*, Natural Resources Defense Council Issue Brief IB-12-03-A (April, 2012), http://www.nrdc.org/water/readiness/files/Water-Readiness-issue-brief.pdf.

45. Matthew Heberger et al., "Impacts of Sea-Level Rise on the California Coast," California Climate Change Center/Pacific Institute, California Energy Commission Report no. CEC-500-2009-024-F (August, 2009), http://www.energy.ca.gov/2009publications/CEC-500-2009-024/CEC-500-2009-024-F.PDF.

46. Richard Gonzales, "After Years of Huge Deficits, California Starts To See a Fiscal Turnaround," NPR News, January 17, 2013, http://www.npr.org/2013/01/17/169644086/after-years-of-huge-deficits-california-starts-to-see-a-fiscal-turnaround.

47. Sir Nicholas Stern, *The Economics of Climate Change: The Stern Review* (Cambridge: Cambridge University Press, 2007).

48. "Climate Change Act 2008," Official Home of UK Legislation, http://www.legislation.gov.uk/ukpga/2008/27/contents.

49. Jim Giles, "Low-Carbon Future: We Can Afford to Go Green," *New Scientist* (December 2, 2009).

50. Stern, *Economics of Climate Change*.

51. R. J. Goettle and A. A. Fawcett, "The Structural Effects of Cap and Trade Climate Policy," *Energy Economics* 31 (2009).

52. *Pledges and Actions—A Scenario Analysis of Mitigation Costs and Carbon Market Impacts for Developed and Developing Countries*, PBL Netherlands Environmental Assessment Agency, http://www.pbl.nl/en/publications/2009/Pledges-and-actions.html.

53. Yvonne Deng et al., *The Ecofys Energy Scenario* (Utrecht, Netherlands: Ecofys and World Wildlife Fund, 2011).

54. The Climate Group, "Carbon Pricing," May 2013.

55. Ibid.

56. Kevin Kennedy, "California's Cap-and-Trade Program Makes Encouraging Headway," World Resources Institute website, August 23, 2013, http://www.wri.org.

57–58. Climate Group, "Carbon Pricing."

59. David Roland-Holst, *Energy Efficiency, Innovation, and Job Creation in California*, Next 10, San Francisco, CA, October 2008.

60. Monika Bauerlein, "Stop Global Warming and Get Rich TOO? Only in California," *Mother Jones* website, August 31, 2006, http://www.motherjones.com.

61. Central Intelligence Agency, *The World Factbook*, aka *CIA World Factbook* (Washington, DC: Central Intelligence Agency, 2007).

62. Center for Continuing Study on the California Economy, *2010 California Economy Rankings*, January 2012, http://www.ccsce.com/PDF/Numbers-Jan-2012-CA-Economy-Rankings-2010.pdf.

63. Duncan Austin, *Climate Protection Policies: Can We Afford to Delay?* (Washington, DC: World Resources Institute, 1997).

64. Climate Group, "Carbon Pricing."

65. Kevin Anderson and Alice Bows, "A New Paradigm for Climate Change," commentary published in *Nature Climate Change* 2 (September 2012; published online August 28, 2013): 639–640, http://www.nature.com/nclimate/journal/v2/n9/full/nclimate1641.html?WT.ec_id=NCLIMATE-201209.

66. Office of National Drug Control Policy, What America's Users Spend on Illegal Drugs, 2000–2006 (Washington, DC: Executive Office, 2012), http://www.whitehouse.gov/sites/default/files/page/files/wausid_report_final_1.pdf.

67. American Cancer Society, "Cigarette Smoking," http://www.cancer.org/cancer/cancercauses/tobaccocancer/cigarettesmoking/cigarette-smoking-who-and-how-affects-health.

68. Campaign for Tobacco-Free Kids, *Toll of Tobacco in the United States*, http://www.tobaccofreekids.org/research/factsheets/pdf/0072.pdf

69. Office of National Drug Control Policy, *What America's Users Spend on Illegal Drugs*, 2012.

70. Office of National Drug Control Policy, "Study Shows Illicit Drug Use Costs US Economy More Than $193 Billion," *ONDCP Update* 2, no. 5 (June 2011).

71. Richard Allen Johnson, "Some Key Gambling Statistics," http://www.Richard AllenJohnson.com; and James Quinn, "American's Gambling $100 Billion in Casinos Like Rats in a Cage," October 11, 2009, http://www.marketoracle.co.uk/Article14128.html.

72. Reuters, "As America's Waistline Expands, Costs Soar," April 30, 2012, http://www.reuters.com/article/2012/04/30/us-obesity-idUSBRE83T0C820120430.

73. US Center for Disease Control, "CDC Reports Excessive Alcohol Consumption Cost the US $224 billion in 2006," news release, October 17, 2011, accessed March 19, 2013, http://www.cdc.gov/media/releases/2011/p1017_alcohol_consumption.html.

74. Robert Scheer, "If Corporations Don't Pay Taxes, Why Should You?" *Nation of Change*, March 13, 2013, http://www.nationofchange.org/if-corporations-don-t-pay-taxes-why-should-you-1363186669.

75. Bruce Bartlett, "The Growing Corporate Cash Hoard," *The New York Times*, February 12, 2013.

76. David Cay Johnston, "Idle Corporate Cash Piles Up," Reuters, July 16, 2012.

77. Ibid.

78. "Wealth Inequality in America," accessed March 13, 2013, http://mashable.com/2013/03/02/wealth-inequality/#.

79–83. Ibid.

84. Office of Management and Budget, *Budget of the United States* (Washington, DC: USGPO, 2012), Tables 3.2 and 7.1; see also: http://www.usfederalbudget.us/us_defense_spending_30.html.

85. Stockholm International Peace Research Institute, "The 15 Countries with the Highest Military Expenditures in 2011," http://www.sipri.org/research/arma ments/milex/resultoutput/milex_15/the-15-countries-with-the-highest-military -expenditure-in-2011-table/view.

86. Melissa Gray, "Blix: US Was 'High on Military' over Iraq," CNN World, July 27, 2010, accessed March 19, 2013, http://articles.cnn.com/2010-07-27/world/uk.iraq .inquiry.blix_1_john-chilcot-iraq-war-hans-blix?_s=PM:WORLD.

87. Joseph E. Stiglitz and Linda J. Bilmes, "The True Cost of the Iraq War: $3 Trillion and Beyond," *Washington Post*, September 5, 2010.

88. The World Bank, "Stocks Traded, Total Value (Current US$)," Data, http:// data.worldbank.org/indicator/CM.MKT.TRAD.CD.

89. Sustainable Energy Act of 2013, S.329, 113th Congress (2013); introduced in the US Senate on February 14, 2013, by senators Bernie Sanders of Vermont and Barbara Boxer of California.

90. US Census Bureau, "Table 939. Petroleum and Coal Products Corporations—Sales, Net Profit, and Profit Per Dollar of Sales: 1990 to 2010," *Statistical Abstract of the United States: 2012*, 592.

91. Robert H. Frank, "A Ripe Moment to Speak Up on Climate Change?" August 15, 2012, http://www.samefacts.com/2012/08/everything-else/a-ripe-mo ment-to-speak-up-on-climate-change/?utm_source=feedburner&utm_medium= feed&utm_campaign=Feed%3A%20RealityBasedCommunity%20%28The%20 RBC%29.

92. Robert H. Frank, "A Small Price for a Large Benefit," *New York Times*, February 21, 2010.

93. Nomi Prins, *It Takes A Pillage: Behind the Bailouts, Bonuses, and Backroom Deals from Washington to Wall Street* (Hoboken, NJ: John Wiley & Sons, Inc., 2009). This book is a tour de force in the tradition of Upton Sinclair's *The Jungle* (a novel about corruption in the turn-of-the century meat packing industry); see also Jacob Riis, *How the Other Half Lives* (an expose on poverty in the slums); Ida Tarbell, *The History of the Standard Oil Company: The Oil Wars of 187*; and Lincoln Steffins, *The Shame of the Cities* (an expose of public corruption).

94. See http://www.nomiprins.com/bailout.html for a report analyzing the $14 trillion in assistance provided by various arms of the federal government, including the Federal Reserve, the Treasury, the FDIC, and the other lenders and guarantors. The data include major corporate recipients and how much they received from specific programs.

95. At the December 2009 Copenhagen talks, the developed nations offered the developing nations only $10 billion per year from 2009 to 2012 for climate adaptation—rather a paltry sum to be spread over many impoverished nations.

96. Neil Barofsky, *Bailout: An Inside Account of How Washington Abandoned Main Street While Rescuing Wall Street* (New York, NY: Free Press/Simon and Shuster, 2012).

97. Stern, *The Economics of Climate Change*.

98. John O. Niles, "Tropical Forests and Climate Change," in *Climate Change Policy: A Survey* (Covelo, CA, and Washington, DC: Island Press, 2002), 355.

99. Duncan MacQueen, *Review of Funds Which Aim to Protect Tropical Forests* (Edinburgh: International Institute, 2010).

100. Ibid.

101. What needs to be done to better protect tropical forests is clearly outlined in the study above by Duncan MacQueen of the nonprofit International Institute for Environment and Development: "The majority of large finance is for the design, implementation and monitoring of REDD [reduced emissions from degradation and deforestation] and also for climate change adaptation—with little channeled towards enhancing agricultural and forest tenure security, building representative institutions in forest areas to absorb climate change payments and reforming governance to combat corruption (except small projects, for example in the CBFF). *In terms of existing fund recipients, there is a heavy weighting towards Government agencies—with almost no funding reaching forest right-holders (poor agro-forest communities, family forest owners and indigenous peoples who live at the forest frontier). There is therefore a major gap in funds to secure rights, build investible institutions and pursue sustainable income generation free from corruption and vested interest—and it is this sort of funding that will ultimately protect forests* [emphasis added]." For additional policy proposals on protecting and restoring tropical forests, see John J. Berger, *Forests Forever: Their Ecology, Restoration, and Protection* (San Francisco and Chicago: Forests Forever Foundation and Center for American Places, 2008).

102. Anna Gibbney, "Longer Roots Equal Less Carbon Dioxide," *Geographical* (October 2011): 13, http://www.geographical.co.uk.

103. D. B. Kell, "Breeding Crop Plants with Deep Roots: Their Role in Sustainable Carbon, Nutrient and Water Sequestration," *Annals of Botany* 108 (2011): 407–418.

104. Catherine Lutz, "Rising from the Ashes," *Stanford Magazine*, March/April, 2013.

105. United Nations Framework Convention on Climate Change, *Report of the Conference of the Parties on Its Sixteenth Session, Held in Cancun from 29 November to 10 December 2010* (United Nations, 2010); and UNFCC, "The Cancun Agreements: Financial, Technology and Capacity-Building support," Key Steps of the United Nations Climate Change Conference, http://cancun.unfccc.int/financial-technology-and-capacity-building-support/fast-start-finance-up-to-2012/#c281.

106. IPCC, 2007: B. Metz, O. R. Davidson, P. R. Bosch, R. Dave, L A. Meyer, eds., *Climate Change 2007: Mitigation of Climate Change. Contribution of Working Group III to the Fourth Assessment Report of the Intergovernmental Panel on Climate Change* (Cambridge and New York: Cambridge University Press, 2007), 785. Hereafter: IPCC, *Climate Change 2007*, vol 3.

107. Export Import Bank of the United States, *2011 Annual Report*, April 2011.

108. The World Bank, "Table 3: World Bank Group Energy Portfolio by Sector, FY2007–FY2011," Energy—Data, accessed April 11, 2012, http://go.worldbank.org/ERF9QNT660; and The World Bank, "The World Bank—Climate Change—Renewables Almost a Quarter of World Bank's Energy Lending," news release, November 7, 2011.

109. IPCC, *Climate Change 2007*, vol. 3: 786.

110. Global Environmental Facility, *Report of the Global Environment Facility to the Seventeenth Session of the Conference of the Parties to the United Nations Framework Convention on Climate Change*, (Washington, DC: Global Environmental Facility, 2011), http://www.thegef.org/gef/sites/thegef.org/files/documents/C.41.Inf_.11_Report_of_the_GEF_to_the_17th_Session_of_the_COP_to_the_UNFCCC.pdf.

111. IPCC, *Climate Change 2007*, vol. 3: 786.

112. Stern, *The Economics of Climate Change*.

113. IPCC, *Climate Change 2007*, vol. 2: 35.

114. National Research Council, *Hidden Costs of Energy: Unpriced Consequences of Energy Production and Use* (Washington, DC: National Academies Press, 2010).

115. Working Group on Public Health and Fossil-Fuel Combustion, "Short-Term Improvements in Public Health from Global-Climate Policies on Fossil-Fuel Combustion: An Interim Report," *Lancet* 350, no. 9088 (November 8, 1997): 1341–1348; and World Resources Institute, "The Hidden Benefits of Climate Policy: Reducing Fossil Fuel Use Saves Lives Now," *Environmental Health Notes* (December, 1997).

Chapter 8

1. IPCC, *Climate Change 2007*, vol. 2.

2. "Climate Change," World Health Organization and United Nations Environment Programme, http://www.who.int/heli/risks/climate/climatechange/en.

3. Ibid.

4. IPCC, *Climate Change 2007*, vol. 2: 407.

5. Ibid.

6. S. M. Hsiang et al., "Quantifying the Influence of Climate on Human Conflict," *Science* 341, no. 6151 (September 13, 2013), published online August 1, 2013, https://www.sciencemag.org/content/341/6151/1235367.abstract.

7. Ibid.

8. Tim McConnell, "Global Warming Could Cause 50 Percent Increase in Violent Conflict," *Mother Jones*, August 1, 2013.

9. Federal Bureau of Investigation, "Violent Crime Offense Figure," *Crime in the United States, 2011, Uniform Crime Reports*, http://www.fbi.gov/about-us/cjis/ucr/crime-in-the-u.s/2011/crime-in-the-US-2011/violent-crime/violent-crime.

10. "Direct Conflict Deaths," *Wikipedia*, http://en.wikipedia.org/wiki/Direct_conflict_deaths.

11. S. M. Hsiang, "Quantifying the Influence."

12. IPCC, *Climate Change 2007*, vol. 2: 398.

13. Ibid., 397.

14. "NASA AIRS Movies Show Evolution of US 2011 Heat Wave," NASA, Jet Propulsion Laboratory, California Institute of Technology, http://photojournal.jpl.nasa.gov/catalog/PIA14480.

15. Steven C. Sherwood and Matthew Huber, "An Adaptability Limit to Climate Change Due to Heat Stress," *Proceedings of the National Academy of Sciences* (Approved March 24, 2010), http://www.pnas.org/content/early/2010/04/26/0913352107.full.pdf; Bob Beale, "Humans Fail in Rising Heat," *Science Alert*, May 4, 2010, http://www.sciencealert.com.au/news/20100505-20918-3.html.

16. Ibid.

17. IPCC, *Climate Change 2007*, vol. 2: 400.

18. Colin Price, "Thunderstorms, Lightning and Climate Change," paper presented at the 29th International Conference on Lightning Protection, Tel Aviv, Israel, June, 2008, http://www.iclp-centre.org/pdf/Invited-Lecture-1.pdf; NASA/ Goddard Space Flight Center, "Global Warming Will Bring Violent Storms And Tornadoes, NASA Predicts," *Science Daily*, http://www.sciencedaily.com/releases /2007/08/070830105911.htm.

19. "Scientific Facts on Air Pollution," Green Facts based on WHO studies, http:// www.greenfacts.org/en/ozone-o3.

20. IPCC, *Climate Change 2007*, vol. 2: 402.

21. Elizabeth Martin Perera and Todd Sanford, *Climate Change and Your Health: Rising Temperatures, Worsening Ozone Pollution* (Cambridge, Massachusetts: Union of Concerned Scientists, June 2011), as cited in "2020-Climate Change Costs," http:// www.global-warming-forecasts.com/cost-climate-change-costs-global-warming .php.

22–24. IPCC, *Climate Change 2007*, vol. 2: 398.

25–26. "Floods in the United States: 2001–Present," *Wikipedia*, http://en.wikipe dia.org/wiki/Floods_in_the_United_States:_2001-present#Decade_of_the_2010s.

27. Kenneth Todar, "Vibrio Vulnificus," *Todar's Online Textbook of Bacteriology*, http://www.textbookofbacteriology.net/V.vulnificus.html.

28. "Disease Fact Sheet IDH-IDCM Vibriosis," Ohio Department of Health (section 3:7, revised January 2009), http://www.odh.ohio.gov/pdf/IDCM/vibrio.pdf, and http://www.tcbh.org/pdfs/Vibriosis.pdf.

29. "Chikungunya Symptoms and Treatment," Centers for Disease Control and Prevention, http://www.cdc.gov/ncidod/dvbid/Chikungunya/CH_SymptomsTreat ment.html.

30. Elisabeth Rosenthal, "As Earth Warms Up, Tropical Virus Moves to Italy," *The New York Times*, December 23, 2007.

31. "Transmission of the Dengue Virus," Centers for Disease Control and Prevention, http://www.cdc.gov/dengue/epidemiology/index.html.

32. "Dengue Frequently Asked Questions: What is Dengue?" Centers for Disease Control and Prevention, http://www.cdc.gov/dengue/fAQFacts/index.html; IPCC, *Climate Change 2007*, vol. 2: 403–404.

33. "Malaria Worldwide," Centers for Disease Control and Prevention, http:// www.cdc.gov/malaria/about/facts.html.

34. Eric Chivian, "Losing It," *World Conservation*, April 2009.

35. Danna Leaman, "Easing the Pressure," *World Conservation*, April 2009.

36. Ibid.

37. Courtney Cavaliere, "Bracing for Change," *World Conservation*, April 2009.

38–40. Eric Chivian, "Losing It."

41. Claire Groden, "From Poison to Potion: Toxins Turned Into Life-Saving Drugs," Time.com/*CNN Health*, July 18, 2013, http://www.cnn.com/2013/07/18/ health/toxin-treatments-time.

42. IPCC, *Climate Change 2007*, vol. 2: 414.

43. "Over 460 Dead, Missing in China Floods," Press TV, http://www.presstv.ir/detail/189447.html.

Chapter 9

1. His Royal Highness Charles, Prince of Wales, speech at the Copenhagen Climate Change Conference, December 15, 2009, http://www.theguardian.com/environment/2009/dec/15/prince-charles-speech-copenhagen-climate.

2. Thomas R. Knutson, "Global Warming and Hurricanes: An Overview of Current Research Results," Geophysical Fluid Dynamics Laboratory/NOAA (September 3, 2008; revised December 30, 2013), http://www.gfdl.noaa.gov/global-warming-and-hurricanes.

3. Jeff Masters, "Haiyan's True Intensity and Death Toll Still Unknown," *Dr. Jeff Masters' WunderBlog*, http://www.wunderground.com/blog/JeffMasters/show.html.

4. John Vidal, "Typhoon Haiyan the Biggest Yet as World's Tropical Storms Gather Force," *The Guardian*, November 8, 2013, http://www.theguardian.com/world/2013/nov/08/typhoon-haiyan-biggest-storms.

5. Ibid.

6. Higher sea surface temperatures and atmospheric humidity are not the only factors that influence trends in hurricane frequency and stability. Hurricane formation depends on convective available potential energy, which is influenced by wind shear as well as atmospheric wind speeds, atmospheric pressure, and atmospheric stability. Hurricane frequency is also a function of natural variability on interannual and multidecadal scales, making attribution of increased frequency harder. See Kevin Trenberth, "Climate: Uncertainty in Hurricanes and Global Warming," *Science* 308, no. 5729 (June 17, 2005):1753-1754.

7. IPCC, *Climate Change 2007*, vol. 1: 312.

8. Ibid., 316.

9. Seth Borenstein, "US in for Perilous Weather as World Warms, NASA Says," Associated Press, *San Francisco Chronicle*, August 31, 2007.

10–11. IPCC, *Climate Change 2007*, vol. 1: 788.

12. Ibid., 789.

13. Office of Public Safety and Emergency Preparedness, State of Louisiana, http://gohsep.la.gov/factsheets/definitionofahurricane.htm.

14. US Department of State, "Bangladesh Country-Specific Information–Natural Disasters," accessed March 28, 2013, http://travel.state.gov/travel/cis_pa_tw/cis/cis_1011.html.

15. Elizabeth Shogren, "Trees Lost to Katrina May Present Climate Challenge," *NPR Special Series: Causes*, January 21, 2008, http://www.npr.org/templates/story/story.php?storyId=17814049.

16. Sonja N. Oswalt et al., "Hurricane Katrina Impacts on Mississippi Forests," *Southern Journal of Applied Forestry* 32, no. 3 (2008): 139–141.

17. Adam B. Smith and Richard W. Katz, *US Billion-Dollar Weather and Climate Disasters: Data Sources, Trends, Accuracy and Biases*, National Climate Data Center and

National Oceanic and Atmospheric Administration, http://www1.ncdc.noaa.gov/
pub/data/papers/smith-and-katz-2013.pdf. This is an unformatted version of an article that has been accepted for publication in the journal *Natural Hazards*. The pre-
publication version is available at http://link.springer.com/article/10.1007/s11069
-013-0566-5.

18. Molly O'Toole,"Weather Disasters Seen Costly Sign of Things to Come," Reuters, July 29, 2011, http://www.reuters.com/article/2011/07/29/us-climate-disasters
-hearing-idUSTRE76S0UC20110729.

19 . Thomas R. Karl et al., eds., *Weather and Climate Extremes in a Changing Climate. Regions of Focus: North America, Hawaii, Caribbean, and US Pacific Islands, Final
Report, Synthesis and Assessment Product 3.3*, US Climate Change Science Program and
the Subcommittee on Global Change Research, Department of Commerce, NOAA's
National Climatic Data Center, 2008, http://www.ssec.wisc.edu/~kossin/articles/
sap3-3-final-all.pdf.

20–22. "2020 Climate Change Costs," Global Warming Forecasts, http://www
.global-warming-forecasts.com/cost-climate-change-costs-global-warming.php.

23. Elizabeth Bunn,"Most Insurers Lack Plans for Climate Change, Survey Finds,"
Bloomberg News, March 7, 2013.

24. IPCC, *Climate Change 2007*, vol. 1: 25.

25. Bunn, "Most Insurers."

26. Chad Hemenway,"Two New States to Require Climate Risk Survey; More Companies Must Now Respond," PropertyCasualty360.com, July 18, 2013, http://www
.propertycasualty360.com.

27. Bryan Walsh, "The Costs of Climate Change and Extreme Weather Are Passing the High-Water Mark," Time.com, July 17, 2013, http://science.time.com.

28. NASA, "NASA Research Finds Last Decade Was Warmest on Record, 2009
One of Warmest Years," news release, January 21, 2010.

29. World Meteorological Organization, *WMO Statement on the Status of the Global
Climate in 2009*, WMO-No. 1055, 2010.

30. Bryan Walsh, "Costs of Climate Change."

31. Thomas R. Karl, *Weather and Climate Extremes*, 1–9.

32. IPCC, *Climate Change 2007*, vol. 1: 312–313.

33. Frances C. Moore, "2007 Second Warmest Year on Record: Northern Hemisphere Temperature Highest Ever." Earth Policy Institute, Eco-Economy Indicators,
January 2008, http://www.earthpolicy.org/Indicators/Temp/2008.htm

34. World Meteorological Organization, "WMO Statement," 5.

35–37. Ibid., 10.

38. Brian K. Sullivan, "Historic Flooding Threatens to Swamp Huge Swath of
Nation," Bloomberg News, *San Francisco Chronicle*, March 17, 2010.

39. V. Lakshmi and K. Schaaf, "Analysis of the 1993 Midwestern Flood Using Satellite and Ground Data," *IEEE Transactions on Geoscience and Remote Sensing* 39, no. 8
(August, 2001): 1736–1743.

40. National Oceanic and Atmospheric Administration, *2008 Midwestern US
Floods*, National Climate Data Center, http://www.ncdc.noaa.gov/special-reports/
2008-floods.html.

41. IPCC, *Climate Change 2007*, vol. 1: 863.

42. World Meteorological Organization, *WMO Statement*, 5.

43. IPCC, *Climate Change 2007*, vol. 1: 260–263.

44. World Meteorological Organization, *WMO Statement*, 5.

45. Joel Brinkley, "Environmental Crisis Gives Iran, US a Common Enemy," *San Francisco Chronicle*, November 17, 2013.

46. IPCC, *Climate Change 2007*, vol. 1: 260–263.

47. Ibid., 265.

48. Ibid., 863.

49. Daniel Pepper, "India Faces Crisis as Groundwater is Used Up," *San Francisco Chronicle*, May 9, 2008.

50. Ibid.

51. Sid Perkins, "Groundwater Use Adds CO_2 to the Air," *Science News*, November 18, 2007.

52. NASA, "NASA Research Finds."

53–55. World Meteorological Organization, *WMO Statement*, 5.

56. IPCC, *Climate Change 2007*, vol. 1: 698.

57. Ibid., 627–629.

58. Ibid., 785–787.

59. Frances C. Moore, "2007 Second Warmest Year."

60. Steve Connor, "Record 22°C Temperatures in Arctic Heatwave," *Independent*, October 3, 2007.

61. NASA, "Impact of Arctic Heat Wave Stuns Climate Change Researchers," *Earth Observatory News*, September 26, 2007, http://earthobservatory.nasa.gov/News room/MediaAlerts/2007/2007092625668.html.

62. Steve Connor, "Record Temperatures."

63–64. IPCC, *Climate Change 2007*, vol. 1: 527.

65. James Brooke, "Heat Wave, Smog Double Moscow's Daily Death Rate," *Voice of America*, August 9, 2010.

66. Vladimir Isachenkov, "Moscow Deaths Double amid Smog to 700 People a Day," Associated Press, August 9, 2010.

67. Fred Weir, "Russia Wildfires: Thick, Toxic Smog Chokes Moscow Residents," *Christian Science Monitor* online, August 8, 2010.

68. Khristina Narizhnaya, "Russia: Wildfire Smog Envelops Moscow," *San Francisco Chronicle*, August 5, 2010.

69. Isachenkov, "Moscow Deaths."

70. Fred Weir, "Russia Wildfires."

71. Stefan Nicola, "Rússian Fires not to Launch Nuclear Cloud," August 12, 2010, UPI.com, http://www.upi.com.

72. Andrew E Kramer, "Russia, Crippled by Drought, Bans Grain Exports, *The New York Times*, August 5, 2010.

73. Anna Smolchenko, "Russia Admits Fires Burned on Chernobyl-Hit Land," Agence France Presse, August 11, 2010.

74. Ibid.

75. Associated Press, "Russian Fires Threaten to Stir and Spread Radioactive Particles Left from Chernobyl Disaster," August 11, 2010.

76. Howard LaFranchi, "Pakistan Floods, Haiti Earthquake: Unprecedented 1-2 Punch for US Aid," *Christian Science Monitor* online, August 13, 2010, http://www. csmonitor.com/USA/Foreign-Policy/2010/0813/Pakistan-floods-Haiti-earthquake -unprecedented-1-2-punch-for-US-aid.

77. Munir Ahmed, "1,000 Marines, Copters Arrive to Help Flood Relief," Associated Press in *San Francisco Chronicle*, August 13, 2010.

78. Mail Foreign Service, "An Island Nation: Millions of Villagers Marooned on Tiny Patches of Land as Floods Devastate Swathes of Pakistan," *Daily Mail* online, August 10, 2010.

79. Ibid.

80. UPI, "Thai Flood Death Toll Surpasses 700," December 14, 2011.

81. Jonathan Watts, "Thailand Seeks Flood Prevention Plan as Bangkok Clean-Up Operation Continues," *The Guardian*, December 26, 2011.

82. Ibid.

Chapter 10

1. Anthony D. Barnosky et al., "Approaching a State Shift in Earth's Biosphere," *Nature* 486, no. 7401 (June 7, 2012): 52–58.

2. Paul R. Ehrlich and Robert M. Pringle, "Where Does Biodiversity Go From Here? A Grim Business-as-Usual Forecast and a Hopeful Portfolio of Partial Solutions," *Proceedings of the National Academy of Sciences*, 105, suppl. 1 (August 12, 2008): 11579–11586; and International Union for the Conservation of Nature, *Red List of Threatened Species*,™ http://www.iucnredlist.org.

3. Michael McCarthy, "Nature Laid Waste: The Destruction of Africa," *Independent*, June 11, 2008.

4. Dave Foreman, "More Immigration=More Americans=Less Wilderness," *Earth Island Journal* (Autumn 2013).

5. Noah S. Diffenbaugh and Christopher B. Field, "Changes in Ecologically Critical Terrestrial Climate Conditions," *Science* 341, no. 6145 (August 2, 2013).

6. Ehrlich, "Where Does Biodiversity Go?"

7. Anthony D. Barnosky et al., "Approaching a State Shift."

8. Abigail E. Cahill et al., "How Does Climate Change Cause Extinction?" *Proceedings of the Royal Society B* 280 (August 13, 2013).

9. John Bongaarts, "Human Population Growth and the Demographic Transition," *Philosophic Transactions of the Royal Society B* 364, no. 1532 (October 2009): 2985–2990.

10. R. Warren et al., "Quantifying the Benefit of Early Climate Change Mitigation in Avoiding Biodiversity Loss," *Nature Climate Change* 3 (May 12, 2013).

11. Jennifer B. Hughes et al., "Population Diversity: Its Extent and Extinction," *Science* 278, no. 5338 (October 24, 1997): 689–692.

12. Ibid.

13. Ehrlich, "Where Does Biodiversity Go?"

14. Hughes, "Population Diversity"; Ehrlich, "Where Does Biodiversity Go?"

15. Paul R. Ehrlich and Anne Erhlich, "The Invisible Ruin," International Human Dimensions Programme on Global Environmental Change, January 9, 2013.

16. "Passenger Pigeon," Chipper Woods Bird Observatory, http://www.wbu .com/chipperwoods/photos/passpigeon.htm.

17. International Union for the Conservation of Nature, *Red List*, and International Union for Conservation of Nature, "More Critically Endangered birds on IUCN Red List," May 2009, http://www.iucn.org/?3159/More-Critically-Endangered -birds-on-IUCN-Red-List-than-ever.

18. International Union for the Conservation of Nature, *Red List*.

19. Wendy B. Foden et al., "Identifying the World's Most Climate Change Vulnerable Species: A Systematic Trait-Based Assessment of All Birds, Amphibians and Corals," *PLOS ONE* 8, no. 6 (June 2013), http://www.plosone.org.

20. Wilson, *The Creation*, 29–30.

21. Andrew Beattie and Paul R. Ehrlich, "De-extinction: Moral Hazard Writ Large," Millennium Alliance for Humanity and the Biosphere (Stanford University and University of Technology, Sydney), November 14, 2013, http://mahb.stanford. edu/blog/deextinction.

22. Paul R. Ehrlich, "The Case Against De-Extinction: It's a Fascinating but Dumb Idea," *Yale Environment 360*, January 13, 2014, http://bit.ly/1gAIuJF.

23. Jason Mark, "Back From the Dead," *Earth Island Journal*, Summer 2013.

24. Beattie, "De-extinction: Moral Hazard."

25. Ehrlich, "Case Against De-Extinction."

26. Cahill, "How Does Climate Change?"

27. Scott R. Loarie et al., "The Velocity of Climate Change," *Nature* 462 (December 24, 2009): 1052–1055

28–30. Noah S. Diffenbaugh and Christopher B. Field, "Changes in Ecologically Critical Terrestrial Climate Conditions," *Science* 341, no. 6145 (August 2, 2013).

31. Janet Marinelli, "Guardian Angles," *Audubon*, May–June 2010; Peter Fimrite, "Animals, Plants Must Move or Die," *San Francisco Chronicle*, December 24, 2009.

32. R. Warren et. al., "Quantifying the Benefit of Early Climate Change Mitigation in Avoiding Biodiversity Loss," *Nature Climate Change* 3 (May 12, 2013).

33. Nathaniel Johnson, "Refining Past Research," *California Magazine*, Spring 2012.

34. Adapted from an article by Sarah Yang, "Yosemite's Alpine Chipmunks Take Genetic Hit From Climate Change," *California Magazine*, Spring 2012.

35–37. Warren, "Quantifying the Benefit."

38. Bradley J. Cardinal et al., "Biodiversity Loss and Its Impact on Humanity," *Nature* 486 (June 7, 2012).

39. Ibid.

40. IPCC, *Climate Change 2007*, vol. 2: 242–245, 512.

41. Ibid.

42–43. Foden, "Identifying Vulnerable Species."

44. James Brooke, "Heat Wave, Smog Double Moscow's Daily Death Rate," Voice of America News, http://www1.voanews.com/english/news/Moscows-Mortality -rate-doubles-during heat-wave-100270414.html.

45. Fred Weir, "Russia Wildfires: Thick, Toxic Smog Chokes Moscow Residents," *Christian Science Monitor* online, August 8, 2010.

46. International Union for Conservation of Nature, *Red List*.

47. Tim Flannery, *The Weather Makers: How Man is Changing the Climate and What it Means for Life on Earth* (New York: Atlantic Monthly Press, 2005).

48. Ibid., 114–117.

49. IPCC, *Climate Change 2007*, vol. 2: 597.

50. Richard Stone, "A Rescue Mission for Amphibians at the Brink of Extinction," *Science* 339, no. 6126 (March 22, 2013).

51. Flannery, *Weather Makers*, 114–117.

52. Barry Sinervo et al., "Erosion of Lizard Diversity by Climate Change and Altered Thermal Niches, *Science* 328, no. 5980 (May 2010): 894–899.

53. Raymond B. Huey et al., "Are Lizards Toast?" *Science* 328, no. 5980: 832–833; David Perlman, "World's Lizards Face Extinction Amid Warming," *San Francisco Chronicle*, May 14, 2010.

54. Sinervo et al., "Erosion of Lizard Diversity," 899.

55. Microdocs Project, "Species on Coral Reefs," http://www.stanford.edu/group/microdocs/species.html.

56. National Oceanic and Atmospheric Administration Coral Reef Conservation Program (NOAA), http://coralreef.noaa.gov/aboutcorals/values/fisheries.

57–58. Microdocs Project, "4 Kinds of Coral Reefs," http://www.stanford.edu/group/microdocs/typesofreefs.html.

59. National Oceanic and Atmospheric Administration Coral Reef Information System (NOAA), http://oceanservice.noaa.gov/facts/coral_species.html.

60. John M. Pandolfi et al., "Projecting Coral Reef Futures Under Global Warming and Ocean Acidification," *Science* 333, no. 6041 (July 22, 2011).

61. Ibid.

62. W. Broadgate et al., eds., *Ocean Acidification Summary for Policymakers*, Third Symposium on the Ocean in a High-CO_2 World, International Geosphere-Biosphere Programme, Intergovernmental Oceanographic Commission, Scientific Committee on Oceanic Research, November 2013.

63. NOAA Coral Reef Conservation Program.

64. Ibid.

65. Coral Reef Alliance, "Coral Reef Overview," http://www.coral.org/resources/about_coral_reefs/coral_overview.

66. Pandolfi, "Projecting Coral Reef Futures."

67–68. Broadgate, *Ocean Acidification Summary*.

69. Wilson, *The Creation*.

70. Flannery, *Weather Makers*, 109.

71. Paul Kvinta, "Rescue Coral Reefs," *Popular Science*, May 2011.

72. Broadgate, *Ocean Acidification Summary*.

73. Stephen Leahy, "Biodiversity: The Twilight of Coral Reefs," Inter Press Service, May 22, 2010.

74. United Nations, *Report of the United Nations Conference on Environment and Development, Annex I, Rio Declaration on Environment and Development;* and *Annex III. Non-Legally Binding Authoritative State of Principles for a Global Consensus on the Management, Conservation, and Sustainable Development of All Types of Forests,* Rio de Janeiro, June 3–14, 1992.

75. Agence France Presse, "UN Fears 'Irreversible' Damage to Natural Environment," May 10, 2010.

76. Jeff Tollefson and Natasha Gilbert, "Earth Summit: Rio Report Card," *Nature* 486, no. 7401 (June 2012).

77. World Resources Institute, "STATEMENT: Rio+20 Wraps Up with 'More of a Whimper Than a Roar,'" news release, June 22, 2010, http://www.wri.org/content/statement-rio20-wraps-more-whimper-roar.

78. Worldwatch Institute, "Rapid Biodiversity Loss Continues in Absence of Political Action and Accurate Assessments of Ecosystem Values," news release, May 22, 2012, http://www.worldwatch.org/rapid-biodiversity-loss-continues-absence-political-action-and-accurate-assessments-ecosystem-values.

79. Convention on Biological Diversity, "Strategic Plan for the Convention on Biological Diversity," http://www.cbd.int/sp.

80. Secretariat of the Convention on Biological Diversity, *Global Biodiversity Outlook 3* (Montreal, Quebec: 2010). The document is compiled by the Secretariat of the Convention on Biological Diversity and the United Nations Environment Programme's World Conservation Monitoring Centre; see http://www.cbd.int/doc/publications/gbo/gbo3-final-en.pdf.

81. United Nations Environment Programme and GRID-Arendal (a foundation established by the Government of Norway to communicate environmental information to policy makers and facilitate environmental decision making), http://www.grida.no/news/press/4221.aspx.

Chapter 11

1. Dorothée Herr and Grantly R. Galland, *The Ocean and Climate Change: Tools and Guidelines for Action* (Gland, Switzerland: International Union for the Conservation of Nature, IUCN US Multilateral Office, Gland, Switzerland, 2009), 36.

2. Ibid.

3. Pew Environment Group, "Marine Fisheries and the World Economy: A Summary of a New Scientific Analysis," *Research Summary, Pew Ocean Science Series,* http://www.pewtrusts.org/uploadedFiles/wwwpewtrustsorg/News/Press_Releases/Protecting_ocean_life/Pew%20OSS%20World%20Economy%20FINAL.pdf, based on A. J. Dyck and U. R. Sumaila, "Economic Impact of Ocean Fish Populations in the Global Fishery," *Journal of Bioeconomics* 12, no. 3 (October 2010), http://link.springer.com/article/10.1007%2Fs10818-010-9088-3.

4. Robert J. Nichols and Anny Cazenave, "Sea-Level Rise and Its Impact on Coastal Zones, *Science* 328, no. 5985 (June 18, 2010).

5. Anil Ananthaswamy, "Sea-Level Rise, It's Worse than We Thought," *New Scientist* 2715 (July 1, 2009).

6. Jonathan T. Overpeck et al., "Paleoclimatic Evidence for Future Ice-Sheet Instability and Rapid Sea-level Rise," *Science* 311, no. 5768 (March 24, 2006).

7. Ibid.

8. Nichols, "Sea-Level Rise."

9. IPCC, *Climate Change 2007*, vol. 2: 334.

10. IPCC, *Climate Change 2007*, vol. 2: 326–328.

11. Justin Gillis, "Reading Earth's Future in Glacial Ice," *The New York Times*, November 14, 2010.

12. Ibid., and Justin Gillis, "As Glaciers Melt, Science Seeks Data on Rising Seas," *The New York Times*, November 13, 2010.

13. "Island States Fear Rising Seas Will Drown Their Beaches and Storm Surges Wash Away Their Homes Unless World Finds Way to Address Climate Change," *Climate Alert* 11, no. 3 (September 1998).

14. Ibid., and "AOSIS Chairman Slade Details Vulnerabilities of Small Island States, *Climate Alert* 11, no. 3 (September 1998).

15. IPCC, *Climate Change 2007*, vol. 2: 317.

16. Jonathan T. Overpeck et al., "Paleoclimatic Evidence."

17. IPCC, *Climate Change 2007*, vol. 2: 334-335.

18. Ibid., 344.

19. "Topographical Shifts at the Urban Waterfront," San Francisco Bay Conservation and Development Commission Rising Tides Competition winning entry, http://www.risingtidescompetition.com/risingtides/Winners_files/13.202523_LR.pdf.

20. IPCC, *Climate Change 2007*, vol. 2: 319.

21. US Department of the Interior Fish and Wildlife Service and National Oceanic and Atmospheric Administration National Marine Fisheries Service, *Status and Trends of Wetlands in the Coastal Watersheds of the Conterminous United States 2004 to 2009* (Washington, DC: US Government Printing Office, 2013).

22. IPCC, *Climate Change 2007*, vol. 2: 339.

23. Ibid., 328.

24. Herr, *Ocean and Climate Change*, 31.

25. IPCC, *Climate Change 2007*, vol. 2: 331.

26. Ibid., 323.

27. Ibid., 324.

28. Ibid., 339.

29. Ulf Riebesell et al., "Sensitivities of Marine Carbon Fluxes to Ocean Change," *Proceedings of the National Academy of Sciences* 106, no. 49 (December 8, 2009): 20602-20609.

30. Ahmed Djoghlaf, statement, Peoples' World Conference on Climate Change and Mother Earth, Cochabamba, Bolivia, April 20, 2010.

31. Royal Society, *Ocean Acidification Due to Increasing Atmospheric Carbon Dioxide*, policy document 12/05, June 2005, http://www.royalsoc.ac.uk; and "Acid Oceans," Weather Underground, http://www.wunderground.com/climate/acidoceans.as.

32. IPCC, *Climate Change 2007*, vol. 2: 529.

33–35. W. Broadgate et al., eds., *Ocean Acidification Summary for Policymakers*, Third Symposium on the Ocean in a High-CO_2 World, International Geosphere-Biosphere

Programme, Intergovernmental Oceanographic Commission, Scientific Committee on Oceanic Research, November 2013.

36. Daniel G. Boyce et al., "Global Phytoplankton Decline Over the Past Century," *Nature* 466, no. 7306 (July 29, 2010): 591–596.

37. Seth Borenstein, "In Hot Water: World Sets Ocean Temperature Record," *Associated Press,* August 20, 2009.

38. Kenneth R. Weiss, "Scientists Blame Ocean Dead Zones on Climate Change," *Los Angeles Times,* February 20, 2008, as reprinted in *San Francisco Chronicle.*

39. Carolyn Lochhead, "'Dead Zone' in Gulf Tied to Ethanol Subsidies," *San Francisco Chronicle,* July 6, 2010.

40. Ibid.

41. USDA, Natural Resource Conservation Service, updated June 23, 2009, http://www.nrcs.usda.gov/programs/crp.

42. Lochhead, "'Dead Zone' in Gulf."

43. Herr, *Ocean and Climate Change,* 56.

44. "As Oceans Warm, Problems From Viruses and Bacteria Mount, *The New York Times,* January 24, 1999; Cervinco et al., "Coral Diseases," *Science* 280 (1998): 99–500; James W. Porter et al., "The Effect of Multiple Stressors on the Florida Keys Coral Reef Ecosystem: A Landscape Hypothesis and a Physiological Test," *Limnology and Oceanography* 44, no. 3, part 2 (May, 1999): 941–949, http://links.jstor.org/sici?sici=00 24-3590%28199905%2944%3A3%3C941%3ATEOMSO%3E2.0.CO%3B2-0.

45. James A. Screen and Ian Simonds, "The Central Role of Diminishing Sea Ice in Recent Arctic Temperature Amplification," *Nature* 464, no. 7239 (April 29, 2010): 1334–1337.

46. Randolph E. Schmid, "Expert Says Arctic 'May Soon Be A Thing Of The Past,'" *San Francisco Chronicle,* September 11, 2009.

47. Eric Post et al., "Ecological Dynamics Across the Arctic Associated with Recent Climate Change," *Science* 325, no. 5946 (September 11, 2009): 1355–1358.

48. Marlowe Hood, "Sea Ice Loss Major Cause of Arctic Warming: Study," *Agence France Presse,* April 28, 2010.

49. David Perlman, "Arctic Ice Getting Thinner, Fading Fast," *San Francisco Chronicle,* April 7, 2009; James Overland and Muyin Wang, "When Will the Summer Arctic Be Nearly Sea Ice Free?" *Geophysical Research Letters* 40, no. 10 (May 28, 2013): 2097–2101.

50. Jane Kay, "Arctic Ice Melt Not Quite A Record This Summer," *San Francisco Chronicle,* September 17, 2008.

51. Perlman, "Arctic Ice Getting Thinner."

52. Post, "Ecological Dynamics."

53. Associated Press, "Much Less Ice for Polar Bears," November 27, 2009.

54–55. Perlman, "Arctic Ice Getting Thinner."

56. Associated Press, "Much Less Ice."

57. Steve Connor, "Vast Methane 'Plumes' Seen in Arctic Ocean as Sea Ice Retreats," *Independent,* December 13, 2011.

Conclusions

1. Sir Nicholas Stern, *The Economics of Climate Change: The Stern Review* (Cambridge: Cambridge University Press, 2007).

2. Barack Obama, "Energy Independence and the Safety of Our Planet," speech, April 3, 2006.

3. IPCC, *Climate Change 2007*, vol.1.

4. Bruce Dorminey, "Giant Sequoias Face Looming threat from Shifting Climate, *Yale Environment 360*, March 21, 2013, http://e360.yale.edu/feature/giant_sequoias_face_looming_threat_from_shifting_climate/2631.

5. Frank Ackerman et al., "The Economics of 350: The Benefits and Costs of Stabilization," Stockholm Environment Institute, Economics for Equity and the Environment Network, 2009, http://e360.yale.edu/images/features/Economics_of_350.pdf.

6. "Military Budget of the United States," *Wikipedia*, http://en.wikipedia.org/wiki/Military_budget_of_the_United_States.

7. Ibid.

8. European Nuclear Society, "Nuclear Power Plants, Worldwide," http://www.euronuclear.org/info/encyclopedia/n/nuclear-power-plant-world-wide.htm.

9. World Economic Forum, *Green Investing 2010: Policy Mechanisms to Bridge the Financing Gap* (January, 2010), http://www3.weforum.org/docs/WEF_IV_Green Investing_Report_2010.pdf.

10. Renewable Energy Network for the 21st Century (REN 21), *Renewables 2011 Global Status Report*, http://www.ren21.net/REN21Activities/GlobalStatusReport .aspx.

11. International Energy Agency, "Investment: the Essence of Energy," *World Energy Outlook 2011*, http://www.worldenergyoutlook.org/publications/weo-2011/#d.en.25173.

12. Richard Heede, "Tracing Anthropogenic Carbon Dioxide and Methane Emissions to Fossil Fuel and Cement Producers, 1854–2010," *Cimatic Change* 122, no. 1–2 (January 2014), http://link.springer.com/article/10.1007/s10584-013-0986-y; and Suzanne Goldenberg, "Just 90 Companies Caused Two-thirds of Man-made Global Warming Emissions," *The Guardian*, November 20, 2013.

13. Dr. Lester R. Brown, "Governments Spend $1.4 Billion Per Day to Subsidize Fossil Fuels," adapted from "World on the Edge," EcoWatch, January 19, 2012, http://ecowatch.org/2012/governments-spend-1-4-billion-per-day-to-destabilize-climate.

14. Robert Reich, *Aftershock: What Has Gone Wrong With Our Economy and Our Democracy, and How to Fix It* (New York: Vintage Books, 2012).

15. Betsy Rosenberg, e-mail message to author, October 15, 2013.

Appendix to Chapter 1

1. Reuters, "Polar Regions Found Warming Fast, Raising Sea Levels," February 25, 2009.

2. Ibid.

3. Jane Kay, "Sea Ice Melt Sets Chilling Record," *San Francisco Chronicle*, September 21, 2007.

4. Ibid.

5. Walter Gibbs, "Research Predicts Summer Doom for Northern Icecap," *The New York Times,* July 11, 2000.

6. Eric Post et al., "Ecological Dynamics Across the Arctic Associated with Recent Climate Change," *Science* 325, no. 5946 (September 11, 2009): 1355–1358.

7. Ibid.

Appendix A to Chapter 4

1. J. M. Holloway and R. A. Dahlgren, "Nitrogen in Rock: Occurrences and Bio-geochemical Implications," *Global Biogeochemical Cycles* 16, no. 4 (2002): 1118.

2. IPCC, *Climate Change 2007*, vol. 1: 141.

3. Ibid., 33.

4. Whereas the Intergovernmental Panel on Climate Change's *Fourth Assessment Report* (Working Group I, Chapter 2), estimated that methane was 25 times more potent than carbon dioxide over 100 years on a per-molecule basis and 72 times as potent over a 20-year period, the Fifth Assessment increased the estimate to 28 times over 100 years and 84 times as potent over 20 years. Older EPA estimates put the per-molecule effect of methane at 20 times the potency of carbon dioxide over 100 years. See Environmental Protection Agency "Methane" EPA (April 1, 2011), accessed March 20, 2012, http://www.epa.gov/methane. See also D. T. Shindell et al., "Improved Attribution of Climate Forcing to Emissions," *Science* 326, no. 5953 (October 2009): 716–718, http://www.sciencemag.org/content/326/5953/716.

5. Stacey C. Jackson, "Parallel Pursuit of Near-Term and Long-Term Climate Mitigation," *Science* 326, no. 5953 (October 2009): 526–527.

Appendix to Chapter 8

1. Jerry Hatfield, "Changing Climate in North America: Implications for Crops," chapter 3.2 in Shyam S. Yadav et al., eds., *Crop Adaptation to Climate Change* (Hoboken, NJ: John Wiley & Sons, Ltd., 2011).

2. Ibid.

3. Josef Schmidhuber and Francesco N. Tubiello, "Global Food Security Under Climate Change," *Proceedings of the National Academy of Sciences* 104, no. 50 (December 11, 2007): 19703–19708.

4. Ibid.

5. Fred Pearce, "Huge Methane Belch in Arctic Could Cost $60 Trillion," *New Scientist* 2927 (July 24, 2013); and Gail Whiteman et al., "Vast Costs of Arctic Change," *Nature* 499, no. 7459 (July 25, 2013).

6. Rex Wyler, "Worldwide Honeybee Population Collapse: A Lesson in Ecology," EcoWatch (June 11, 2013), http://ecowatch.com/2013/worldwide-honey-bee-col lapse-a-lesson-in-ecology.

7. Mark Rowe, "The Slow Burn," *Geographical*, December 2011.

8. Schmidhuber, "Global Food Security."

Appendix to Chapter 10

1. Juliet Eilperin, "Most of Earth's 8.7 Million Species Still Undiscovered," *Washington Post*, August 24, 2011.

2. Christine Dell'Amore, "30 Amphibian Species Wiped Out in Panama Forest," *National Geographic News*, July 20, 2010, http://news.nationalgeographic.com/news/2010/07/100720-amphibians-lost-species-extinct-panama-science-environment.

3. E. O. Wilson, *The Creation: An Appeal To Save Life on Earth* (New York and London: W.W. Norton & Co., 2006), 85.

4. Ibid., 117

5. Eilperin, "Most of Earth's Species."

6–7. Wilson, *The Creation*, 32.

8. Wilson, *The Creation*.

9. Grid Arendal, "New Vision Required to Stave Off Dramatic Biodiversity Loss, Says UN Report," press release, May 10, 2010, http://www.grida.no/news/press/4221.aspx.

10. USDA., Agricultural Research Service, "Honey Bees and Colony Collapse Disorder," http://www.ars.usda.gov/News/docs.htm?docid=15572, last modified April 2, 2013.

11. Richard Black, "Nature Loss 'To Damage' Economies," BBC News, May 10, 2010, http://www.bbc.co.uk/news/10103179.

12. Wilson, *The Creation*, 98.

13. Ibid., 99.

14. "Humans Change the World," Smithsonian National Museum of Natural History, http://humanorigins.si.edu/human-characteristics/change.

15. "Paleolithic," *Wikipedia*, http://en.wikipedia.org/wiki/Paleolithic.

16. Stephen P. Hinshaw, *Origins of the Human Mind, Course Guidebook* (Chantilly, VA: The Teaching Company, 2010).

17. "Conservation of Biological Diversity," United Nations Environment Programme, http://www.unep.org/documents.multilingual/default.asp?DocumentID=52&ArticleID=63&l=en.

18. International Union for Conservation of Nature, "Saving Biodiversity—An Economic Approach," *World Conservation*, July 2010.

19. Grid Arendal, "New Vision Required."

Appendix to Chapter 11

1. Ulf Riebesell et al., "Sensitivities of Marine Carbon Fluxes to Ocean Change," *Proceedings of the National Academy of Sciences* 106, no. 49 (December 8, 2009): 20602-20609.

2–3. Ibid.

4. Dorothée Herr and Grantly R. Galland, *The Ocean and Climate Change: Tools and Guidelines for Action* (Gland, Switzerland: International Union for the Conservation of Nature, IUCN US Multilateral Office, Gland, Switzerland, 2009), 29.

5. Richard Stone, "The Invisible Hand Behind a Vast Carbon Reservoir," *Science* 328, no. 5985 (June 18, 2010).

6. Ibid.

7. David Beerling, *The Emerald Planet: How Plants Changed Earth's History* (Oxford: Oxford University Press, 2007).

8–9. Stone, "Invisible Hand."

10. Vladimir V. Yurkov and J. Thomas Beatty, "Anaerobic Anoxygenic Phototropic Bacteria," *Microbiology and Molecular Biology Reviews* 62, no. 3 (September, 1998): 695–724, http://www.ncbi.nlm.nih.gov/pmc/articles/PMC98932.

11. Riebesell, "Sensitivities of Marine Carbon."

12. Royal Society, *Ocean Acidification Due to Increasing Atmospheric Carbon Dioxide,* policy document 12/05, June, 2005, http://www.royalsoc.ac.uk.

13. Committee on the Development of an Integrated Science Strategy for Ocean Research, Monitoring, and Impacts Assessment et al., *Ocean Acidification: A National Strategy to Meet the Challenges of a Changing Ocean* (Washington, DC: National Academies Press, 2010); and Office of News and Public Information, National Research Council, "CO_2 Emissions Causing Ocean Acidification to Progress at Unprecedented Rate," September 7, 2010.

14. Committee on Development of Integrated Science Strategy, *Ocean Acidification.*

15. John A. Church, "The Changing Oceans," *Science* 328, no. 5985 (June 18, 2010).

16. IPCC, *Climate Change 2007*, vol. 2: 317.

17. Ibid., 334.

18. Ibid., 334 and 339.

19. Robert J. Nichols and Anny Cazenave, "Sea-Level Rise and its Impact on Coastal Zones," *Science* 328, no. 5985 (June 18, 2010).

20. Ibid.

21. Peter U. Clark et al., "Rapid Rise of Sea Level 19,000 Years Ago and Its Global Implications," *Science* 304, no. 5674 (May 21, 2004).

22. Daniel Howden, "Shockwaves From Melting Icecaps Are Triggering Earthquakes, Say Scientists," *Independent* (London), September 8, 2007.

23. Justin Gillis, "Reading Earth's Future in Glacial Ice," *The New York Times,* November 14, 2010.

24. Howden, "Shockwaves."

25. Gillis, "Reading Earth's Future."

26. Seth Borenstein, "Petermann Glacier in Greenland Breaks off Iceberg Twice the Size of Manhattan," *Huffington Post Green,* July 17, 2012.

27. Miguel Llanos, "Ice Melt Found across 97 Percent of Greenland, Satellites Show," NBC News, July 25, 2012, 7:59 p.m.

28. Borenstein, "Petermann Glacier."

29. Alexander Robinson et al., "Multistability and Critical Thresholds of the Greenland Ice Sheet," *Nature Climate Change* 2 (published online March 11, 2012), http://www.nature.com/nclimate/journal/v2/n6/full/nclimate1449.html.

30. Llanos, "Ice Melt."

31–32. Anil Ananthaswamy, "Sea-Level Rise, It's Worse than We Thought," *New Scientist* 2715 (July 1, 2009).

33. Jonathan L. Bamber et al., "Reassessment of the Potential Sea-level Rise from a Collapse of the West Antarctic Ice Sheet," *Science* 324, no. 5929 (May 15, 2009).

34. Ananthaswamy, "Sea-Level Rise."

35. Malcolm W. Browne, "Under Antarctica, Clues to an Icecap's Fate," *The New York Times*, October 26, 1999.

36. Donald Blankenship, comments, "Warnings From The Ice," *Nova* 2508, PBS, April 21, 1998.

37. Robert J. Nichols and Anny Cazenave, "Sea-Level Rise and Its Impact on Coastal Zones," *Science* 328, no 5985 (June 18, 2010).

38. "Konrad Steffin: Greenland Melt and the Complexities of Sea level Rise," interview by Jean DePomeru, International Polar Foundation, December 10, 2010, http://www.sciencespoles.org/articles/article_detail/konrad_steffen_greenland_melt_and_the_complexities_of_ sea_ level_rise.

39. Jonathan T. Overpeck et al., "Paleoclimatic Evidence for Future Ice-Sheet Instability and Rapid Sea-level Rise," *Science* 311, no. 5768 (March 24, 2006).

40. Stuart Staniford, "Greenland, or Why You Might Care About Ice Physics," *The Oil Drum*, January 28, 2007, http://theoildrum.com/story/2005/12/9/31522/5910.

41. Llanos, "Ice Melt."

42. Overpeck, "Paleoclimatic Evidence."

43. Nichols, "Sea-level Rise."

44. Jonathan L. Bamber et al., "Reassessment of the Potential Sea-Level Rise."

45. Deborah Zabarenko, "Greenland Ice Could Fuel Severe US Sea Level Rise," Reuters, May 27, 2009.

46. Herr, *The Ocean and Climate Change.*

47. Carolyn Lochhead, "Immediate Shift Needed to Slow Global Warming," *San Francisco Chronicle*, January 28, 2013.

48. J. Silverman et al., "Coral Reefs May Start Dissolving When Atmospheric CO_2 Doubles," *Geophysical Research Letters* 36, no. 5 (March 13, 2009).

49. Ibid.

50. Riebesell, "Sensitivities of Marine Carbon."

51. US Department of Commerce, "NOAA/NSF Cruise Reveals Impacts of Ocean Acidification on Chemistry, Biology of North Pacific Ocean," *NOAA Magazine*, April 5, 2006.

52. Andrew Moy et al., "Reduced Calcification in Southern Ocean Planktonic Foraminifera," letter, *Nature Geoscience* 2 (March 8, 2009): 272–280.

GLOSSARY

Albedo: surface reflectivity of the Earth or another object, often given as a fraction or percentage of incident solar radiation.

Acidification: an increase in the acidity of seawater due to the absorption of carbon dioxide and measured by a decrease in pH (the negative log of the hydronium ion concentration).

Adaptation: generally, action to reduce the vulnerability of the human environment to climate change or mitigate its costs, for example by building seawalls or retreating from the coast in response to sea-level rise. May also refer to natural adjustments by ecosystems or organisms to climate change.

Aerosols: solid or liquid airborne particles that usually remain in the atmosphere for a few hours and are typically 0.01 to 10 millionths of a meter in size.

Algal bloom: a rapid, dense overgrowth of phytoplankton in a body of water.

Aquifer: an underground geological formation consisting of a layer of permeable rock, sand, or gravel saturated with water.

Atlantic Meridional Overturning Circulation (AMOC): an enormous ocean current that transfers heat from the equator to the poles.

Biochar: charcoal produced by heating plant material in a zero- or low-oxygen atmosphere.

Biodiversity: the total number of kinds of all types of organisms found in an ecosystem or area of any size, from the microscopic to the planetary scale.

Bioengineering: the use of an engineered device, including artificially cultured living tissue, to repair biological damage to an organism; hence, also the artificial creation of life from genetic material.

Biomass: the total mass of organisms or organic matter, such as plants, animals, or vegetation, in a given area or habitat. Usually used in reference to plants or vegetation, particularly those suitable for use as fuel for energy production.

Carbonate rock: generally a sedimentary rock containing more than 50 percent carbonate minerals, usually limestone (calcium carbonate) or dolostone (calcium-magnesium carbonate). The carbonate particles in limestone typically incorporate fossilized shells of marine organisms.

Carbon dioxide: the gaseous compound formed of carbon and oxygen that both occurs naturally and is also created when fossil fuels are burned. Because of its long persistence in the atmosphere, its heat-trapping ability, and the billions of tons that have been released, it is the principal greenhouse gas.

Carbon sink: generally a natural system, such as a forest, grassland, wetland, or ocean, that absorbs more carbon than it releases and stores it for a period of time, thus keeping it out of the atmosphere.

Chytrid fungus: a now-widespread fungus, *Batrachochytrium dendrobatidis*, of the class Chytridiomycetes, that causes a lethal disease known as chytridiomycosis among amphibians. Most likely introduced by long-distance human transportation of wildlife, it infects the skin of amphibians and is responsible for a large number of rapid population declines and extinctions. While all 6,000 species of amphibians are potentially susceptible to infection, some species, such as the American bullfrog and the African clawed frog, carry the infection but do not become ill.

Clathrate (methane): See Methane hydrate.

Climate cycles: periodic variations in climate that are beyond the scope of individual weather events and are caused either by processes internal to the climate system, by solar variations and volcanoes, or are induced by human combustion of fossil fuels and land-use changes that together alter the atmosphere.

Climate feedback: a response from the climate system that intensifies or diminishes an initial climate change. When the feedback intensifies the original process, it is called a positive feedback, and when it diminishes the effect, it is known as a negative feedback.

Climate sensitivity: the responsiveness of the climate to a change in the concentration of greenhouse gases in the atmosphere. In reports of the IPCC, climate sensitivity refers specifically to the equilibrium annual mean global surface temperature change that results when the concentration of greenhouse gases is doubled.

Cimate thresholds: a condition on a continuum of climate change that is both brought about by a climate forcing (that is, by an intervention in a normal climate cycle) and beyond which a significant climate change occurs that is irreversible or recoverable only over a very long time span. An example is an increase in average global surface temperature sufficient to cause widespread melting of Arctic permafrost.

Climate tipping element: a part of the climate system that is susceptible to responding to a climate forcing by causing the climate system to abruptly shift to another state. The new state is characterized by a new mean condition outside the range of normal fluctuations seen in the existing state around its mean.

Cognitive dissonance: a disharmony or incompatibility between two or more conflicting beliefs, values, attitudes, behaviors, or emotions concurrently held by the same person, often manifested as a disparity between attitude and behavior or between one's self-interest and ideals. The discrepancy produces psychological discomfort, which the person may try to resolve by changing their belief, their behavior, or more typically, by changing the perception of the action by rationalizing it.

Convection: the transfer of energy by the mass movement of molecules. In meteorology, the rising of air warmed by contact with the Earth's surface to drive the circulation of the Earth's atmosphere. The convection of moist air causes the condensation of moisture into clouds and the release of heat, which can intensify convection.

Coral bleaching: the loss or expulsion of zooxanthellae by coral polyps due to rising ocean temperatures or other adverse conditions, resulting in a whitening of the coral through loss of the more colorful algae.

Coral polyp: a tiny colonial organism with a hollow cylindrical body and stinging tentacles related to sea anemone but with a hard calcium carbonate (limestone) skeleton called a *calicle* at its base. The coral polyp attaches itself to a hard surface of the sea floor, where it buds or divides and forms an interconnected living mat of polyps known as a colony. Over long period of time, the replicating polyps accumulate in large numbers, the annual growth of their calicles gradually accreting to form a reef.

Cyclone: the name for a hurricane in the Indian Ocean and Bay of Bengal.

Dansgaard-Oescher oscillations: Rapid large temperature changes of 10.8 to 18°F that occurred during the last ice age.

Disease vector: that which hosts and then transmits a pest or disease organism.

El Niño/Southern Oscillation (ENSO): a naturally occurring climate cycle consisting of a sequential fluctuation (usually on a four-year cycle) in the location of a vast pool of warm ocean water in the tropical Pacific. Its eastward expansionary phase is known as an El Niño, marked by unusually warm water (that may exceed 80°F) in the equatorial Pacific. Its contraction is called a La Niña and is associated with a strong pullback of the warm pool and an unusually cold equatorial Pacific. The ENSO cycle not only affects ocean temperatures but also winds and upwelling. It also affects rainfall and temperature patterns in various parts of the world.

European Union's Emissions Trading Scheme (ETS): a regional carbon trading scheme.

Evapotranspiration: the evaporation of water from the soil and from the surfaces of plant leaves by transpiration, the process by which water is extracted from the soil by roots and raised by capillary action to tiny pores on the undersides of leaves, from where it vaporizes.

Extratropical: outside the tropics and often used to refer to extratropical cyclones or hurricanes arising in midlatitudes. In an extratropical hurricane, the primary energy source has shifted from the latent heat of condensation to the contrast in temperature between warm and cold fronts.

Foraminifera: single-celled marine protozoa of the phylum *Foraminifera* having perforated shells through which pseudopods protrude for trapping food. Foraminifera have been around for 540 million years and are pervasive in marine sediments. The composition of their shells changes with changes in the chemical composition of seawater, and hence they are valuable proxy for ancient climates.

Forcing agents: determinants of climate that are not interactive with the normal internal workings of the climate system and thus, in that sense, operate outside the system, influencing it without themselves being altered. Examples include human-induced alterations in the composition of the atmosphere and land-use changes as well as volcanic eruptions and variations in the sun's output.

Greenhouse gases: Greenhouse gases are natural and synthetic atmospheric gases that absorb and reemit heat radiation emitted by the Earth, atmosphere, and clouds, thereby reducing the transmission of heat to interplanetary space and warming the Earth by the greenhouse effect. The primary greenhouse gases are water vapor, carbon dioxide, nitrous oxide, methane, and ozone, all of which occur naturally in the atmosphere but are augmented by human activity. Humans are also responsible for releasing synthetic halocarbons, other chlorine and bromine compounds, and sulfur hexafluoride, all of which are greenhouse gases.

Gross Domestic Product: all the goods and services domestically produced by the citizens of a nation, hence a measure of its economic performance.

Groundwater: a subsurface flow of water that causes springs to flow, and moistens the root zones of vegetation, and is often pumped out from wells.

Halocarbons: a class of synthetic gaseous chemical compounds of carbon and one of the halogen elements (chlorine, fluorine, bromine, iodine). Commonly used for propellants and refrigerants until regulated by the Montreal Protocol, they are inert and nontoxic in the lower atmosphere. In the stratosphere, however, they destroy large amounts of ozone that otherwise would shield life from dangerous ultraviolet radiation.

Heat exhaustion: a more serious heat illness than heat stress. In addition to all the symptoms of heat stress, a heat exhaustion victim may also experience chills, pale, moist skin, lightheadedness, fainting, weakness, nausea, cramps, excessive sweating, headache, mood changes, and symptoms of dehydration, such as dry mouth, extreme thirst, and dark-colored urine. As in heat stress, the brain and other internal organs are being deprived of blood flow, and hence oxygen, as the body's heat-regulating mechanisms increase blood flow to the skin to cool the body.

Heat stress: an early-stage heat-induced illness caused by exposure to excessive heat and marked by symptoms that may include dizziness, fatigue, impaired judgment, inability to concentrate, and irritability. Heat stress occurs when—in an effort to maintain a normal core body temperature of 98.6°F—blood flow is diverted from the brain and other internal organs to the skin to cool the body.

Heat stroke: the final stage of a continuum of disorders commencing with heat stress and progressing through heat exhaustion. This life-threatening heat illness may occur suddenly and is marked by extremely high body temperature, lack of sweating, fast pulse, dizziness, and dry, hot, flushed skin. Confusion, aggressive behavior, seizures, convulsions, or coma may occur. The high internal body temperature associated with it may cause brain and other organ damage and death.

Holocene Epoch: the current warm period of geologic time that has lasted since the end of the last ice age 11,700 years ago. It coincides with the dawn of agriculture and with all of recorded human history during which humans evolved from their primitive conditions in the preceding Pleistocene Epoch to inhabitants of modern industrial civilization.

Infrared radiation: heat energy, a form of electromagnetic radiation whose wavelength is longer than that of visible light but shorter than that of radio waves, hence it ranges from 700 nanometers to 1 mm. (A nanometer is equal to a billionth of a meter.) The name *infrared* comes from the fact that its frequency is below that of red light. The Earth's surface, the atmosphere, and the clouds all emit longwave infrared radiation.

Invertebrates: animals without a backbone or bony skeleton. They are found in more than 30 phyla or major taxonomic divisions of animal life and comprise 97 percent of all species on the Earth.

Khystym Disaster: the 1957 explosion of a nuclear waste dump in Chelyabinsk in the Ural Mountains.

Krill: a small plankton-feeding marine crustacean of the family Euphausiidae that is an important part of the marine food web and is found in all the world's oceans, where it is eaten by many marine animals, including baleen whales, other marine mammals, and fish.

Kyoto Protocol: the world's first international climate protection treaty adopted in Kyoto, Japan, in 1997 as a part of the 1992 UN Framework Convention on Climate Change. The Kyoto Protocol called on the world's industrialized nations to voluntarily and collectively reduce their human-caused greenhouse gas emissions by at least 5.2 percent below their 1990 levels between 2008 and 2012, a modest first step toward control of global carbon emissions. The United States signed but did not ratify the agreement after a fossil fuel industry campaign of opposition led to its defeat in the Senate. The treaty went into effect in 2005 after more than 55 nations representing more than 55 percent of the world's greenhouse gas emissions approved the agreement.

La Niña: a periodic cooling in the central and eastern tropical Pacific that usually lasts for 9 to 12 months and may persist as long as two years. Sea surface temperatures in parts of the tropical Pacific may fall by as much as 7°F during a La Niña, accompanied by a strengthening of easterly trade winds. Like El Niño, La Niña influences global and regional weather patterns, often producing opposite climate conditions from El Niño in a given locality. See El Niño/Southern Oscillation.

Market externalities: impacts of economic activities that do not register properly in the marketplace and therefore are not explicitly or properly charged for nor credited in the market. In the context of climate change, externalities inflicted on the public by the combustion of fossil fuels include climate disturbance from the discharge of greenhouse gases, and the production of air, water, and noise pollution and public health effects.

Megadrought: a prolonged, severe drought lasting decades to centuries. The paleoclimatic record reveals that megadroughts have previously occurred in the American Southwest during the Medieval Warm Period of about 900 to 1,300 AD when the Northern Hemisphere warmed.

Methane: the second most important greenhouse gas and the main constituent of natural gas, it only persists for about a dozen years in the atmosphere but is many times more powerful a greenhouse gas than carbon dioxide. It occurs naturally but is also discharged to the atmosphere during the production and transport of natural gas as well as from livestock raising and decomposition of organic matter in landfills.

Methane hydrates: a slushy mixture of methane gas and ice usually found buried in ocean sediments where, due to cold temperatures and high pressure, the gas is trapped in a crystal lattice cage of frozen water.

Milanković cycles: The Earth's tilt, wobble, orbital shape changes, and the changes in timing of the perihelion are collectively known as the Milanković cycles. Their combined effect slightly varies the amount and distribution of radiation reaching the Earth over geologic time, thus initiating the Earth's climate cycles.

Monsoon: a persistent seasonal wind that brings heavy rains to South Asia and elsewhere. The monsoon is caused by a difference in temperature between a land mass and the ocean; it typically blows toward land during summer and away from land in winter.

Negative feedback: See Climate feedback.

Nitrous oxide (N_2O): a powerful and long-lived noncarbon greenhouse gas that occurs naturally and is also released mainly in the course of human agricultural activities, particularly from decomposing manure, as well as from fossil fuel burning and industrial activities.

Ozone: a corrosive allotrope of oxygen composed of three oxygen atoms instead of the two of a normal oxygen molecule. In the lower atmosphere, ozone is formed by a chemical reaction of nitrogen oxides, methane, and volatile organic compounds (VOCs).

Paleocene-Eocene Thermal Maximum (PETM): a period of time about 56 million years ago when the Earth experienced a rapid heating (over a period of only a few thousand years) that raised global average temperatures by more than 9°F. The PETM lasted for about 100,000 years and caused widespread ecological changes, ocean acidification, and mass extinctions, all caused by a massive increase in atmospheric carbon that may have come from the sudden release of methane hydrates from the ocean floor or other sources.

Palmer Drought Severity Index: a measure of soil moisture deficits based on temperature, precipitation, and available water content.

Perihelion: the point on the orbit of a planet, comet, or asteroid when it is nearest to the sun.

Permafrost: perennially frozen ground, most formed during the last glacial period, containing billions of tons of frozen organic carbon that is gradually being released as the permafrost melts. Permafrost occurs in a quarter of the land area of the Northern Hemisphere in distributions ranging from patchy to continuous, and may be anywhere from a yard to almost a mile thick.

Photosynthesis: a biochemical process powered by sunlight in which a plant, alga, or cyanobacterium absorbs solar energy in its chlorophyll (a light-sensitive pigment), takes carbon dioxide from the air, combines it chemically with water, and uses it to produce carbohydrates and oxygen. The oxygen not needed for cellular respiration is released to the atmosphere as a waste product of photosynthesis and comprises the oxygen we breathe.

Phytoplankton: consists of microscopic floating unicellular photosynthetic organisms including algae, bacteria, and protists (an organism that typically has a single nucleated cell and the characteristics of both a plant and an animal but is neither). Phytoplankton form the base of the ocean food web and include green algae, cyanobacteria, diatoms, protozoa, dinoflagellates, and cocolithosphores.

Positive feedback: See Climate feedback.

Pteropods: a small group of transparent marine snails and slugs that propel themselves through the waters of the open ocean and are found in great abundance in Arctic seas and elsewhere. They are an important component of the oceanic food web.

Quaternary Period: the most recent major geologic period, which lasted for about 2 million years and encompasses both the Pleistocene and, during the past 17,000 years, our own Holocene Epoch. The Quartenary consisted of long ice ages in which vast continental ice sheets and mountain glaciers and permafrost accumulated, punctuated by relatively short interglacial periods. The Holocene is one of these relatively brief interglacials.

Residence time: the average length of time a substance persists in some defined area, as in a lake, soil, ocean, atmosphere, body, or cell, or in a particular condition. Thus, the term is used to describe the time that a pollutant might persist in a reservoir before flowing out, or the length of time a greenhouse gas persists in the atmosphere or other part of the climate system.

Respiration: in common usage, the acquisition of oxygen by higher animals through inhalation or diffusion and the release of carbon dioxide and water. However, in biology, respiration is the organism's total physical and chemical oxygen demand, during which oxygen is combined with organic compounds in a series of metabolic steps within cells to release energy, while releasing carbon dioxide and water as byproducts to the environment. Anaerobic organisms perform a series of similar energy-releasing metabolic reactions without oxygen that are also referred to as respiration.

Sahel: an arid and semiarid geographic zone extending in an east–west band about 600 miles wide from the Atlantic to the Red Sea across 12 countries of sub-Saharan Africa, some of which are likely to experience severe drought impacts due to climate change within coming decades.

Saltwater intrusion: generally, the infiltration of saltwater from the ocean or estuary through the ground and into freshwater supplies. Often used to refer to the contamination of coastal freshwater aquifers by seawater caused by the overpumping of groundwater. Rising sea level may increase the rate and extent of saltwater intrusion.

Thermal inertia: is a measure of a body's ability to store and conduct heat and therefore determines the time required for a body to adjust to the temperature of its surroundings. Formally, it is the square root of the product of thermal conductivity multiplied by volumetric specific heat, which is density times specific heat. (Specific heat is the energy required to raise a substance by one degree kelvin.) The larger and denser a body is, the greater its capacity to store heat and the longer it will take to adjust to the temperature of its surroundings. Thus, the temperature change of the ocean lags the temperature changes of the atmosphere.

Thermocline: the shifting subsurface boundary layer in the ocean between the relatively well-mixed, warmer surface waters and the cooler, deeper waters. It is here that water temperature declines most rapidly with increasing depth compared with the surface water above or deeper water below. The depth of the thermocline is affected by surface heating (hence, by the seasons), by winds, and by El Niño/Southern Oscillation cycles.

Thermohaline circulation: oceanic circulation that transports heat from the equatorial regions toward the poles and returns cold dense water at depth from polar to equatorial regions.

Tipping element: a part of the climate system that, in response to a climate forcing, may cause the climate system to abruptly shift to another state outside the range of normal climate fluctuations. Examples include the Amazon forest, the Greenland Ice Sheet, and seabed methane-hydrate deposits.

Tipping point: a point in the condition of a tipping element at which further change would precipitate major irreversible climate change.

Troposphere: Earth's lower atmosphere extending from the planet's surface to a height of 4 to 12 miles, containing 75 to 80 percent of the entire mass of the atmosphere as well as almost all its water vapor (hence almost all clouds) and dust particles. Temperature, density, and atmospheric pressure decline with altitude in the troposphere. Its height varies by season, latitude, and from day to night. Almost all of the world's weather is created in the troposphere.

Troubled Asset Relief Program (TARP): A $700 billion Treasury fund created by Congress during the subprime mortgage financial crisis of 2007 and 2008 as part of the Emergency Economic Stabilization Act of 2008 to rescue troubled banks and other large financial institutions. The fund, eventually reduced by Congress to $475 billion, was used to subsidize banks with public funds by purchasing steeply devalued "toxic" mortgage-backed securities for which no market existed, or far above their market value, and by purchasing shares of affected financial institutions. A major purpose of TARP was to unfreeze credit; however, banks generally used TARP funds to strengthen their balance sheets without significantly increasing credit. Most TARP funds have now been repaid to the Treasury.

Tundra: vast flat treeless plains of the Arctic, Antarctica, and various alpine zones where tree growth is inhibited by cold and by the short growing season.

Typhoon: the term used for hurricanes in the western Pacific.

Standard deviation: the square root of the variance, i.e., the square root of the average of the squared differences of each value in the population (or sample) from the mean.

Stratosphere: the part of the Earth's atmosphere above the lower atmosphere (troposphere) extending from about 11 to 32 miles above the Earth's surface to the height at which the mesosphere begins. Whereas temperature decreases with altitude in the troposphere, it increases slightly with altitude in the stratosphere because of the increasing absorption of ultraviolet radiation there.

Subduction zones: areas in which the enormous plates of the Earth's crust collide with each other causing one plate to plunge beneath the other, resulting in fractures in the Earth's crust, mountain building, and volcanic activity.

Urban heat island effect: a process in which solar heat is absorbed by stone and other building materials and reradiated within the confines of an urban area.

Volatile organic compounds (VOCs): organic chemical compounds, frequently toxic, involved in the manufacture of many everyday products, such as solvents and fuels, that evaporate under normal atmospheric conditions and are widespread in the outdoor and indoor environment. They participate in reactions that cause photochemical smog and are common groundwater contaminants. Examples include acetone, benzene, formaldehyde, methyl chloride, propane, and other gasoline components.

Wet-bulb temperature: the temperature of a thermometer wrapped in a wet porous material, providing a single measurement based both on temperature and humidity.

Zooplankton: floating and drifting marine invertebrates that feed upon phytoplankton and do not perform photosynthesis. Most marine taxa are represented in plankton. Some are microscopic while others range in size up to several inches. They include animals that live their entire lives as plankton as well as the larvae of other marine organisms, such as mollusks, coral, and fish, that eventually change into them. (See Phytoplankton.)

Zooxanthellae: a single-celled algae found in a nutritional partnership with coral polyps to which they provide oxygen and nutrients produced during photosynthesis.

RESOURCES FOR FURTHER
CLIMATE INFORMATION

Recommended Books

Archer, David. *Global Warming: Understanding the Forecast.* 2nd ed. Hoboken, NJ: Wiley-Blackwell, 2011.

Archer, David. *The Global Carbon Cycle.* Princeton: Princeton University Press, 2010.

Archer, David. *The Long Thaw: How Humans are Changing the Next 100,000 Years of Earth's Climate.* Princeton: Princeton University Press, 2009.

Beerling, David. *The Emerald Planet: How Plants Changed Earth's History.* Oxford: Oxford University Press, 2007.

Benestad, Rasmus. *Solar Activity and Earth's Climate.* 2nd ed. Berlin: Springer-Verlag, 2006.

Berger, John J. *Climate Myths: The Campaign Against Climate Science.* Berkeley, CA: Northbrae Books, 2013.

Bowen, Mark. *Thin Ice: Unlocking the Secrets of Climate in the World's Highest Mountains.* New York: Henry Holt and Company, LLC, 2005.

Bradley, Raymond. *Global Warming and Political Intimidation.* Amherst: University of Massachusetts Press, 2011.

Bradley, Raymond. P*aleoclimatology: Reconstructing Climates of the Quaternary.* Amsterdam: Academic Press, 2005.

Brown, Lester R. *Plan B 4.0: Mobilizing to Save Civilization.* New York: W.W. Norton & Company, 2009.

Dauncey, Guy. *The Climate Challenge: 101 Solutions to Global Warming.* Gabriola Island, BC: New Society Publishers, 2009.

Dyer, Gwynne. *Climate Wars: The Fight for Survival as the World Overheats.* Oxford: Oneworld Publications, 2011.

Epstein, Paul R. and Dan Ferber. *Changing Planet, Changing Health: How the Climate Crisis Threatens Our Health and What We Can Do About It.* Berkeley, CA: University of California Press, 2011.

Fagan, Brian M. *The Great Warming: Climate Change and the Rise and Fall of Civilizations.* New York: Bloomsbury Press, 2008.

Flannery, Tim. *The Weather Makers: How Man Is Changing the Climate and What it Means for Life on Earth.* New York: Atlantic Monthly Press, 2005.

Gardiner, Stephen M. *A Perfect Moral Storm: Understanding the Ethical Tragedy of Climate Change.* New York: Oxford University Press, 2011.

Gore, Al. *An Inconvenient Truth: The Planetary Emergency of Global Warming and What We Can Do About It*. Emmaus, PA: Rodale Inc., 2006.

Gore, Al. *Our Choice: A Plan to Solve the Climate Crisis*. Emmaus, PA: Rodale Inc., 2009.

Hansen, James. *Storms of My Grandchildren: The Truth About the Coming Climate Catastrophe and Our Last Chance to Save Humanity*. New York: Bloomsbury, 2009.

Harte, John and Harte, Mary Ellen. *Cool the Earth, Save the Economy: Solving the Climate Crisis is EASY*. 2008. Web, http://cooltheearth.us.

Intergovernmental Panel on Climate Change. *Climate Change 2013: The Physical Science Basis. Working Group 1 Contribution to the Fifth Assessment Report of the Intergovernmental Panel on Climate Change*. New York: Cambridge University Press, 2014.

Koomey, Jonathan. *Cold Cash, Cool Climate: Science-Based Advice for Ecological Entrepreneurs*. Chicago: Analytics Press, 2012.

Linden, Eugene. *The Winds of Change: Climate, Weather, and the Destruction of Civilizations*. New York: Simon & Schuster, 2006.

Lovins, Amory B. and Rocky Mountain Institute. *Reinventing Fire: Bold Business Solutions for the New Energy Era*. White River Junction, VT: Chelsea Green Publishing, 2011.

Lovins, L. Hunter, and Boyd Cohen. *Climate Capitalism: Capitalism and the Age of Climate Change*. New York: Hill and Wang, 2011.

Lynas, Mark. *Six Degrees: Our Future on a Hotter Planet*. Washington, DC: National Geographic, 2008.

Mann, Michael E. *The Hockey Stick and the Climate Wars: Dispatches from the Front Lines*. New York: Columbia University Press, 2012.

Mann, Michael and Lee R. Kump. *Dire Predictions: Understanding Global Warming*. New York: DK Publishing, 2008.

Makhijani, Arjun. *Carbon-Free and Nuclear-Free: A Roadmap for US Energy Policy*. Takoma Park, MD: IEER Press, 2007. Muskegon, MI: RDR Books, 2007.

Metz, Bert. *Controlling Climate Change*. Cambridge: Cambridge University Press, 2010.

National Research Council. *Abrupt Impacts of Climate Change: Anticipating Surprises*. Washington, DC: The National Academies Press, 2013.

Oreskes, Naomi, and Erik Conway. *Merchants of Doubt: How a Handful of Scientists Obscured the Truth on Issues from Tobacco Smoke to Global Warming*. New York: Bloomsbury Press, 2010.

Pearce, Fred. *With Speed and Violence: Why Scientists Fear Tipping Points In Climate Change*. Boston: Beacon Press, 2007.

Pierrehumbert, Ray. *Principles of Planetary Climate*. Cambridge: Cambridge University Press, 2010.

Pierrehumbert, Ray and David Archer. *The Warming Papers: The Scientific Foundation for the Climate Change Forecast*. Hoboken, NJ: Wiley–Blackwell, 2011.

Powell, James Laurence. *The Inquisition of Climate Science*. New York: Columbia University Press, 2011.

Rahmstorf, Stefan and David Archer. *The Climate Crisis: An Introductory Guide to Climate Change*. Cambridge: Cambridge University Press, 2010.

Richardson, Katherine and Stefan Rahmstorf. *Our Threatened Oceans*. London: Haus Publishing, 2008.

Romm, Joseph. *Hell and High Water: Global Warming—the Solution to the Politics—and What We Should Do*. New York: William Morrow, 2007.

Royal Society and US National Academy of Sciences. *Global Climate Change: Evidence and Causes. An Overview from the Royal Society and the US National Academy of Sciences*. London: Royal Society, 2013. Washington, DC: National Academic Press, 2013.

Washington, Hayden and John Cook. *Climate Change Denial: Heads in the Sand*. London: Earthscan, 2011.

Wolfe, Joshua and Gavin Schmidt. *Climate Change: Picturing the Science*. New York: W.W. Norton, 2009.

Sources of Climate Science Information

The most comprehensive scientific reports about climate change are those of the United Nations Intergovernmental Panel on Climate Change, which are available on the web at www.ipcc.ch and in print from Cambridge University Press. The site includes many valuable links for further information.

Climate Action Network: a global network of NGOs working to promote action to limit human-induced climate change.

ClimateArk, www.climateark.org: a customized search and news feed of reviewed, climate news.

Climate Central, www.climatecentral.org: an independent organization of leading scientists and journalists researching and reporting the facts about our changing climate and its impact on the American public. Climate Central surveys and conducts scientific research on climate change and informs the public of key findings.

Climate Change News Feed of the Environmental News Network, www.enn.com/topics/climate.

Climate Science Legal Defense Fund, climatesciencedefensefund.org: a fund providing legal defense for climate scientists targeted for intimidation, so that climate scientists can conduct their research without the threat of politically motivated attacks.

Climate Science Rapid Response Team, www.climaterapidresponse.org: a group advocating for science education that matches climate scientists with lawmakers and the media to provide rapid, high-quality information to media and government officials.

Climate Science Watch, www.climatesciencewatch.org: a nonprofit public interest education and advocacy project dedicated to holding public officials accountable for using climate research effectively and with integrity in dealing with the challenge of global climate disruption. With a focus on US national policy developments, Climate Science Watch investigates the misuse of climate change research and assessments in politics and policy making.

DeSmogblog.com, www.DeSmogblog.com: a blog that provides fact-based information on global warming and exposes climate science disinformation campaigns.

Energy and Environmental Study Institute, www.eesi.org: was founded by a bipartisan congressional caucus in 1984 and now is an independent organization that serves as a trusted source of credible, nonpartisan information on climate and on energy and environmental solutions for Congress and other stakeholders.

Exxonsecrets.org: a Greenpeace research project devoted to exposing the climate disinformation campaign that Exxon has supported for more than a decade in an effort to downplay the urgency of addressing climate change

Indymedia UK, www.indymedia.org.uk/en/topics/climate: a network of individuals and independent and alternative media activists and organizations offering grassroots, noncorporate, noncommercial coverage of important social and political issues.

International Climate Change Partnership, www.iccp.net: a global coalition of companies and trade associations from diverse industries committed to constructive and responsible participation in the international policy process concerning global climate change.

International Council for Local Environmental Initiatives, www.iclei.org: a network of 12 megacities, 100 supercities and urban regions, 450 large cities, as well as 450 medium-sized cities and towns in 86 countries collaborating on lowering greenhouse gas emissions and promoting global sustainability in many other ways.

International Energy Agency, www.iea.org: an autonomous organisation with 28 member countries concerned about reliable affordable energy and economic development. They publish the *World Energy Outlook* annually and other important energy documents.

International Institute for Sustainable Development, www.iisd.org: a Canadian-based, international public policy research institute focusing on sustainable development and issues "ripe for transformation." Provides newsletters and videos for keeping up with international climate negotiations and other issues.

Organization for Economic Cooperation and Development, www.oecd.org: exists to promote policies that will improve economic and social well-being worldwide, and provides a forum for governments to work cooperatively on common problems. Its reports measure productivity and global flows of trade and investment and predict future trends.

RealClimate, www.realclimate.org: provides authoritative climate science from climate scientists.

Reality Drop Project, realitydrop.org: collects important climate news stories, provides related climate science, and invites readers to share the news with their social networks.

Skeptical Science, www.skepticalscience.com: examines the science and arguments of global warming skepticism. Through news and analysis of common skeptical arguments, this blog debunks climate science denial.

SourceWatch, www.sourcewatch.org: a collaborative online wiki operated by the Center for Media and Democracy that profiles the activities of energy industry front groups, industry-friendly experts, industry-funded organizations, and think tanks trying to manipulate public opinion on behalf of corporations or government. SourceWatch also discusses public policies they are trying to affect and provides avenues for citizen involvement.

Target 300, www.Target 300.org: a group working toward stabilizing atmospheric carbon dioxide at 300 ppm or less.

350.org, www.350.org: a group building a global grassroots movement to solve the climate crisis by instituting policies to reduce atmospheric carbon dioxide levels back to 350 ppm.

UN Framework Convention on Climate Change, www.unfccc.de: the website of the international agreement that produced the Kyoto Protocol and coordinates other international climate negotiations. The site has a searchable, country-by-country database on greenhouse gas emissions and extensive other climate information resources.

US Department of Energy, Energy Information Administration (EIA), www.eia.gov: has masses of data on energy-related greenhouse gas emissions and is a great resource on major energy sources and technologies, including data on production of electricity and fuels as well as renewable energy sources. The EIA is host to the National Energy Information Center, which responds to public information requests.

US Department of Energy, National Renewable Energy Laboratory, www.nrel.org: the place to go for scientific and technical information on renewable energy, as are the websites of other national laboratories, such as the Lawrence Berkeley National Laboratory and DOE's 16 other national laboratories.

US Department of Energy, National Center for Atmospheric Research (NCAR) and University Corporation for Atmospheric Research (UCAR), www2.ucar.edu: The University Corporation for Atmospheric Research is a consortium of more than 100 member colleges and universities focused on research and training in the atmospheric and related Earth system sciences. It sets directions and priorities for NCAR. NCAR provides research, observing and computing facilities, and a variety of services for atmospheric and other Earth scientists.

US Environmental Protection Agency, www.epa.gov: has extensive information on climate change and greenhouse gas emissions by emitting facility.

US Federal Highway Administration, Office of Planning, Environment & Realty: Climate Change, www.fhwa.dot.gov/environment/climate_change: supports transportation and climate change research and disseminates the results in the areas of climate change mitigation and adaptation, provides technical assistance to stakeholders, and coordinates its activities within the US Department of Transportation (DOT) and other federal agencies.

US Global Climate Change Research Program, www. globalchange.gov: integrates federal climate research across all federal agencies and provides studies of regional and sectoral climate impacts, other climate research, as well as links to most other federal climate-related programs.

US National Aeronautics and Space Administration, www.nasa.gov: has a master directory of worldwide climate change data holdings, the Global Change Master Directory at www.gcmd.gsfc.nasa.gov, and much other useful information. Its Goddard Institute for Space Studies has extensive climate data sets for professional climate science research, and other useful resources.

Environmental, Climate, and Energy Organizations

Alliance for Climate Protection: www.climateprotect.org

Alliance for Renewable Energy: www.allianceforrenewableenergy.org

Alliance to Save Energy: www.ase.org

American Council for an Energy Efficient Economy: www.aceee.org

American Council on Renewable Energy: www.acore.org

American Forests: www.americanforests.org

American Solar Energy Society: www.ases.org

Business Council for Sustainable Energy: www.bcse.org

Center for Change and Energy Solutions: www.c2es.org

Center for Energy Efficiency and Renewable Technology: and www.ceert.org

Center for Environmental Information, Inc.: www.rochesterenvironment.com/resources.htm

Center for Resource Solutions: www.resource-solutions.org

Clean Power Campaign: www.cleanpower.org

Climate Progress: thinkprogress.org/climate/issue

Climate Action Network: www.climatenetwork.org

Corporate Watch: www.corpwatch.org

Critical Mass Energy Project: www.citizen.org/cmep

Earth Day 2000: www.earthday.net

Earth Island Institute: www.earthisland.org

Environmental Alliance for Senior Involvement: www.easi.org

Environmental Defense Fund: www.edf.org

Environmental Media Services: www.ems.org

The Federation of State Public Interest Research Groups: www.uspirg.org

Friends of the Earth: www.foe.org

Global Climate Change Digest, published in *Global Change*, the electronic edition: www.globalchange.org/default.htm

Greenpeace: www.greenpeace.org

Midwest Renewable Energy Association: www.midwestrenew.org/home

National Audubon Society: www.audubon.org

National Wildlife Federation: www.nwf.org

Natural Resources Defense Council: www.nrdc.org

OSS Foundation: www.ossfoundation.us

Pace Law Energy and Climate Center: www.law.pace.edu/energy-and-climate
    -center

Pacific Institute for Studies in Development, Environment, and Security:
    www.pacinst.org

Physicians for Social Responsibility: www.psr.org

Renewable Energy Policy Project: www.repp.org

The Rocky Mountain Institute: www.rmi.org

The Rural Alliance for Renewable Energy: www.infinitepower.org/rare

Sierra Club: www.sierraclub.org

Solar Century: www.solarcentury.co.uk

Stockholm Environment Institute: www.sei.se

Tata Energy Research Institute: www.teriin.org

Tiempo Climate Newswatch: www.tiempocyberclimate.org/newswatch

Union of Concerned Scientists: www.ucsusa.org

University of East Anglia (England) Climate Research Unit: www.cru.uea.ac.uk

US Climate Action Network (same as CAN): www.usclimatenetwork.org

US Country Studies Program: www.gcrio.org/CSP

Western Clean Energy Campaign: www.westerncec.org

World Energy Council: www.worldenergy.org

World Health Organization: www.who.int/en

World Meteorological Organization: www.wmo.int/pages/index_en.html

Worldwatch Institute: www.worldwatch.org

World Wildlife Federation: www.wwf.panda.org/about_our_earth/aboutcc

Scientific Journals on Climate Change and the Environment

Atmospheric Chemistry and Physics

Atmospheric Environment

Climatic Change

Climate Dynamics

Climate Policy

Climate Research

Energy Policy

Environmental Research Letters

Environmental Resource Economics
Environmental Science and Technology
Geophysical Research Letters
Global Change Biology
Journal of Climate
Journal of Geophysical Research
Nature
Nature Geoscience
Nature Reports Climate Change
Proceedings of the National Academy of Science
Science

Major Scientific Societies That Cover Climate and Related Environmental Issues

American Geophysical Union
American Institute of Professional Geologists
American Meteorological Society
American Sociological Association
American Solar Energy Society
Association of American Geographers
Association of Climate Change Officers
Association of Environmental and Engineering Geologists
Association of Meteorology and Atmospheric Sciences
Development Studies Association
Ecological Society of America
Environmental and Energy Study Institute
Environmental and Engineering Geophysical Society
European Association of Geoscientists and Engineers
European Geosciences Union
European Wind Energy Association
Geochemical Society Geological Society (London, UK)
Geological Society of America
International Association for Urban Climate International
International Society for Ecological Economics
International Society for Environmental Biogeochemistry
The Royal Society Oceanography Society
The World Conservation Union

ACKNOWLEDGMENTS

I AM DEEPLY GRATEFUL to my wife, Nancy Gordon, for keeping the home fires burning and fueling my lunch box every morning, and for many important editorial and managerial suggestions as well as for mining numerous important climate news stories for me from print media and the web.

I'm also grateful to my research and administrative assistants: Wendy Li, Jennifer Millman, Irene Saunders, Maria Terekhov, Ashley Warner, Donna Woo, Ping Wu, and Dennis Martz.

I also want to thank my friend and colleague Benson Lee for keeping me in shape by cycling with me every weekend in the East Bay hills and around San Francisco Bay and for sharing my deep interest in climate science. During our sometimes strenuous outings, we held many hours of thoughtful discussion that influenced my thinking about climate change, and Benson would later share relevant articles and information that often had a major impact on the book. Whenever my natural tendency to view dire problems through rose-colored glasses surfaced, he was rigorously pragmatic and invariably persuaded me to take a more hard-nosed approach.

I am immensely grateful to David King Dunaway for his wise counsel and insightful editing on various chapters of the book. His patient and forceful advice prevailed on me to shorten the book and reduce its scientific complexity. I also benefited from the comments of the members of Real Writing—The East Bay Nonfiction Writers group. Every time I read a few pages to them, it invariably led to hours of rewriting—a good thing.

I'm especially thankful to have had the generous guidance of Dr. Kevin Trenberth of the National Center for Atmospheric Research on questions of climate science, particularly regarding questions of ocean heat transport processes and the role of the ocean in the Earth's long-term climate cycle.

I am grateful for the advice of the following other professional climate scientists who were gracious to offer perspective and information, including: Dr. James Hansen, recently of NASA Goddard Institute for Space Studies; Professor John Harte, University of California; Dr. Stacy C. Jackson, University of California, Berkeley; Dr. Benjamin Santer, Lawrence Livermore National

Laboratory; Dr. Benjamin I. Cook, NASA Goddard Institute for Space and Lamont-Doherty Earth Observatory Studies; Professor John R. Porter, University of Copenhagen; Dr. Mark Howden, Commonwealth Scientific and Industrial Research Organization; and Dr. Andy Challinor, University of Leeds. Drs. Porter, Howden, and Challinor are all members of the IPCC Task Force on Food Production Systems and Food Security for Assessment 5.

I especially appreciate the assistance of Dr. Kevin Kennedy of the World Resources Institute for information and discussion of the intricacies of California's cap-and-trade program;

Dr. Stacy C. Jackson, recently with the Energy and Resources Group, University of California, Berkeley, for sharing her unpublished doctoral dissertation and for valuable assistance comprehending the implications of various emissions cycles, the role of the ocean in determining future warming, and for the discussion of climate models in her dissertation;

Dr. Shaye Wolf of the Center for Biological Diversity for discussions regarding the role of the ocean in long-term climate change and for bringing valuable references on climate and extinctions to my attention. Friends Ron Feldman, Sharon Smith, Dr. David Lenderts, Bob Hall, Virginia Morgan, Maureen Murphy, and Mary Planding also read the manuscript or provided encouragement. Notwithstanding the suggestions received, I am fully responsibile for any remaining errors.

I appreciate the scientific editing provided by my contract editor, Dr. Marta Tanrikulu, and the work of editor Elisabeth Ptak, who reedited the manuscript, raised useful questions, and helped make the book more accessible to nontechnical readers. I especially appreciate the skillful interior book design and layout work of Nancy Austin and the excellent proofreading and copyediting by Molly Woodward. My thanks to Harry Bego of TExtract (www.Texyz.com) for patiently answering my many questions about the software.

I'm particularly grateful to designer Shanon Bodie of Lightbourne, Inc. for her patience and superb book cover design, and to her patient, supportive Lightbourne colleague Jann Armstrong. Thanks as well to all the friends and family who provided valuable early input on the various cover concepts and mockups.

Climate research support was generously provided by the late Newton D. Becker of the Newton D. and Rochelle F. Becker Foundation.

ABOUT THE AUTHOR AND CONTRIBUTORS

About the Author

John J. Berger, PhD is the author and editor of 11 books on climate, energy, and natural resources. He is a graduate of Stanford University and has a master's in energy and natural resources from UC Berkeley and a PhD in ecology from UC Davis. A journalist, professor, and leader of national environmental organizations, he has also served as a consultant on energy and natural resources to government, scientific, academic, and nonprofit organizations, including the US Congress and the National Academy of Sciences.

Contributors' Biographical Information

Ben Santer, PhD is an atmospheric scientist at Lawrence Livermore National Laboratory (LLNL). His research focuses on such topics as climate model evaluation, the use of statistical methods in climate science, and identification of natural and anthropogenic "fingerprints" in observed climate records. Dr. Santer spent much of the last decade addressing the contentious issue of whether model-simulated changes in tropospheric temperature are in accord with satellite-based temperature measurements. His recent work has attempted to identify anthropogenic fingerprints in a number of different climate variables, such as tropopause height, atmospheric water vapor, the temperature of the stratosphere and troposphere, and ocean surface temperatures.

Dr. Santer served as convening lead author of the climate-change detection and attribution chapter of the 1995 IPCC report. More recently, he was the convening lead author of a key chapter of the US Climate Change Science Program's report on "Temperature Trends in the Lower Atmosphere." His awards include the Norbert Gerbier–MUMM International Award (1998), a MacArthur Fellowship (1998), the US Department of Energy's E. O. Lawrence Award (2002), and a Distinguished Scientist Fellowship from the US Department of Energy, Office of Biological and Environmental Research (2005).

PAUL R. EHRLICH is Bing Professor of Population Studies and president of the Center for Conservation Biology, Department of Biology, Stanford University; and adjunct professor, University of Technology, Sydney. He does research in population biology (which includes ecology, evolutionary biology, behavior, and human ecology and cultural evolution). Ehrlich has carried out field, laboratory, and theoretical research on a wide array of problems ranging from the dynamics and genetics of insect populations, studies of the ecological and evolutionary interactions of plants and herbivores, and the behavioral ecology of birds and reef fishes, to experimental studies of the effects of crowding on human beings and studies of cultural evolution. He is heavily involved in the Millennium Alliance for Humanity and the Biosphere (MAHB—http://mahb.stanford.edu/) and is author and coauthor of more than 1000 scientific papers and articles in the popular press and over 40 books. Ehrlich is a Fellow of the American Academy of Arts and Sciences and the Beijer Institute of Ecological Economics, and a member of the United States National Academy of Sciences and the American Philosophical Society. He is a foreign member of the Royal Society and an honorary member of the British Ecological Society. Among his many other honors are the Royal Swedish Academy of Sciences' Crafoord Prize in Population Biology and the Conservation of Biological Diversity (an explicit replacement for the Nobel Prize); a MacArthur Prize Fellowship; the Volvo Environment Prize; UNEP Sasakawa Environment Prize; the Heinz Award for the Environment; the Tyler Prize for Environmental Achievement; the Heineken Prize for Environmental Sciences; the Blue Planet Prize; the Eminent Ecologist award of the Ecological Society of America; and the Margalef Prize in Ecology and Environmental Sciences. Dr. Ehrlich has appeared as a guest on hundreds of TV and radio programs; he also was a correspondent for NBC News. He has given many public lectures in the past 40 years.

ANNE E. EHRLICH is a senior research scientist and associate director for policy of the Center for Conservation Biology at Stanford University. She has carried out research and coauthored many technical articles on population biology and ecology and has taught seminar courses on environmental policy at Stanford since 1981. She also has written extensively on issues of public concern such as population, environmental protection, and the environmental consequences of nuclear war, and has coauthored a dozen books, including *The Population Explosion* (1990), *Betrayal of Science and Reason* (1996), *One with Nineveh* (2004), and *The Dominant Animal* (2008).

Anne served as one of seven outside consultants to the White House Council on Environmental Quality's Global 2000 Report (1977–80) and on a task group for the President's Commission on Sustainable Development (1994–95). She has served on the boards of directors of Friends of the Earth, the Rocky Mountain Biological Laboratory, the Sierra Club, and the Ploughshares Fund, and currently serves on the boards of the Pacific Institute for Studies in Environment, Development, and Security and the New-Land Foundation. Anne is a fellow of the American Academy of Arts and Sciences, and her honors include two honorary doctorate degrees, the United Nations Environmental Programme/Sasakawa Prize, the Heinz Award for Environmental Achievement, and the Tyler Prize (all shared with Paul Ehrlich).

INDEX

AVAILABLE NOW IN PAPERBACK & ALL EBOOK FORMATS

Climate Myths: The Campaign Against Climate Science

Scientists have been warning the world for decades about the climate dangers—extreme weather in all its forms—linked to our continued heavy reliance on fossil fuels. Why haven't policy makers heeded their warnings and acted long ago?

Climate Myths explains why. It exposes the fossil fuel industries' successful 20-year-long campaign to mislead the public and legislators about climate change. It shows how this campaign managed to sow doubt and confusion about climate change through a well-paid network of prominent proxy organizations and special interest groups.

Beyond describing how valuable years slipped away during which needed US climate legislation and global climate policy were stalemated, the book carefully dissects and rebuts the fossil fuel industry's main myths and misconceptions about climate change—in everyday language ordinary readers can understand.

Anyone curious about the politics of climate change and the strategy and tactics of the disinformation campaign need look no farther. *Climate Myths* will be of particular interest to the intellectually curious and to college faculty and students, environmentalists, activists, renewable energy advocates, and entrepreneurs as well as legislators and their staffs.

With a preface and foreword by two eminent climate scientists and an introduction by John H. Adams, winner of the 2010 Presidential Medal of Freedom.

Paperback available through Amazon, Barnes & Noble, Ingram, and your local bookstore: 5¼ x 8 inches, 132 pages, ISBN 978-0-98590-920-8, US $9.95

eBook available in all major formats, including Kindle, iBooks, Nook and PDF. ISBN 978-0-98590-921-5, US $6.99

Bulk orders and all other inquiries, contact: info@northbraepublishers.com
More information available at www.johnjberger.com